INTRODUCTION TO CONTROL THEORY, INCLUDING OPTIMAL CONTROL

ELLIS HORWOOD SERIES IN MATHEMATICS & ITS APPLICATIONS

Series Editor: Professor G. M. BELL

Chelsea College, University of London

The works in this series will survey recent research, and introduce new areas and up-to-date mathematical methods. Undergraduate texts on established topics will stimulate student interest by including present-day applications, and the series can also include selected volumes of lecture notes on important topics which need quick and early publication.

In all three ways it is hoped to render a valuable service to those who learn, teach, develop and use mathematics.

MATHEMATICAL THEORY OF WAVE MOTION
G. R. BALDOCK and T. BRIDGEMAN, University of Liverpool.

MATHEMATICAL MODELS IN SOCIAL MANAGEMENT AND LIFE SCIENCES
D. N. BURGHES and A. D. WOOD, Cranfield Institute of Technology.

MODERN INTRODUCTION TO CLASSICAL MECHANICS AND CONTROL
D. N. BURGHES, Cranfield Institute of Technology and A. DOWNS, Sheffield University.

CONTROL AND OPTIMAL CONTROL
D. N. BURGHES, Cranfield Institute of Technology and A. GRAHAM, The Open University, Milton Keynes.

TEXTBOOK OF DYNAMICS
F. CHORLTON, University of Aston, Birmingham.

VECTOR AND TENSOR METHODS
F. CHORLTON, University of Aston, Birmingham.

ADVANCED TOPICS IN OPERATIONAL RESEARCH
B. CONNOLLY, Chelsea College, University of London.

LECTURE NOTES ON QUEUEING SYSTEMS
B. CONNOLLY, Chelsea College, University of London.

MATHEMATICS FOR THE BIOSCIENCES
G. EASON, C. W. COLES, G. GETTINBY, University of Strathclyde.

HANDBOOK OF HYPERGEOMETRIC INTEGRALS: Theory, Applications, Tables, Computer Programs
H. EXTON, The Polytechnic, Preston.

MULTIPLE HYPERGEOMETRIC FUNCTIONS
H. EXTON, The Polytechnic, Preston

COMPUTATIONAL GEOMETRY FOR DESIGN AND MANUFACTURE
I. D. FAUX and M. J. PRATT, Cranfield Institute of Technology.

APPLIED LINEAR ALGEBRA
R. J. GOULT, Cranfield Institute of Technology.

MATRIX THEORY AND APPLICATIONS FOR ENGINEERS AND MATHEMATICIANS
A. GRAHAM, The Open University, Milton Keynes.

APPLIED FUNCTIONAL ANALYSIS
D. H. GRIFFEL, University of Bristol.

GENERALISED FUNCTIONS: Theory, Applications
R. F. HOSKINS, Cranfield Institute of Technology.

MECHANICS OF CONTINUOUS MEDIA
S. C. HUNTER, University of Sheffield.

GAME THEORY: Mathematical Models of Conflict
A. J. JONES, Royal Holloway College, University of London.

USING COMPUTERS
B. L. MEEK and S. FAIRTHORNE, Queen Elizabeth College, University of London.

SPECTRAL THEORY OF ORDINARY DIFFERENTIAL OPERATORS
E. MULLER-PFEIFFER, Technical High School, Ergurt.

SIMULATION CONCEPTS IN MATHEMATICAL MODELLING
F. OLIVEIRA-PINTO, Chelsea College, University of London.

ENVIRONMENTAL AERODYNAMICS
R. S. SCORER, Imperial College of Science and Technology, University of London.

APPLIED STATISTICAL TECHNIQUES
K. D. C. STOODLEY, T. LEWIS and C. L. S. STAINTON, University of Bradford.

LIQUIDS AND THEIR PROPERTIES: A Molecular and Macroscopic Treatise with Applications
H. N. V. TEMPERLEY, University College of Swansea, University of Wales and D. H. TREVENA, University of Wales, Aberystwyth.

GRAPH THEORY AND APPLICATIONS
H. N. V. TEMPERLEY, University College of Swansea.

INTRODUCTION TO CONTROL THEORY, INCLUDING OPTIMAL CONTROL

D. N. BURGHES, B.Sc., Ph.D., F.R.A.S., F.I.M.A.
Department of Mathematics
Cranfield Institute of Technology

and

A. GRAHAM, M.A., M.Sc., Ph.D., C.Eng., M.I.E.E.
Department of Mathematics
The Open University
Milton Keynes

ELLIS HORWOOD LIMITED
Publishers Chichester

Halsted Press: a division of
JOHN WILEY & SONS
New York - Chichester - Brisbane - Toronto

First published in 1980 and
Reprinted in 1986 by

ELLIS HORWOOD LIMITED
Market Cross House, Cooper Street,
Chichester, West Sussex, PO19 1EB, England
The publisher's colophon is reproduced from James Gillison's drawing of the ancient Market Cross, Chichester.

Distributors:
Australia and New Zealand:
JACARANDA WILEY LIMITED
GPO Box 859, Brisbane, Queensland 4001, Australia
Canada:
JOHN WILEY & SONS CANADA LIMITED
22 Worcester Road, Rexdale, Ontario, Canada
Europe and Africa:
JOHN WILEY & SONS LIMITED
Baffins Lane, Chichester, West Sussex, England
North and South America and the rest of the world:
Halsted Press: a division of
JOHN WILEY & SONS
605 Third Avenue, New York, NY 10158, USA

British Library Cataloguing in Publication Data
Burghes, David Noel
Introduction to control theory, including Optimal Control. –
(Ellis Horwood series in mathematics and its applications).
1. Control theory
I. Title II. Graham, Alexander
629.8'312 QA402.3 80–40386
ISBN 0–85312–181–8 (Ellis Horwood Ltd., Publishers – Library Edition)
ISBN 0–470–26998–7 (Halsted Press)

Typeset in Press Roman by Ellis Horwood Ltd.
Printed in Great Britain by R. J. Acford Ltd., Chichester

Table of Contents

Preface 9

PART I - CONTROL

Chapter 1 System Dynamics and Differential Equations

1.1 Introduction 13
1.2 Some System Equations 13
1.3 System Control 19
1.4 Mathematical Models and Differential Equations 21
1.5 The Classical and Modern Control Theory 23

Chapter 2 Transfer Functions and Block Diagrams

2.1 Introduction 27
2.2 Review of Laplace Transforms 27
2.3 Applications to Differential Equations 38
2.4 Transfer Functions 39
2.5 Block Diagrams 41

Chapter 3 State-Space Formulation

3.1 Introduction 49
3.2 State-Space Forms 49
3.3 Using the Transfer Function to Define State Variables 57
3.4 Direct Solution of the State Equation 62
3.5 Solution of the State Equation by Laplace Transforms 70
3.6 The Transformation from the Companion to the Diagonal State Form 73
3.7 The Transfer Function from the State Equation 78

Chapter 4 Transient and Steady State Response Analysis

4.1 Introduction 83
4.2 Response of First Order Systems 84

4.3 Response of Second Order Systems. 86
4.4 Response of Higher Order Systems 92
4.5 Steady State Error. 95
4.6 Feedback Control 99

Chapter 5 Stability
5.1 Introduction. 107
5.2 The concept of Stability. 107
5.3 Routh Stability Criterion 110
5.4 Introduction to Liapunov's Method 116
5.5 Quadratic Forms 126
5.6 Determination of Liapunov's Functions. 128
5.7 The Nyquist Stability Criterion 131
5.8 The Frequency Response 131
5.9 An Introduction to Conformal Mappings 138
5.10 Applications of Conformal Mappings to the Frequency Response . 142

Chapter 6 Controllability and Observability
6.1 Introduction. 149
6.2 Controllability 149
6.3 Observability 155
6.4 Decomposition of the System State 157
6.5 A Transformation into the Companion Form 165

Chapter 7 Multivariable Feedback and Pole Location
7.1 Introduction. 171
7.2 State Feedback of a SISO System. 171
7.3 Multivariable Systems. 175
7.4 Observers. 188

PART II - OPTIMAL CONTROL

Chapter 8 Introduction to Optimal Control
8.1 Control and Optimal Control. 197
8.2 Examples. 197
8.2.1 Economic Growth. 197
8.2.2 Resource Depletion 200
8.2.3 Exploited Populations 200
8.2.4 Advertising Policies 201
8.2.5 Rocket Trajectories. 203
8.2.6 Servo Problem 204
8.3 Functionals 204
8.4 The Basic Optimal Control Problem 206

Chapter 9 Variational Calculus
9.1 The Brachistochrone Problem 209
9.2 Euler Equation 210
9.3 Free End Conditions 219
9.4 Constraints 222

Chapter 10 Optimal Control with Unbounded Continuous Controls
10.1 Introduction 231
10.2 The Hamiltonian 233
10.3 Extension to Higher Order Systems 235
10.4 General Problem 238

Chapter 11 Bang-Bang Control
11.1 Introduction 245
11.2 Pontryagin's Principle 249
11.3 Switching Curves 254
11.4 Transversality Conditions 258
11.5 Extension to the Boltza Problem 261

Chapter 12 Applications of Optimal Control
12.1 Introduction 267
12.2 Economic Growth 267
12.3 Resource Depletion 270
12.4 Exploited Populations 274
12.5 Advertising Policies 276
12.6 Rocket Trajectories 279
12.7 Servo Problem 281

Chapter 13 Dynamic Programming
13.1 Introduction 289
13.2 Routing Problem 289
13.3 D.P. Notation 291
13.4 Examples 295
13.4.1 Resources Allocation 295
13.4.2 Production Planning 297
13.5 Bellman's Equation 300
13.6 The Maximum Principle 307

Appendix 1 Partial Fractions 315

Appendix 2 Notes on Determinants and Matrices
A2.1 Determinants 319
A2.2 Partitioned Matrices 322

A2.3 Eigenvectors and Eigenvalues 323
A2.4 The Companion Matrix 325
A2.5 The Cayley–Hamilton Theorem 326
A2.6 Linear Dependence and the Rank of a Matrix 328

Solutions to Problems 331

Bibliography 393

Index 397

Preface

Our aim in writing this book is to introduce a number of important concepts both in linear control and in optimal control at a level suitable for advanced undergraduate and postgraduate students in applied mathematics, engineering and related subjects. The book is divided into two parts; the first part dealing with the control of continuous systems and the second part dealing with optimal control.

In the first part, we present concepts in both classical and modern control theory, using many illustrative examples in order to develop the theory in as simple a manner as possible. We concentrate on concpets rather than associated mathematical ramifications, and so, for example, we deal mainly with system matrices which can be diagonalised.

To avoid giving the impression that there exists some rigid division between classical and modern control theories, we have ordered the chapters so as to deal with the concepts as they naturally arise.

In considering system stability, we discuss three methods, including one based on Liapunov's work. Although this method has been developed for non-linear systems, we feel that our simplified presentation will be a useful introduction to this topic.

In the second part, we again concentrate on concepts rather than unnecessary mathematical complexity. An introduction to calculus of variations is given, which leads neatly into continuous optimal control problems. A verification is given for Pontryagin's principle, but emphasis is placed on its use in practical problems.

We end the book with an introductory chapter on dynamic programming, illustrating its connections with both calculus of variations and optimal control.

We hope that this book will provide a valuable introduction at an elementary level to the study of control and optimal control, both for students who will be taking the subject no further, and for others who will be moving on to more advanced studies in continuous, discrete and stochastic control.

D. Burghes
A. Graham

Part I – Control

CHAPTER 1

System dynamics and differential equations

1.1 INTRODUCTION

This book is intended for the student who has acquired certain mathematical skills, but who does not necessarily have an engineering background. We shall therefore begin by showing that investigation of dynamic behaviour of (idealised) components, and the collection of interacting components (or systems), are equivalent to studying differential equations. We shall see that the dynamic behaviour of very different systems can be expressed by differential equations which, when appropriate scaling has taken place, may be identical. Once this point has been made, we shall be in a position to go forward in discussing various concepts in control theory.

1.2 SOME SYSTEM EQUATIONS

1. In electromagnetic theory it is known that, for a coil (Fig. 1.1) having an inductance L the electromotive force (e.m.f) E is proportional to the rate of change of the current I, at the instant considered, that is,

$$E = L\frac{\mathrm{d}I}{\mathrm{d}t} \quad \text{or} \quad I = \frac{1}{L}\int E\,\mathrm{d}t. \tag{1.1}$$

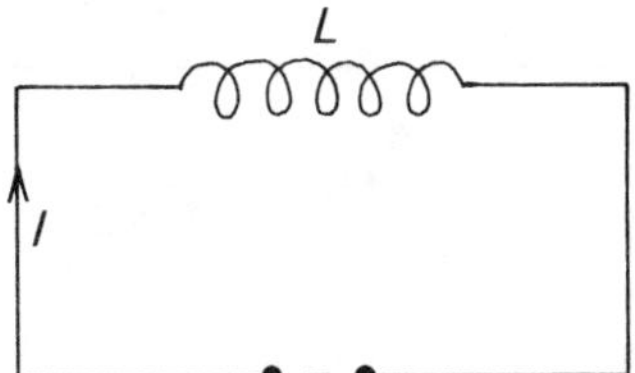

Fig. 1.1

2. Similarly, from electrostatics theory we know (Fig. 1.2) that the voltage V and current I through a capacitor of capacitance C are related at the instant t by

$$E = \frac{1}{C}\int I\,\mathrm{d}t \quad \text{or} \quad I = C\frac{\mathrm{d}E}{\mathrm{d}t}. \tag{1.2}$$

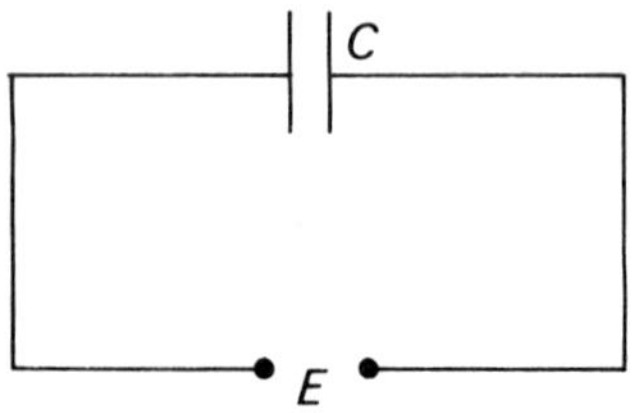

Fig. 1.2

3. Fig. 1.3 illustrates a dashpot device which consists of a piston sliding in an oil filled cylinder. The motion of the piston relative to the cylinder is resisted by the oil, and this viscous drag can be assumed to be proportional to the velocity of the piston.

   If the applied force is $f(t)$ and the corresponding displacement is $y(t)$, then Newton's Law of motion is

$$f = \mu\frac{\mathrm{d}y}{\mathrm{d}t} \tag{1.3}$$

   where the mass of the piston is considered negligible, and μ is the viscous damping coefficient.

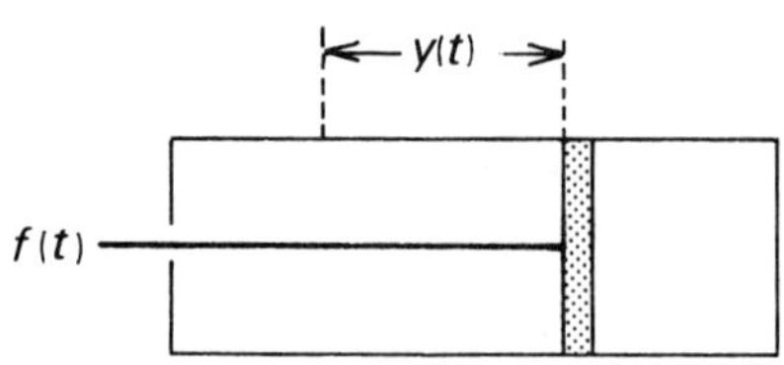

Fig. 1.3

4. To analyse the control of the level of a liquid in a tank we must consider the input and output regulations. Fig. 1.4 shows a tank with:

   inflow rate q_i
   outflow rate q_o
   head level h
   the cross-section area of the tank is A.

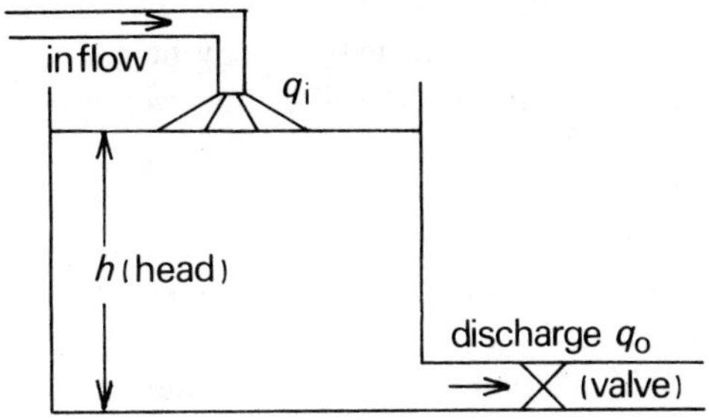

Fig. 1.4

If $q = q_0 - q_i$ is the net rate of inflow into the tank, over a period δt, and δh is the corresponding change in the head level, then $q\delta t = A\delta h$ and in the limit as $\delta t \to 0$

$$q = A\frac{\mathrm{d}h}{\mathrm{d}t} \quad \text{or} \quad h = \frac{1}{A}\int q\,\mathrm{d}t. \tag{1.4}$$

We could produce many more examples of very diverse situations in economics, physiology, etc. in which the dynamic behaviour of the system can be shown to be characterised by a differential equation similar to the above.

All these equations (1.1)-(1.4) have something in common - they can all be written in the form

$$a\frac{\mathrm{d}y}{\mathrm{d}t} = x, \quad \text{that is,} \tag{1.5}$$

in (1.1) $x = E, y = I, a = L$

in (1.2) $x = I, y = E, a = C$

in (1.3) $x = f, a = \mu$

in (1.4) $x = q, y = h, a = A.$

Equation (1.5) is interesting not only because its solution is also the solution to any one of the systems considered, but also because it shows the direct analogies which can be formulated between quite different types of components and systems. This has very important implications in mathematical modelling because solving a differential equation leads to the solution of a vast number of problems in different disciplines, all of which are modelled by the same equation.

For a more complicated system, consisting of (say) two interconnected tanks, we must obtain explicitly an equation for the outflow rate q_0 (see Fig. 1.4) in terms of the resistance offered by the discharge valve. This in fact is an interesting problem which uses a 'linearisation' technique - a mathematical simplification of a type frequently used to obtain an approximate solution.

Over small ranges of the variables involved the loss of accuracy may be very small and the simplification of calculations may be great. It is known that the flow rate through a restriction such as a discharge valve is of the form

$$q = VP^{\frac{1}{2}}$$

where P is the pressure across the valve and V is a coefficient dependent on the properties of the liquid and the geometry of the valve.

Fig. 1.5 shows the relation between the pressure and the flow rate and also indicates the assumption we make, that is, that in the neighbourhood of the pressure $P = P_1$

$$\frac{\text{change in pressure}}{\text{change in flow rate}} \simeq R'$$

where R' is a constant called the resistance of the valve at the point (of pressure) considered.

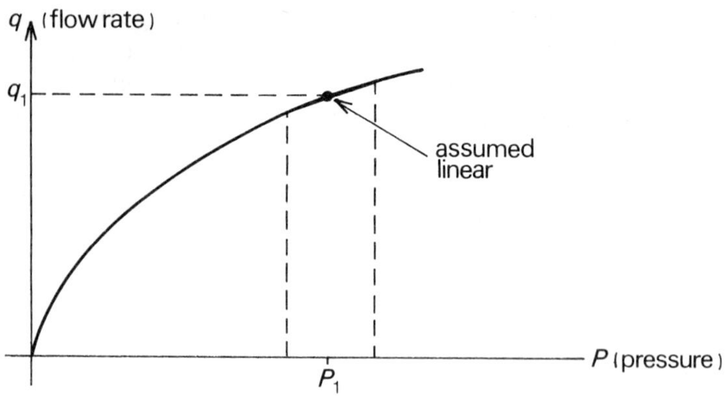

Fig. 1.5

This type of assumption, which assumes that an inherently nonlinear situation can be approximated by a linear one albeit in a restricted range of the variable, is fundamental to many applications of control theory.

Since, at a given point, the pressure

$$P = h\rho g$$

(ρ is the density of the liquid and $g = 9.8\ \text{m/s}^2$) we can write the above relation as

$$R = h/q_o \tag{1.6}$$

where $$R = \frac{R'}{\rho g}.$$

A two tank system with one inflow (q) and two discharge valves (R_1 and R_2) is shown in Fig. 1.6.

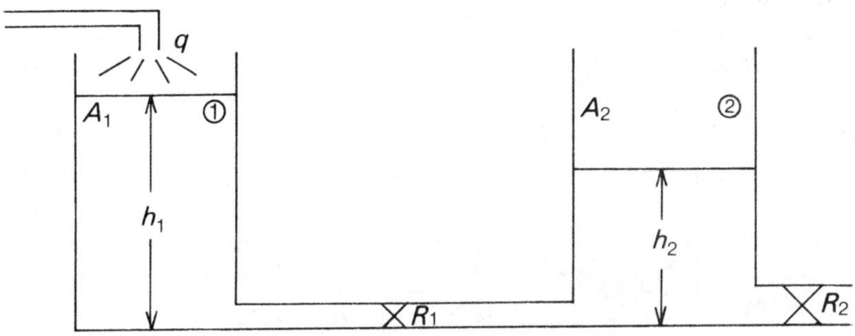

Fig. 1.6

For tank 1

$$A_1 \frac{\mathrm{d}h_1}{\mathrm{d}t} = q - \frac{1}{R_1}(h_1 - h_2)$$

or $$\dot{h}_1 = \frac{1}{A_1} q - \frac{1}{A_1 R_1} h_1 + \frac{1}{A_1 R_1} h_2 \tag{1.7}$$

where $\dot{h}_1 = \mathrm{d}h_1/\mathrm{d}t$.

For tank 2

$$A_2 \frac{\mathrm{d}h_2}{\mathrm{d}t} = \frac{1}{R_1}(h_1 - h_2) - \frac{1}{R_2} h_2$$

or $$\dot{h}_2 = \frac{1}{A_2 R_1} h_1 - \frac{1}{A_2}\left(\frac{1}{R_1} + \frac{1}{R_2}\right) h_2 \tag{1.8}$$

where $$\dot{h}_2 = \frac{\mathrm{d}h_2}{\mathrm{d}t}.$$

We can write equations (1.7) and (1.8) in matrix form as

$$\begin{bmatrix} \dot{h}_1 \\ \dot{h}_2 \end{bmatrix} = \begin{bmatrix} -\dfrac{1}{A_1R_1} & \dfrac{1}{A_1R_1} \\ \dfrac{1}{A_2R_1} & -\dfrac{1}{A_2}\left(\dfrac{1}{R_1}+\dfrac{1}{R_2}\right) \end{bmatrix} \begin{bmatrix} h_1 \\ h_2 \end{bmatrix} + \begin{bmatrix} \dfrac{1}{A_1} \\ 0 \end{bmatrix} q$$

that is, in the form

$$\dot{\underline{h}} = A\underline{h} + Bq \tag{1.9}$$

where

$$\underline{h} = \begin{bmatrix} h_1 \\ h_2 \end{bmatrix}, \quad A = \begin{bmatrix} -\dfrac{1}{A_1R_1} & \dfrac{1}{A_1R_1} \\ \dfrac{1}{A_2R_1} & -\dfrac{1}{A_2}\left(\dfrac{1}{R_1}-\dfrac{1}{R_2}\right) \end{bmatrix}$$

and

$$B = \begin{bmatrix} \dfrac{1}{A_1} \\ 0 \end{bmatrix}.$$

We can also write the set of simultaenous first order equations ((1.7) and (1.8)) as one differential equation of the second order.

On differentiating equation (1.8), we obtain

$$\ddot{h}_2 = \frac{1}{A_2R_1}\dot{h}_1 - \frac{1}{A_2}\left(\frac{1}{R_1}+\frac{1}{R_2}\right)\dot{h}_2. \tag{1.10}$$

Substituting $\dot{h}_1$ from equation (1.7) and h_1 from equation (1.8) in the above equation and rearranging we finally obtain

$$\ddot{h}_2 + \left[\frac{1}{A_2}\left(\frac{1}{R_1}+\frac{1}{R_2}\right) + \frac{1}{A_1R_1}\right]\dot{h}_2 - \left[\frac{1}{A_1A_2R_1^2} - \frac{1}{A_1A_2R_1}\left(\frac{1}{R_1}+\frac{1}{R_2}\right)\right]h_2 = \frac{1}{A_1A_2R_1}q. \tag{1.11}$$

Equation (1.11) is a differential equation of order 2 of the form

$$\ddot{y} + a_1\dot{y} + a_2 y = b_0 x.$$

In general it is possible to write n simultaneous first order differential equations, each representing the dynamic behaviour of a 'simple' system as an nth order differential equation representing the dynamic behaviour of the resulting 'complex' system which is a collection of the interacting subsystems.

Conversely, as we shall discuss in Chapter 3, it is possible to express an nth order differential equation as a set of n simultaneous first order differential equations. This is in fact fundamental in modern control theory.

1.3 SYSTEM CONTROL

In the previous section we considered a rather vague definition of a system as one or a collection of interacting components or **subsystems**. This definition is purposely vague so as to allow for the consideration of a very wide variety of systems from that of a complex nuclear reactor to the economy of a nation.

In control terminology the word **plant** is used for that part of a system which is to be controlled.

For example, we have already considered (Fig. 1.4) a tank with an inflow q_i and a discharge q_o. This can be regarded as a **system**, but as yet it is not **controlled** in any way. But by an addition of a float and a regulator valve we can quite easily convert the tank to obtain the automatic control of the liquid level - as shown in Fig. 1.7.

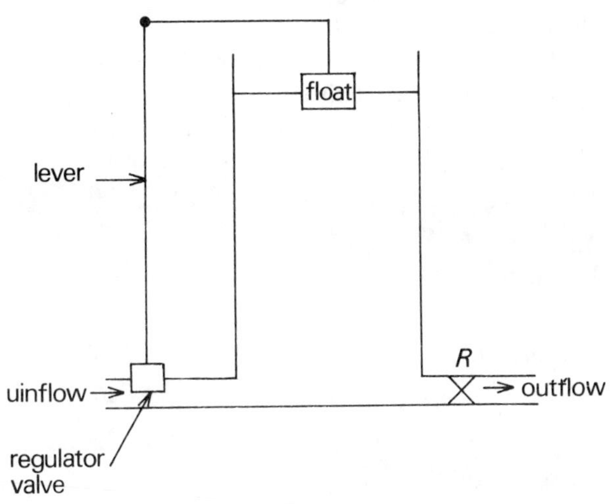

Fig. 1.7

We now have a control system - a combination of components that act together to maintain the head at some desired level h. The actual tank would now be referred to as the plant.

The control system works very simply. Suppose for one reason or another, that is, owing to some form of an external disturbance, the level of the liquid in the tank rises above h. The float senses this rise, and communicates it via the lever to the regulating valve which reduces (or stops) the inflow rate. This situation will continue until the level of the liquid again stabilises at its steady state h.

The above illustration is a simple example of a **feedback control system.** We can illustrate the situation schematically by the use of a block diagram as in Fig. 1.8.

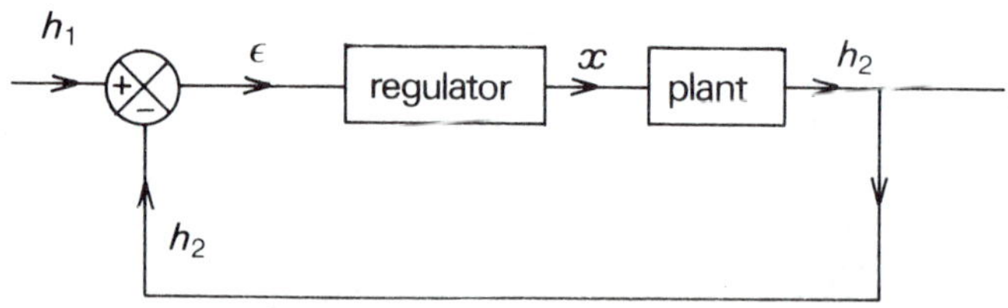

Fig. 1.8

h_1 represents the desired value of the head
h_2 represents the actual value of the head
$\epsilon = h_1 - h_2$ is called the **error**.

The regulator receives the signal ϵ which it transforms into a movement x of the lever which in turn influences the position of the regulator valve.

To obtain the dynamic characteristics of this system we consider a deviation ϵ from the desired level h_1 over a short time interval (small t).

If q_i is the change in the inflow and q_o is the change in the outflow, (see equation (1.4)),

$$A\frac{d\epsilon}{dt} = q_i - q_o$$

where (see equation (1.6))

$$q_o = \epsilon/R.$$

So that the equation can be written as

$$AR\frac{d\epsilon}{dt} + \epsilon = Rq_i$$

q_i, the change in the inflow, depends on the characteristics of the regulator valve and may be of the form

$$q_i = K\epsilon \quad \text{where } K \text{ is a constant.}$$

In many feedback systems there is a component in the feedback loop, known as a **transducer**, which is a device transforming one type of signal into another. For example, to control the speed of a motor (the plant) the actual speed (h_2 in Fig. 1.8) is generally transformed by the transducer (in this case, a tachometer) into a voltage. So in general a **feedback** or a **closed loop control system** is one which compares the output (or some function of the output) with the input and forms an error actuating signal from their difference. This signal produces inputs to the system which will reduce the error and bring the output towards the desired value.

Another type of control system is known as an **open loop control system**. It is one in which the output of the system is not involved in its control. Basically the control on the plant is exercised by a **controller** as shown in Fig. 1.9. Examples of open-loop control systems are traffic lights operating on a fixed time changeover basis, automatic washing machines, toasters, etc.

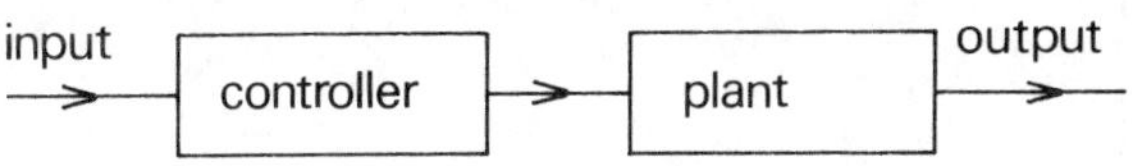

Fig. 1.9

There are advantages and disadvantages in using either type of control system. Among the advantages of a feedback system are that (as we have seen) for a relatively cheap and fairly crude measuring device we can obtain an accurate control of a plant which is relatively insensitive to system parameter variations or external disturbances.

An inherent disadvantage of such systems is the problem of stability. Every engineer who has designed a feedback amplifier will appreciate this fact! Even in our tank control system we can appreciate that, depending on the time constants involved, by the time the signal from the float has moved the regulating valve, too much liquid may have been discharged and the float would again indicate a difference between h_1 and h_2, thus setting up an oscillatory situation for the liquid level in the tank. We shall discuss system stability problems in some detail in Chapter 5.

In general, when the input signals are predictable, open loop control is preferable, but in a complex system a judicious combination of the two types of control is usually found to give the most satisfactory overall performance.

1.4 MATHEMATICAL MODELS AND DIFFERENTIAL EQUATIONS

As was pointed out in Section 1.1, many dynamic systems are characterised by differential equations. The process involved, that is, the use of physical laws

(Newton's, Kirchhoff's, etc.) together with various assumptions of linearity, etc. is known as **mathematical modelling**. This process can be vary complicated and is a topic of great importance in its own right. In most cases we shall assume that someone has very kindly done the job for us, that is provided us with the differential equation which characterises the dynamic behaviour of the system being investigated. Since to us the differential equation will characterise the system, we first discuss the type of equations we shall consider.

We discuss only **ordinary differential equations**. This implies that the systems we consider are either lump parameter ones or approximately so. For example, we consider a liquid heated in a tank to be isothermal, otherwise we would need to use complicated partial differential equations to represent the heat diffusion in the tank. Again this is an approximation to the true state of affairs, but in most situations, an acceptable one.

Of greatest interest will be **linear differential equations**, that is equations (linear in y and its derivatives) whose coefficients are either constants or functions of the independent variable. For example

$$\frac{d^2y}{dt^2} + 2\frac{dy}{dt} - 3y = u$$

and

$$\frac{d^2y}{dt^2} - 2t\frac{dy}{dt} + y = u$$

are both linear differential equations. The first equation is **time-invariant** (or **autonomous**) whereas the second is a **time-varying** equation, since at least one of the coefficients is a function of time (the independent variable).

Of all the types of differential equations, it is the linear ones which are most amenable to mathematical treatment and hence best understood and most used in control theory. The most important property of such equations (and hence of the systems they model) is the **principle of superposition**. It states that the response of a linear system to the simultaneous application of two different forcing functions, is the sum of the two individual responses and conversely.

For example, let $y_1(t)$ and $y_2(t)$ be the responses to $Au_1(t)$ and $Bu_2(t)$ (A and B are constants) in the second of the above equations, that is,

$$\frac{d^2y_1}{dt^2} - 2t\frac{dy_1}{dt} + y_1 = Au_1, \quad \text{and}$$

$$\frac{d^2y_2}{dt^2} - 2t\frac{dy_2}{dt} + y_2 = Bu_2.$$

On addition, we obtain

$$\frac{d^2}{dt^2}(y_1+y_2) - 2t\frac{d}{dt}(y_1+y_2) + (y_1+y_2) = Au_1 + Bu_2$$

which shows that $[y_1(t)+y_2(t)]$ is the system response to the input function $[Au_1(t)+Bu_2(t)]$.

Nonlinear differential equations are ones which are not linear. For example

$$\frac{d^2y}{dt^2} + \left(\frac{dy}{dt}\right)^2 - 2y = u$$

and

$$\frac{d^2y}{dt^2} - y^2\frac{dy}{dt} + y = u$$

are both nonlinear and as can be easily verified, the superposition principle does not apply to them.

Solutions of nonlinear equations are generally very complicated. In the classical control theory, based on Laplace transform methods, one becomes involved with convolution integrals which more often than not cannot be solved analytically. The state-state approach of modern control theory offers much greater flexibility, but even with this approach the difficulties encountered are enormous.

Fortunately many physical systems are nearly linear over a fairly wide range of the variables involved. For many others, a linearised form of their equations is a valid approximation over a specified range of operation. It is for these reasons that we shall study the control theory of linear systems.

1.5 THE CLASSICAL AND MODERN CONTROL THEORY

Classical control theory is based on Laplace transforms (which implies a transformation to a complex frequency domain) and applies to linear autonomous systems having scalar (that is, single) input and scalar output. A function called the **transfer function** relating the input-output relationship of the system is defined. One of the objectives of control theory is to design - in terms of the transfer function - a system which satisfies certain assigned specifications. This objective is primarily achieved by a trial and error approach.

Modern control theory is not only applicable to linear autonomous systems but also to time-varying systems and is useful when dealing with nonlinear systems. In particularly it is applicable to multiple-input and multiple-output systems - in contrast to the classical control theory. The approach is based on the concept of **state**. It is the consequence of an important characteristic of a dynamical system, namely that its instantaneous behaviour is dependent on its past history, so that the behaviour of the system at time $t > t_0$ can be determined given

(1) the forcing function (that is, the input), and

(2) the state of the system at $t = t_0$.

If an nth order differential equation describes the behaviour of a system it can be expressed by the introduction of auxiliary variables as a set of n first order simultaneous equations which is written as a vector matrix differential equation (as equation (1.9)).

We shall see that the choice of these auxiliary variables is in general by no means unique; but having selected a particular set, it uniquely determines the system behaviour so that it can be taken as the set of state variables for the system.

The vector-matrix differential equation can be solved by a procedure which is similar to the one used for a first order (scalar) differential equation.

In contrast to the trial and error approach of classical control, it is often possible in modern control theory to make use of optimal control methods, some of which are discussed in Part II of this book.

PROBLEMS

Note When a mass moves in a horizontal plane, it is assumed that it does so on a frictionless support.

(1) for each of the elements shown below, write down the differential equations relating the input force $f(t)$ and the displacement $y(t)$. (M is a mass, k is a spring constant and μ is a damping coefficient.)

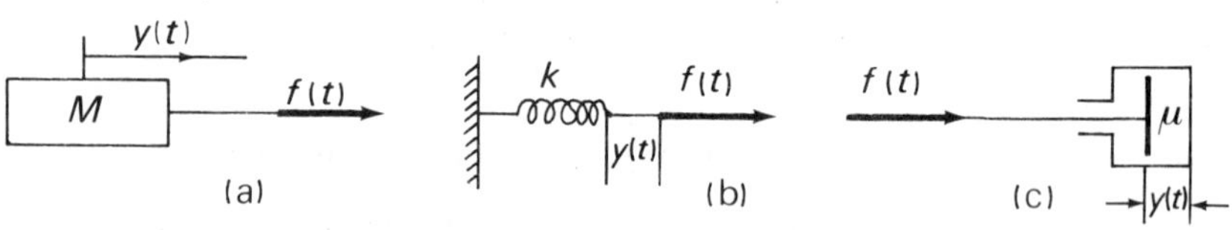

(2) Write down the differential equations which governs the displacements of each mass for the following systems.

(a)

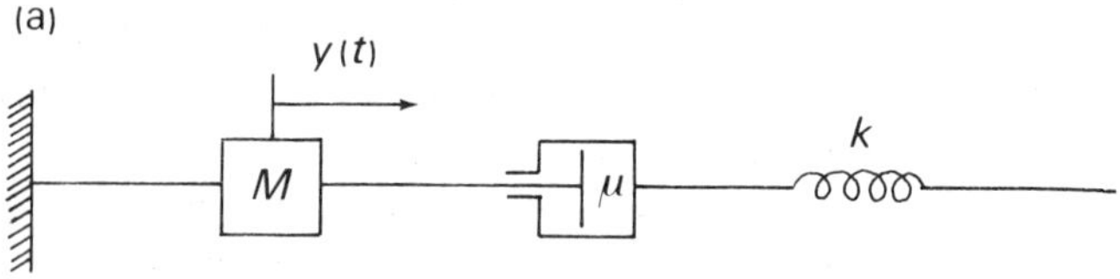

(b)

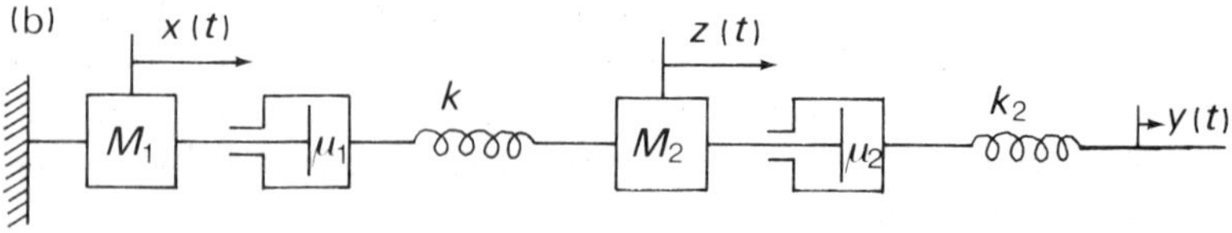

(3) Obtain the differential equation for the displacement $y(t)$ for the system shown below when the force $f(t)$ is applied as indicated.

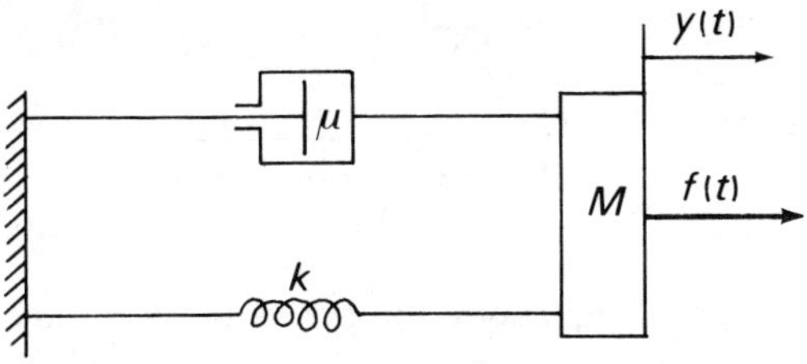

(4) (i) Determine the stiffness K of the spring equivalent to the systems shown below.

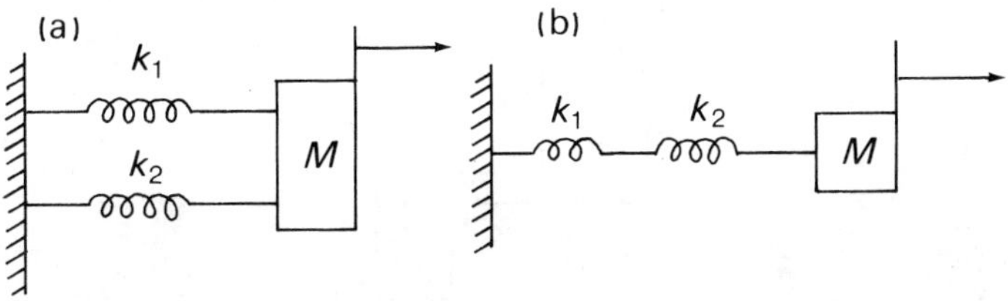

(ii) Determine the equations of motion of the mass in each of the following cases.

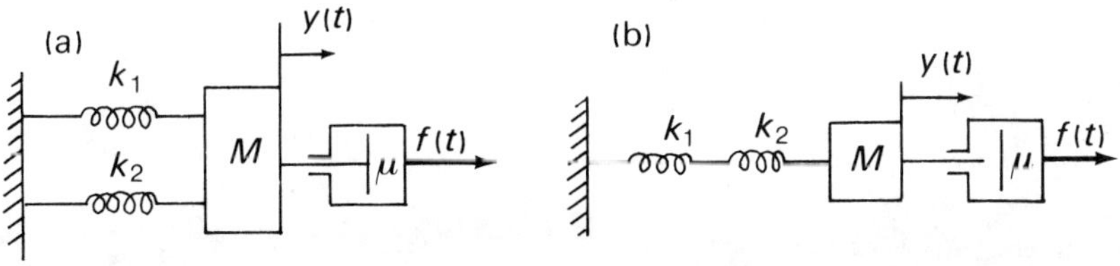

(5) Write down the differential equation which determines the voltage across the capacitor C in the following circuit. (Hint, use the results in section 1.1).

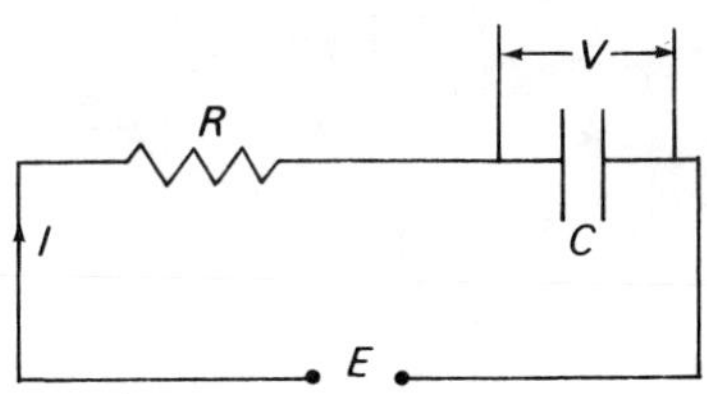

(6) Write down the differential equation which specifies the head $h(t)$ for the system shown in the figure below.

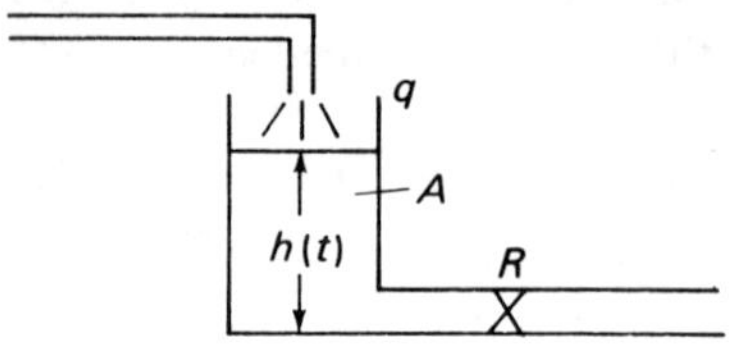

(7) Find the differential equations for the heads $h_1(t)$ and $h_2(t)$ for the system shown in the figure below. Both the tanks have cross-sectional area A.

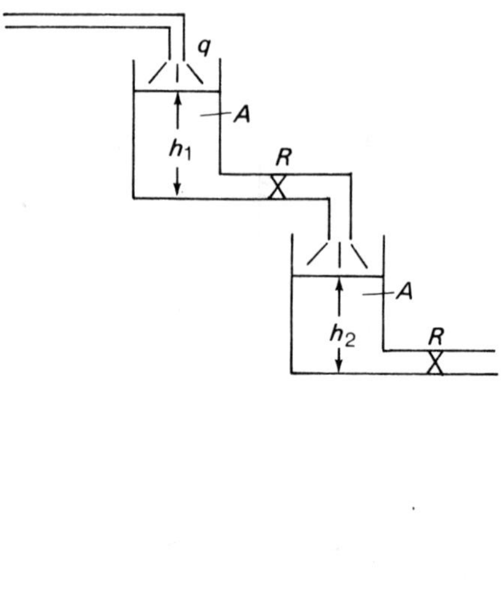

CHAPTER 2

Transfer functions and block diagrams

2.1 INTRODUCTION

It is assumed that the reader is familiar with the classical methods for the solution of ordinary differential equations in the time domain, and therefore these methods will not be reviewed at this stage although there is a very brief discussion in section 3.4. On the other hand, because of the importance of Laplace transform method in control theory, it will be reviewed briefly here, although for detailed proofs and a more rigorous approach the reader is advised to consult specialised references, for example [10] and [38].

Taking the Laplace transform of a differential equation, transforms it from a function of time into an algebraic equation, a function of the complex frequency s. The method is particularly useful for the solution of high order equations.

2.2 REVIEW OF LAPLACE TRANSFORMS

When dealing with certain types of calculations, we sometimes find it much easier to use a transformation to simplify the work.

For example, it is easier to add numbers than to multiply them, hence if

$$AB = C \quad \text{(say)}$$

then by taking the logarithm of both sides, we find

$$\log A + \log B = \log C,$$

so that we have transformed a multiplication into an addition.

To find C we look at a table of inverse transformations to obtain

$$C = \log^{-1}(\log A + \log B).$$

Using Laplace transforms leads to a similar simplification.

Definition

The Laplace transform of $f(t)$, a sectionally continuous function of time, denoted by $\mathcal{L}[f(t)]$, is defined as

$$\mathcal{L}[f(t)] = \int_0^\infty f(t)\mathrm{e}^{-st}\,\mathrm{d}t = F(s) \tag{2.1}$$

where $s = \sigma + i\omega$, σ and ω are real variables.

A few words now about the existence of the integral in equation (2.1). A sufficient condition for

$$\lim_{R\to\infty} \int_0^R f(t)\mathrm{e}^{-st}\,\mathrm{d}t \tag{A}$$

to exist for $\mathrm{Re}(s) > k$ (Re stands for 'the real part of') is

$$\lim_{t\to\infty} |f(t)|\mathrm{e}^{-kt} = 0. \tag{B}$$

If the limit (A) exists and is equal to I, we say that $\int_0^\infty f(t)\mathrm{e}^{-st}\,\mathrm{d}t$ exists and its value is also I. In other words, we write

$$\int_0^\infty f(t)\mathrm{e}^{-st}\,\mathrm{d}t = \lim_{R\to\infty} \int_0^R f(t)\mathrm{e}^{-st}\,\mathrm{d}t.$$

The lower bound of $k = \sigma_c$ (say) for which (B) is true is called the **abscissa of convergence**.

Hence the Laplace transform is defined for any sectionally continuous function $f(t)$, satisfying (B), provided that $\mathrm{Re}(s) > \sigma_c$.

For example, if $f(t) = \mathrm{e}^{3t}$, then

$$|f(t)|\mathrm{e}^{-kt} = \mathrm{e}^{(3-k)t} \to 0 \text{ as } t\to\infty \text{ when } k > 3.$$

In this case $\sigma_c = 3$ (for calculation of $F(s)$ see example 2.3).

In equation (2.1), $f(t)$ is called the **inverse transform** of $F(s)$; this is written as

$$\mathcal{L}^{-1}[F(s)] = f(t).$$

Notice that the definition of $F(s)$ makes no use of the values of $f(t)$ for $t < 0$, hence it is natural to expect that in general $\mathcal{L}^{-1}[F(s)]$ only coincides with $f(t)$ when $t > 0$. An important exception to this statement is a class of functions $\{f(t)\}$ which are defined for all t by their values for positive t. Functions analytic for $t > 0$ form such a class.

We shall therefore generally consider functions $f(t)$ which are zero for $t<0$.

Example 2.1

Consider the step functions $f(t)$

$$f(t) = 0 \quad \text{for} \quad t<0$$

$$f(t) = c\,(\text{constant}) \quad \text{for} \quad t \geqslant 0.$$

Find the Laplace transform of $f(t)$.

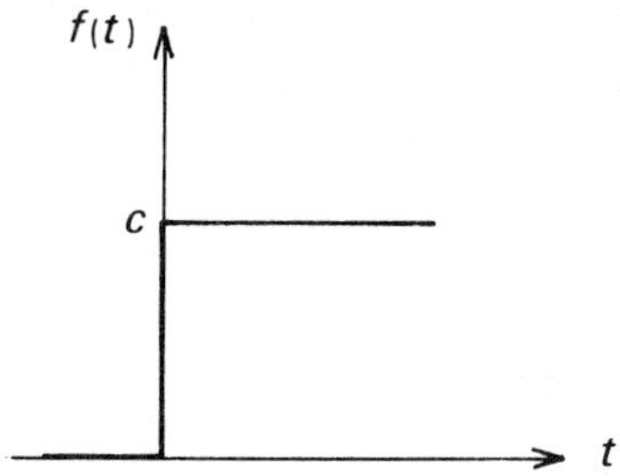

Fig. 2.1

Note: The step function of unit height ($c = 1$) or the **unit step function** is often denoted by $u(t)$. The above function is written as $cu(t)$.

Solution

$$F(s) = \int_0^\infty ce^{-st}\,\mathrm{d}t = -c\left[\frac{1}{s}\,\mathrm{e}^{-st}\right]_0^\infty =$$

$$= -\frac{c}{s}\lim_{t\to\infty}\mathrm{e}^{-st} + \frac{c}{s}.$$

So long as σ (= real part of s) is positive

$$\lim_{t\to\infty}\mathrm{e}^{-st} = 0, \quad \text{hence}$$

$$F(s) = \frac{c}{s} \quad (\text{so long as } \sigma > 0).$$

Some Properties of Laplace Transforms

A is a constant and $\mathcal{L}[f(t)] = F(s)$.

(1) $\mathcal{L}[Af(t)] = A\mathcal{L}[f(t)] = AF(s)$

(2) $\mathcal{L}[f_1(t) + f_2(t)] = \mathcal{L}[f_1(t)] + \mathcal{L}[f_2(t)] = F_1(s) + F_2(s)$.

(1) and (2) prove that the operator $\mathcal{L}$ is linear.

(3) **Differentiation**

On integration by parts, we obtain

$$F(s) = \int_0^\infty f(t)e^{-st}\,dt = -\left[f(t)\frac{1}{s}e^{-st}\right]_0^\infty + \frac{1}{s}\int_0^\infty \frac{d}{dt}[f(t)]e^{-st}\,dt$$

$$= \frac{f(0)}{s} + \frac{1}{s}\mathcal{L}[\frac{d}{dt}f(t)].$$

Hence

$$\mathcal{L}[\frac{d}{dt}f(t)] = sF(s) - f(0). \tag{2.2}$$

Example 2.2

Find $\mathcal{L}[\frac{d^2}{dt^2}f(t)]$, given $[\frac{d}{dt}f(t)]$.

Solution

Assume that $df/dt = g(t)$, then

$$\mathcal{L}[\frac{d^2}{dt^2}f(t)] = \mathcal{L}[\frac{d}{dt}g(t)] = sG(s) - g(0) \quad \text{(from equation (2.2))}$$

$$= s\mathcal{L}[\frac{d}{dt}f(t)] - f'(0)$$

$$= s[sF(s) - f(0)] - f'(0)$$

$$= s^2F(s) - sf(0) - f'(0).$$

In generalising the above result, we obtain

$$\mathcal{L}[\frac{d^nf}{dt^n}] = s^nF(s) - s^{n-1}f(0) - s^{n-2}f'(0) - \ldots - f^{(n-1)}(0). \tag{2.3}$$

In particular if at $t = 0$, f and all its derivatives are zero, equation (2.3) becomes

$$\mathcal{L}[\frac{d^nf}{dt^n}] = s^nF(s) \tag{2.4}$$

(4) **Integration**

On integration by parts, we obtain

$$\mathcal{L}\left[\int_0^t f(x)\mathrm{d}x\right] = \int_0^\infty \left(\int_0^t f(x)\mathrm{d}x\right)\mathrm{e}^{-st}\,\mathrm{d}t$$

$$= -\left[\frac{1}{s}\mathrm{e}^{-st}\int_0^t f(x)\mathrm{d}x\right]_0^\infty + \frac{1}{s}\int_0^\infty f(t)\mathrm{e}^{-st}\,\mathrm{d}t$$

$$= \frac{F(s)}{s}. \qquad (2.5)$$

Example 2.3

Find

(i) $\mathcal{L}[\mathrm{e}^{-\alpha t}]$,

(ii) $\mathcal{L}[\sin\alpha t]$,

(iii) $\mathcal{L}[\cos\alpha t]$.

Solution

(i) $$\mathcal{L}[\mathrm{e}^{-\alpha t}] = \int_0^\infty \mathrm{e}^{-\alpha t}\,\mathrm{e}^{-st}\,\mathrm{d}t = \int_0^\infty \mathrm{e}^{-(\alpha+s)t}\,\mathrm{d}t = \frac{1}{s+\alpha}$$

(ii) (a) *Direct method:*

Since $\mathrm{e}^{i\alpha t} = \cos\alpha t + i\sin\alpha t$

and $\mathrm{e}^{-i\alpha t} = \cos\alpha t - i\sin\alpha t$

it follows that

$$\sin\alpha t = \frac{1}{2i}(\mathrm{e}^{i\alpha t} - \mathrm{e}^{-i\alpha t}).$$

Hence $$\mathcal{L}[\sin\alpha t] = \frac{1}{2i}\int_0^\infty (\mathrm{e}^{i\alpha t} - \mathrm{e}^{-i\alpha t})\mathrm{e}^{-st}\,\mathrm{d}t$$

$$= \frac{1}{2i}\left[\frac{1}{(s-i\alpha)} - \frac{1}{(s+i\alpha)}\right], \quad \text{using (i) above}$$

$$= \frac{\alpha}{s^2+\alpha^2}.$$

(b) *Indirect method* using the rule for differentiation. We define functions

$$f(t) = 0 \quad \text{for} \quad t<0$$
$$= \sin\alpha t \quad \text{for} \quad t \geqslant 0, \quad \text{then for} \quad t \geqslant 0$$

$$f(t) = \sin\alpha t \qquad \text{hence,} \quad f(0) = 0$$

$$f'(t) = \alpha\cos\alpha t \quad \text{hence,} \quad f'(0) = \alpha$$

and $$f''(t) = -\alpha^2\sin\alpha t.$$

Using equation (2.3) and property (1) of Laplace transforms, that is,

$$\mathcal{L}[-\alpha^2\sin\alpha t] = -\alpha^2\mathcal{L}[\sin\alpha t], \quad \text{we obtain}$$

$$-\alpha^2\mathcal{L}[\sin\alpha t] = -\alpha^2 F(s) = s^2F(s) - 0 - \alpha$$

that is, $$(s^2 + \alpha^2)F(s) = \alpha.$$

So that $$F(s) = \frac{\alpha}{s^2 + \alpha^2}.$$

(iii) $$\mathcal{L}[\cos\alpha t] = [\frac{\mathrm{d}}{\mathrm{d}t}(\frac{1}{\alpha}\sin\alpha t)]$$

$$= \frac{1}{\alpha}\mathcal{L}[\frac{\mathrm{d}}{\mathrm{d}t}(\sin\alpha t)] = \frac{1}{\alpha}[sF(s) - f(0)] \quad \text{(by equation (2.2))}$$

where $f(t) = \sin\alpha t$ and $F(s) = \dfrac{\alpha}{s^2 + \alpha^2}$ by (ii) above. Hence

$$\mathcal{L}[\cos\alpha t] = \frac{1}{\alpha}\cdot s\frac{\alpha}{s^2 + \alpha^2} = \frac{s}{s^2 + \alpha^2}$$

(5) **The Delay Function**

Given the Laplace transform $F(s)$ of a function $f(t)$ such that

$$f(t) = 0 \quad \text{for} \quad t < 0$$

we wish to find the Laplace transform of $f(t-\alpha)$ where $f(t-\alpha) = 0$ for $t < \alpha$ (see Fig. 2.2)

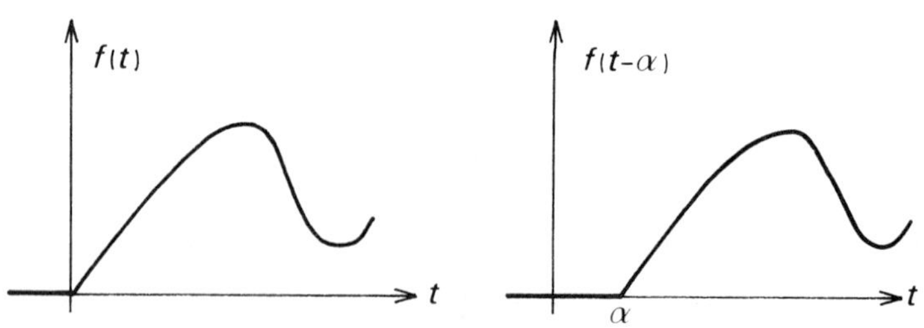

Fig. 2.2

$$F(s) = \int_0^\infty f(x)e^{-sx}\,dx.$$

Let $x = t - \alpha$, then

$$F(s) = \int_\alpha^\infty f(t-\alpha)e^{-s(t-\alpha)}\,dt = e^{\alpha s}\int_0^\infty f(t-\alpha)e^{-st}\,dt$$

$$= e^{\alpha s}\mathcal{L}[f(t-\alpha)].$$

Hence $\mathcal{L}[f(t-\alpha)] = e^{-\alpha s}F(s)$. (2.6)

(6) **The Unit Impulse Function**

Consider the function shown in Fig. 2.3

$$f(t) = \frac{1}{c} \qquad 0 \leqslant t \leqslant c$$

$$= 0 \qquad t > c \quad .$$

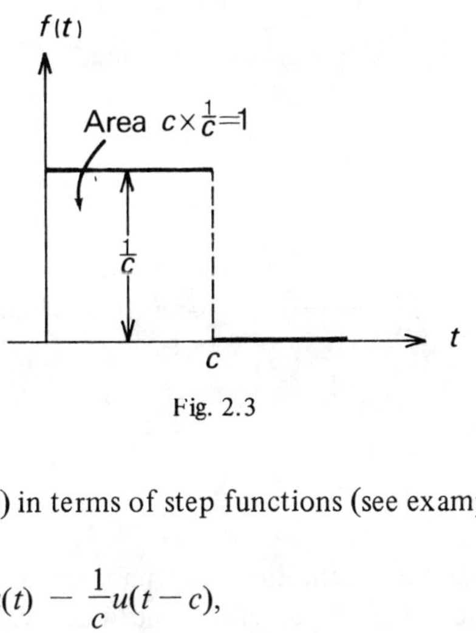

Fig. 2.3

We can express $f(t)$ in terms of step functions (see example 2.1) as

$$f(t) = \frac{1}{c}u(t) - \frac{1}{c}u(t-c),$$

that is, a step function beginning at $t = 0$ minus a step function beginning at $t = c$. This leads to the definition of the **unit impulse function** or the **Dirac delta function** denoted by $\delta(t)$ defined by

$$\delta(t) = \lim_{c \to 0} \frac{u(t) - u(t-c)}{c}.$$

Fig. 2.4 represents the difference of the step functions

$$\frac{1}{c_j}[u(t)-u(t-c_j)]$$

as c_j $(j = 1, 2, 3)$ become progressively smaller.

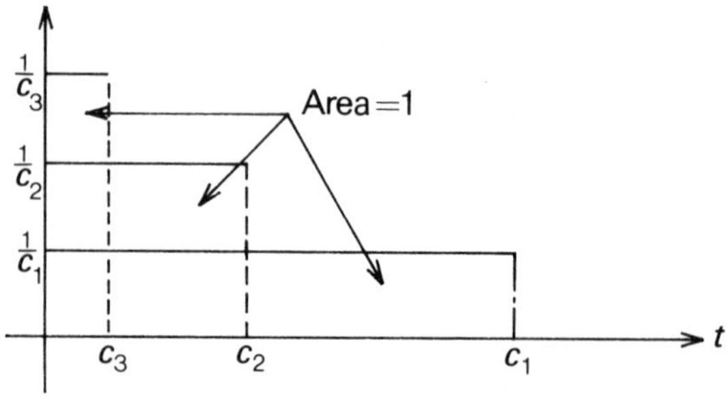

Fig. 2.4

Using the results of example 2.1 and equation 2.6

$$\begin{aligned}\mathcal{L}[\delta(t)] &= \lim_{c\to 0}\frac{1}{cs}(1-e^{-cs})\\ &= \lim_{c\to 0}\frac{\frac{\mathrm{d}}{\mathrm{d}c}(1-\mathrm{e}^{-cs})}{\frac{\mathrm{d}}{\mathrm{d}c}(cs)} = \text{(using l'Hopital's rule)}\\ &\lim_{c\to 0}\frac{s\mathrm{e}^{-cs}}{s} = 1.\end{aligned} \tag{2.7}$$

Notice that from the definition, the unit impulse function has infinite height, zero width (that is duration) and unit area. Although such a function connot exist, it is used as an approximation to a pulse input (such as a blow by a hammer). Mathematical manipulations with this 'non-existing' function can be made quite rigorous by using the following properties:

$$\delta(t-t_0) = \begin{cases} 0 & \text{for } t \neq t_0 \\ \infty & \text{for } t = t_0 \end{cases}$$

$$\int_0^{\infty} \delta(t)\mathrm{d}t = 1$$

$$\int_0^{\infty} \delta(t-a)g(t)\mathrm{d}t = g(a) \quad \text{for a continuous function } g(t).$$

(7) **Time scale change**

It is frequently convenient to normalise a time function so that the results are applicable to systems having similar properties. Suppose that $\mathcal{L}[f(t)] = F(s)$, and we wish to find $\mathcal{L}[f(\frac{t}{\alpha})]$ $(\alpha > 0)$.

$$\begin{aligned}\mathcal{L}[f(\tfrac{t}{\alpha})] &= \int_0^{\infty} f(\tfrac{t}{\alpha})\mathrm{e}^{-st}\,\mathrm{d}t \\ &= \int_0^{\infty} f(x)\mathrm{e}^{-s\alpha x}\,\alpha\mathrm{d}x \quad \text{where } x = \frac{t}{\alpha}. \\ &= \alpha F(\alpha s). \end{aligned} \tag{2.8}$$

Example 2.4

Find the Laplace transform of

(1) $f(t) = \mathrm{e}^{-3t}$

(2) $g(t) = \mathrm{e}^{-0.3t}$.

Solution

(1) $\mathcal{L}[\mathrm{e}^{-3t}] = \dfrac{1}{s+3} = F(s)$ (say) (see example 2.3).

(2) Since $g(t) = f(\frac{t}{10})$, $\mathcal{L}[\mathrm{e}^{-.3t}] = 10\,F(10s) = \dfrac{10}{10s+3} = \dfrac{1}{s+0.3}$ (see Fig. 2.5).

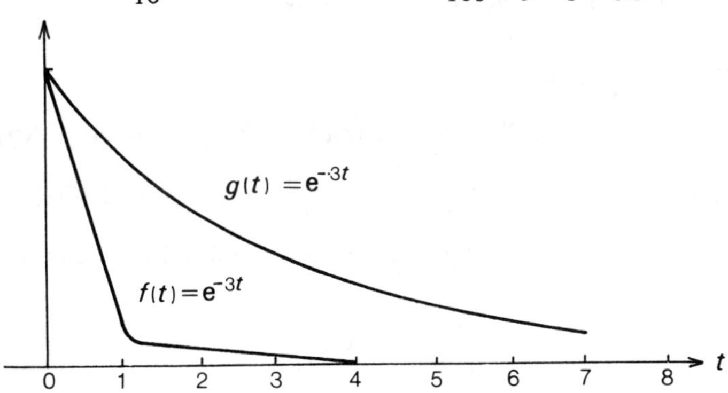

Fig. 2.5

Two theorems very useful for some practical calculations follow.

(8) **Initial Value Theorem**

This theorem makes it possible to find $f(0)$ directly from $F(s)$, without actually evaluating the inverse Laplace transform. The theorem states that provided

$$\lim_{s\to\infty} sF(s) \text{ exists, then}$$

$$f(0) = \lim_{s\to\infty} sF(s). \tag{2.9}$$

Since

$$\mathcal{L}[f'(t)] = sF(s) - f(0) \quad \text{(see equation (2.2))}$$

and $$\lim_{s\to\infty} \int_0^\infty f'(t)e^{-st}\,dt = 0$$

hence $0 = \lim_{s\to\infty} sF(s) - f(0)$, and (2.9) follows.

(9) **Final Value Theorem**

This theorem makes it possible to obtain $f(\infty)$ directly from $F(s)$. Provided the indicated limits exist

$$\lim_{t\to\infty} f(t) = f(\infty) = \lim_{s\to 0} sF(s). \tag{2.10}$$

Indeed, as above

$$\mathcal{L}[f'(t)] = \int_0^\infty f'(t)e^{-st}\,dt = sF(s) - f(0).$$

Now

$$\lim_{s\to 0} \int_0^\infty f'(t)e^{-st}\,dt = \int_0^\infty f'(t)dt = f(t)\Big|_0^\infty = f(\infty) - f(0).$$

Hence

$$f(\infty) - f(0) = \lim_{s\to 0} sF(s) - f(0), \quad \text{and (2.10) follows.}$$

Example 2.5

Verify the Initial and Final value theorems for the function $f(t) = 5e^{-3t}$.

Solution

$$F(s) = \frac{5}{s+3}.$$

(a) *Initial value theorem:*

$$f(0) = \lim_{s \to \infty} sF(s)$$

$$\lim_{s \to \infty} sF(s) = \lim_{s \to \infty} \frac{5s}{s+3} = 5$$

and $\lim_{t \to 0} f(t) = 5.$

(b) *Final value theorem*

$$f(\infty) = \lim_{s \to 0} sF(s) = \lim_{s \to 0} \frac{5s}{s+3} = 0$$

and $f(\infty) = \lim_{t \to \infty} f(t) = 0.$

These values of $f(0)$ and $f(\infty)$ are as expected by direct calculation from the function.

We are now in a position to compile a table of Laplace transform pairs. The reader may wish to prove some of the results included for himself. This table may be used in a similar way to a table of standard derivatives or integrals. It is used to obtain the Laplace transform of a time function, and conversely it can be used to obtain the time function corresponding to a Laplace transform, that is, the inverse Laplace transform.

$f(t)$	$F(s)$	$f(t)$	$F(s)$
$\delta(t)$ (unit impulse)	1	$\cos \alpha t$	$\frac{s}{s^2+\alpha^2}$
$u(t)$ (unit step)	$\frac{1}{s}$	$e^{-bt} \sin \alpha t$	$\frac{\alpha}{(s+b)^2+\alpha^2}$
t^n $(n = 1, 2, 3, \ldots)$	$\frac{n!}{s^{n+1}}$	$e^{-bt} \cos \alpha t$	$\frac{s+b}{(s+b)^2+\alpha^2}$
$e^{-\alpha t}$	$\frac{1}{s+\alpha}$	$\sinh \alpha t$	$\frac{\alpha}{s^2-\alpha^2}$
$t^n e^{-\alpha t}$ $(n = 1, 2, \ldots)$	$\frac{n!}{(s+\alpha)^{n+1}}$	$\cosh \alpha t$	$\frac{s}{s^2-\alpha^2}$
$\sin \alpha t$	$\frac{\alpha}{s^2+\alpha^2}$		

2.3 APPLICATIONS TO DIFFERENTIAL EQUATIONS

By taking the Laplace transform of a differential equation it is transformed into an algebraic one in the variable s. This equation is rearranged so that all the terms involving the dependent variable are on one side and the solution is obtained by taking the inverse transform as the following example illustrates. (We make use of a standard notation when differential equations are discussed

$$\mathrm{D}y = \frac{\mathrm{d}y}{\mathrm{d}t}, \quad \mathrm{D}^2y = \frac{\mathrm{d}^2y}{\mathrm{d}t^2}, \quad \text{etc.})$$

Example 2.6

Solve the following equation

$$\mathrm{D}^2y - 2\mathrm{D}y + y = \mathrm{e}^{3t}$$

subject to the initial conditions:

$$y(0) = 1 \quad \text{and} \quad y'(0) = 2.$$

Solution

Taking Laplace transform throughout the equation gives

$$\mathcal{L}[\mathrm{D}^2y] - 2\mathcal{L}[\mathrm{D}y] + \mathcal{L}[y] = \mathcal{L}[\mathrm{e}^{3t}], \quad \text{so that}$$

$$[s^2Y(s) - sy(0) - y'(0)] - 2[sY(s) - y(0)] + Y(s) = \frac{1}{s-3}$$

(where $Y(s) = \mathcal{L}[y(t)]$.)

Since $y(0) = 1$ and $y'(0) = 2$, we obtain

$$(s^2 - 2s + 1)Y(s) - s = \frac{1}{s-3}$$

giving

$$Y(s) = \frac{1}{(s-3)(s-1)^2} + \frac{s}{(s-1)^2}$$

$$= \frac{s^2 - 3s + 1}{(s-3)(s-1)^2}.$$

This expression for $Y(s)$ does not correspond to a form found in the table of Laplace transforms. On the other hand we can expand $Y(s)$ as a sum of partial fractions (see Appendix 1), that is as

$$Y(s) = \frac{\frac{3}{4}}{s-1} + \frac{\frac{1}{2}}{(s-1)^2} + \frac{\frac{1}{4}}{s-3}.$$

Each term on the right-hand side does correspond to a form in the table, so that we can now find the inverse transform term by term, that is

$$\mathcal{L}^{-1}[Y(s)] = \tfrac{3}{4}\mathcal{L}^{-1}\left[\frac{1}{s-1}\right] + \tfrac{1}{2}\mathcal{L}^{-1}\left[\frac{1}{(s-1)^2}\right] + \tfrac{1}{4}\mathcal{L}^{-1}\left[\frac{1}{s-3}\right]$$

where $\mathcal{L}^{-1}[f(s)]$ denotes the inverse Laplace transform of $f(s)$. Hence

$$y(t) = \tfrac{3}{4}e^t + \tfrac{1}{2}te^t + \tfrac{1}{4}e^{3t}.$$

2.4 TRANSFER FUNCTIONS

We have agreed in Chapter 1 that with approptiate approximations and using idealised components, we can often describe the dynamic behaviour of a system by a linear n^{th} order differential equation.

In general form the nth order system having a single input and a single output (Fig. 2.6) may have an associated differential equation

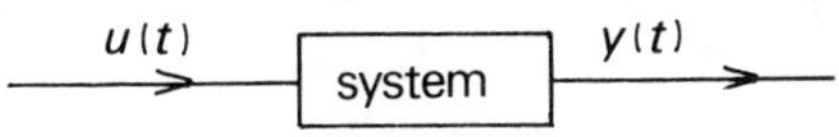

Fig. 2.6

$$a_0D^n y + a_1D^{n-1}y + \ldots + a_n y = b_0D^m u + b_1D^{m-1}u + \ldots + b_m u \qquad (2.11)$$

where $u(t)$ is the input and $y(t)$ is the output. In general, $n \geqslant m$, in which case the system is said to be **proper**, otherwise, if $m > n$, it is said to be **improper**. Once the input and the initial conditions $y(0)$, $y'(0), \ldots, y^{(n)}(0)$ are given, the output response $y(t)$ is found by solving (2.11).

In the previous section we have used Laplace transforms to solve differential equations. If we make the assumption that *all initial conditions are zero*, we obtain a great simplification in the calculations. With this assumption, the Laplace transform of (2.11) becomes

$$(a_0s^n + a_1s^{n-1} + \ldots + a_n)Y(s) = (b_0s^m + b_1s^{m-1} + \ldots + b_m)U(s).$$

Definition

The **transfer function** $G(s)$ of a linear autonomous system is the ratio of the Laplace transform of the output to the Laplace transform of the input, with the restriction that all initial conditions are zero.

From the above equation it follows that

$$G(s) = \frac{Y(s)}{U(s)} = \frac{b_0 s^m + b_1 s^{m-1} + \ldots + b_m}{a_0 s^m + a_1 s^{n-1} + \ldots + a_n}. \qquad (2.12)$$

The transfer function $G(s)$ depends only on the system parameters and is independent of the input.

Equation (2.12) can be written as

$$Y(s) = G(s)U(s) \qquad (2.13)$$

which shows that the transfer function $G(s)$ is always a factor of the output transform $Y(s)$, whatever input $u(t)$ is applied.

Example 2.7

Obtain the transfer function for a fluid tank having a cross-sectional area A with one inflow $q(t)$, a head $h(t)$ and with an outflow valve offering a resistance R.

Solution

The system can be described by the differential equation (1.7)

$$A\frac{\mathrm{d}h}{\mathrm{d}t} = q - \frac{1}{R}h$$

or

$$AR\frac{\mathrm{d}h}{\mathrm{d}t} + h = qR \quad \text{(where } A \text{ and } R \text{ are constants).}$$

On taking the Laplace transforms (with zero initial conditions)

$$ARsH(s) + H(s) = Q(s)R.$$

Hence

$$G(s) = \frac{H(s)}{Q(s)} = \frac{R}{1 + ARs}.$$

Notice that $G(s)$ is dependent only on the system characteristics (that is, on A and R).

Example 2.8

Find the transfer function of the system described by the equation

$$\mathrm{D}^2 y + 3\mathrm{D}y + 2y = u$$

where $u(t)$ is the input and $y(t)$ is the output.

Solution

Taking Laplace transforms of both sides (assuming zero initial conditions) we obtain

$$(s^2 + 3s + 2)Y(s) = U(s), \text{ hence}$$

$$G(s) = \frac{Y(s)}{U(s)} = \frac{1}{(s+1)(s+2)}.$$

Notice from equation (2.13) that if the input to the system is $u(t) = \delta(t)$, the unit impulse function, then

$$Y(s) = G(s) \quad (\text{since } U(s) = 1 \text{ in this case}).$$

On taking the inverse Laplace transform, we obtain

$$y(t) = g(t) \tag{2.14}$$

where $g(t) = \mathcal{L}^{-1}[G(s)]$ is called the **impulse response** or the **weighting function** for the system. In other words we can (at least theoretically) obtain the transfer function $G(s)$ of a system by taking the Laplace transform of its response to a unit impulse (the initial conditions are assumed to be zero).

We shall consider a system's response to other inputs in Chapter 4.

It is helpful to have a pictorial representation of the functions performed by each component of a system. One such representation, particularly useful when transfer functions are used, is called a block diagram (also see Fig. 1.8).

2.5 BLOCK DIAGRAMS

The basic block diagram usually indicates the transfer function of the component or the system in a **block**. Several blocks can be connected by arrows indicating the direction of signal flow.

The equation $Y(s) = G(s)U(s)$ is represented by Fig. 2.7. It shows the transform $U(s)$ of the input $u(t)$ flowing into the system having a transfer function $G(s)$. The Laplace transform $Y(s)$ of the output $y(t)$ is shown as the system outflow.

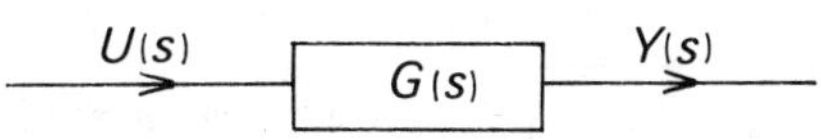

Fig. 2.7

Besides blocks, such diagrams also make use of **sensing** and **identity** devices. Typical sensing devices are shown in Fig. 2.8 (a) and (b).

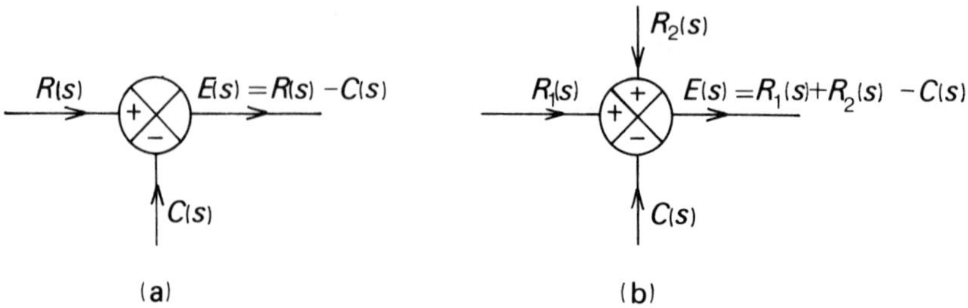

Fig. 2.8

For linear systems the sensing devices are used for the addition and/or subtraction of signals. Fig. 2.8 (a) shows a very common device - it is an **error detector**, which produces the **error signal** $E(s)$, the difference between a **reference signal** $R(s)$ and the **feedback signal** $C(s)$.

Fig. 2.9 shows an identity relation

$$Y(s) = Y_1(s) = Y_2(s)$$

by an **identity** diagram.

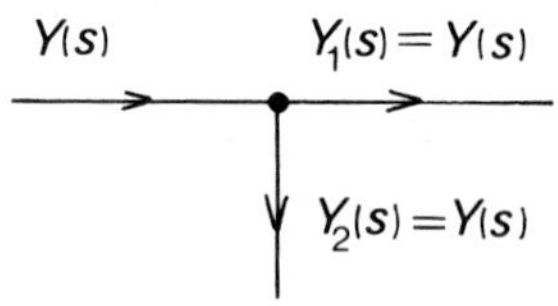

Fig. 2.9

Reduction of Block Diagrams

When a system consists of several components, each represented by a transfer function, we can in general reduce the corresponding block diagram to one represented in Fig. 2.7, which is known as the **system transfer function**.

In Fig. 2.10, the output from block **A** is $X(s) = A(s)U(s)$ where $X(s)$ is the input to block B.

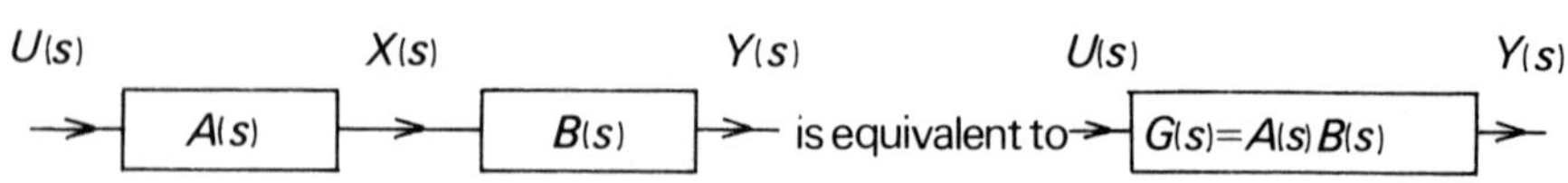

Fig. 2.10

The output from block B is $Y(s) = B(s)X(s) = B(s)A(s)U(s)$. Hence

$$\frac{Y(s)}{U(s)} = G(s) = A(s)B(s). \tag{2.15}$$

Blocks in Parallel

Fig. 2.11 shows a **negative** feedback loop. We have:

$$E(s) = R(s) - B(s)C(s)$$

$$C(s) = A(s)E(s)$$

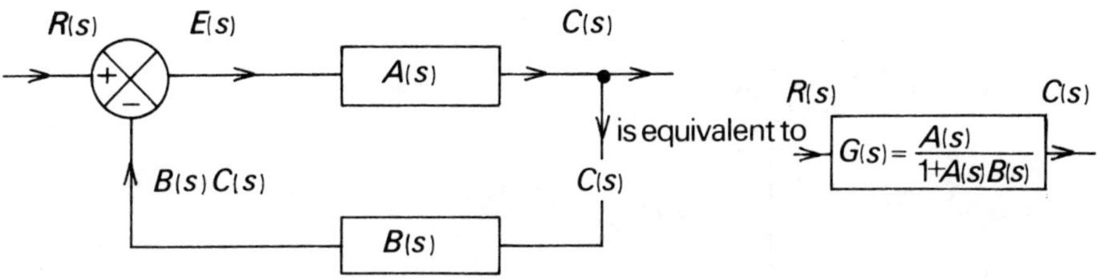

Fig. 2.11

On eliminating $E(s)$ between these two equations, we obtain

$$C(s) = A(s)R(s) - A(s)B(s)C(s)$$

so that

$$G(s) = \frac{C(s)}{R(s)} = \frac{A(s)}{1 + A(s)B(s)} \tag{2.16}$$

Example 2.9

Obtain the system transfer function for the positive feedback system shown in Fig. 2.12.

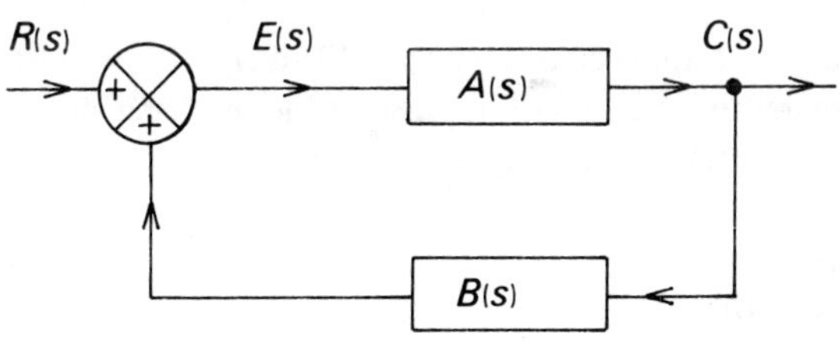

Fig. 2.12

Solution

$$E(s) = R(s) + B(s)C(s)$$

$$C(s) = A(s)E(s)$$

therefore

$$C(s)(1 - A(s)B(s)) = A(s)R(s) \quad \text{and}$$

$$G(s) = \frac{C(s)}{R(s)} = \frac{A(s)}{1 - A(s)B(s)}$$

A special and an important case of a feedback system is the **unity feedback system** shown in Fig. 2.13.

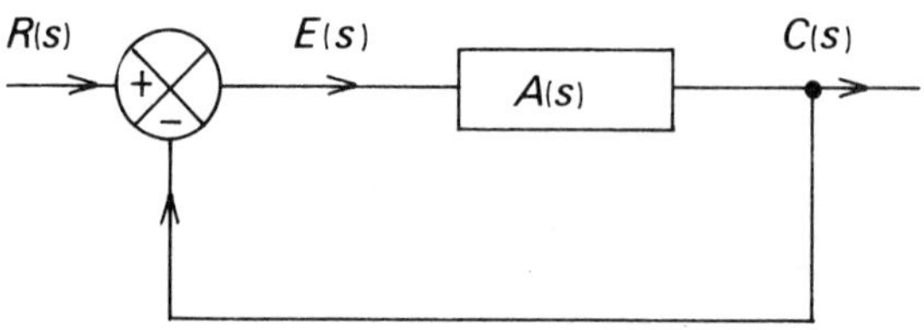

Fig. 2.13

From equation (2.16) with $B(s) = 1$, the system transfer function in this case is

$$G(s) = \frac{A(s)}{1 + A(s)}. \tag{2.17}$$

Example 2.10

Consider a system having the transfer function

$$A(s) = \frac{K}{s-2} \quad (K \text{ is constant}).$$

Find the time response of this system to a unit impulse.

A unity feedback is applied to $A(s)$, as in Fig. 2.13.
Find the time response to a unit impulse of the resulting closed loop system.
Comment on the stability of the two systems.

Solutions

Use equation (2.12), that is,

$$Y(s) = A(s)U(s), \quad U(s) = 1$$

hence

$$y(t) = Ke^{2t}.$$

For the closed loop system

$$G(s) = \frac{A(s)}{1 + A(s)} = \frac{K}{s + (K-2)}.$$

This time

$$y(t) = Ke^{-(K-2)t}.$$

Comment

In the first case it is seen that $y(t)$ increases rapidly as t increases (independently of K). The system having the transfer function $A(s)$ is unstable.

When the feedback is applied, it is seen that $y(t)$ will in fact decrease as t increases, so long as $K > 2$.

This example shows an interesting possible affect of feedback. Here we start with an unstable system, and on applying feedback, the resulting system is stable for an appropriate choice of the constant K. On the other hand, it is possible for $A(s)$ to be the transfer function of a stable system and yet the resulting closed loop system may be unstable. We in fact consider such a situation in Chapter 5 (see example 5.11). Indeed this is the reason why the analysis of system stability is so important in control theory.

PROBLEMS

(1) (i) Find (for $t \geqslant 0$) the Laplace transform of the ramp function $f(t) = t$.
(ii) Given that $\mathcal{L}\{f(t)\} = F(s)$, find $\mathcal{L}\{f(t)e^{-\alpha t}\}$
(iii) Find $\mathcal{L}\{te^{-\alpha t}\}$.

(2) (i) Given that $\mathcal{L}\{f(t)\} = F(s)$, find $\mathcal{L}\{tf(t)\}$
(ii) Find $\mathcal{L}\{t^n\}$.

(3) If $f(t) = \sin 2t$, find
(i) $\mathcal{L}\{f(t)\}$,
(ii) $\mathcal{L}\{f'(t)\}$ and
(iii) $\mathcal{L}\{\int_0^t f(u)du\}$.

(4) Find
(a) $\mathcal{L}\{t^3 e^{2t}\}$
(b) $\mathcal{L}\{e^{-3t} \sin 2t\}$
(c) $\mathcal{L}\{e^{-2t}(\cos 2t - 3 \sin 2t)\}$.

(5) Find
(i) $\mathcal{L}^{-1}\{\frac{1}{s^3}\}$

(ii) $\mathcal{L}^{-1}\{\frac{s}{s^2+4}\}$;

(iii) $\mathcal{L}^{-1}\{\frac{2s+5}{s^2-4s+5}\}$

(iv) $\mathcal{L}^{-1}\{\frac{1}{(s-2)^3}\}$

(v) $\mathcal{L}^{-1}\{\frac{e^{-3s}}{(s-2)^3}\}$.

(6) Use Laplace transforms to solve

(i) $$\frac{d^2y}{dt^2} - 2\frac{dy}{dt} + 2y = 3e^{2t}$$

given that when $t = 0$, $y = 1$ and $dy/dt = -1$.

(ii) $$\frac{d^2y}{dt^2} - \frac{dy}{dt} - 2y = \cos 2t$$

given that when $t = 0$, $y = 1$ and $dy/dt = 1$.

(7) The differential equation governing the behaviour of a system is

$$D^2y - 2Dy + y = Du + 3u$$

where $Dy \equiv dy/dt$.

Given that when $t = 0$, $y = y_0$ and $Dy = y'_0$, find the Laplace transform $Y(s)$ of the solution $y(t)$, indicating respectively the contributions due to the initial conditions and the input.

What is the system transfer function?

(8) (i) Find $\frac{E(s)}{R(s)}$ for the two systems shown in Fig. 2.14.

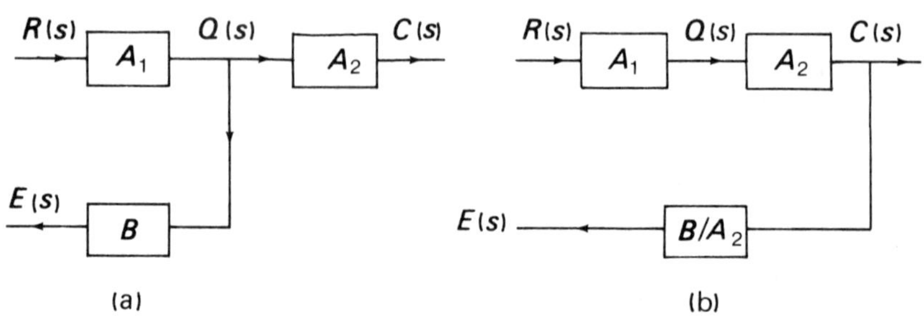

Fig. 2.14

(ii) Find the transfer function of the system shown in Fig. 2.15.

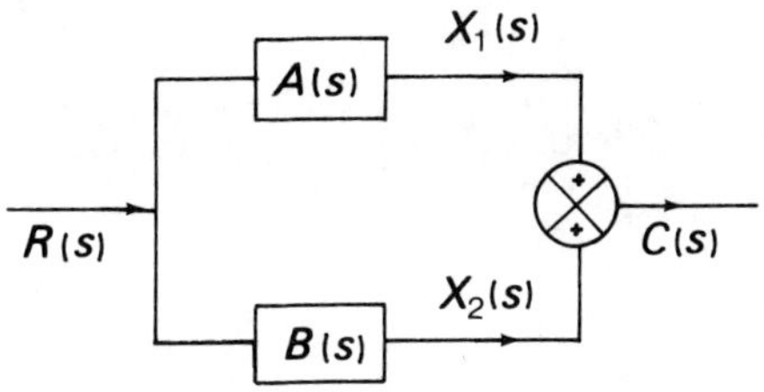

Fig. 2.15

(9) Find the transfer function of the systems shown in Figs 2.16 and 2.17

(i)

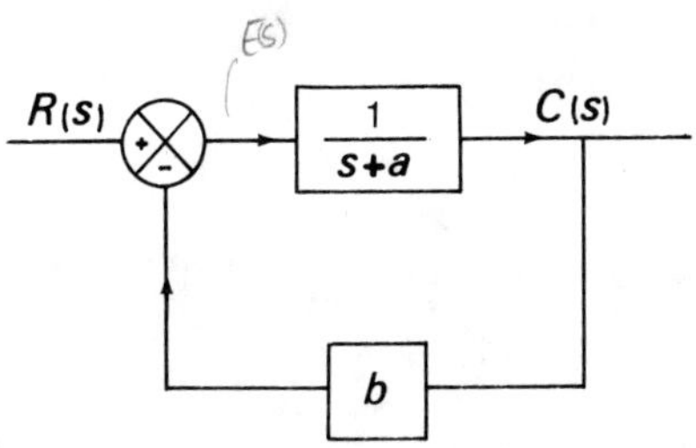

Fig. 2.16

(ii)

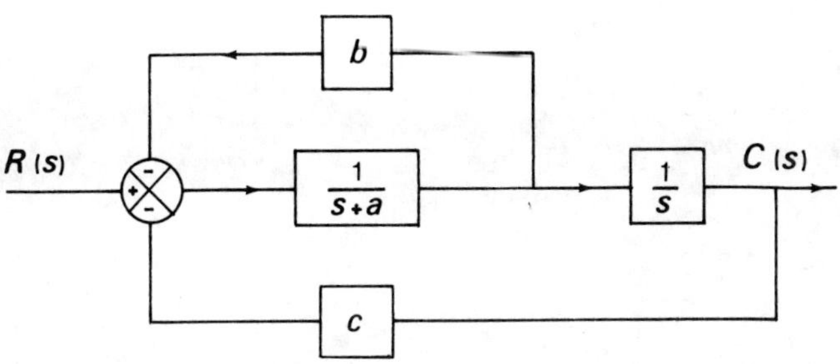

Fig. 2.17

CHAPTER 3

State-space formulation

3.1 INTRODUCTION

In the previous chapter we have discussed the characterisation of a system by its transfer function. It has already been mentioned (Chapter 1) that such a description is very convenient but that it suffers from several disadvantages; among them is the fact that all initial conditions are assumed to be zero. Since the behaviour of a system in the time domain is dependent on its past history, the transfer function description is not always adequate and we need to use the state-space characterisation.

Another advantage of the state-space approach is that it is applicable to multivariable systems (that is, multi-input, multi-output systems) in contrast to the transfer function techniques which are generally used for single-input, single-output systems.

For simplicity we begin by considering single-input single-output systems.

We have already considered a state-space representation of system equations in Chapter 1 (see equation 1.9).

3.2 STATE-SPACE FORMS

We now consider a more general case - a system characterised by an nth order differential equation. We shall also assume that the systems we are dealing with are **autonomous**, which implies that the *free* system (where the input is zero) does not depend explicitly on time.

Note: It is customary in modern control theory to use the notation

$$\dot{y} = \frac{\mathrm{d}y}{\mathrm{d}t},\ \ddot{y} = \frac{\mathrm{d}^2 y}{\mathrm{d}t^2}, \ldots, \overset{(n)}{y} = \frac{\mathrm{d}^n y}{\mathrm{d}t^n}.$$

The system equation has the form:

$$\overset{(n)}{y} + a_1 \overset{(n-1)}{y} + a_2 \overset{(n-2)}{y} + \ldots + a_{n-1}\dot{y} + a_n y = u; \tag{3.1}$$

it is assumed that

$$y(0), \dot{y}(0), \ldots, \overset{(n-1)}{y}(0) \quad \text{are known.}$$

If we define $x_1 = y$, $x_2 = \dot{y}, \ldots, x_n = \overset{(n-1)}{y}$, then we can write equation (3.1) as a system of n simultaneous differential equations, each of order 1, namely

$$\begin{aligned}
\dot{x}_1 &= x_2 \\
\dot{x}_2 &= x_3 \\
&\vdots \\
\dot{x}_{n-1} &= x_n
\end{aligned}$$

and from (3.1) $\dot{x}_n = -a_n x_1 - a_{n-1} x_2 - \ldots - a_1 x_n + u$

which can be written as a vector-matrix differential equation

$$\begin{bmatrix} \dot{x}_1 \\ \dot{x}_2 \\ \vdots \\ \dot{x}_{n-1} \\ \dot{x}_n \end{bmatrix} = \begin{bmatrix} 0 & 1 & \ldots & 0 & 0 \\ 0 & 0 & \ldots & 0 & 0 \\ \vdots & & & & \\ 0 & 0 & & 0 & 1 \\ -a_n & -a_{n-1} & & -a_2 & -a_1 \end{bmatrix} \begin{bmatrix} x_1 \\ x_2 \\ \vdots \\ x_{n-1} \\ x_n \end{bmatrix} + \begin{bmatrix} 0 \\ 0 \\ \vdots \\ 0 \\ 1 \end{bmatrix} u \qquad (3.2)$$

that is, as

$$\dot{\mathbf{x}} = A\mathbf{x} + Bu$$

where $\mathbf{x}$, A, and B are defined in equation (3.2). The output of the system is y, which was defined as x_1 above, and is written in matrix form as

$$y = [1 \;\; 0 \;\; 0 \;\ldots\; 0] \begin{bmatrix} x_1 \\ x_2 \\ \vdots \\ x_n \end{bmatrix} \qquad (3.3)$$

that is, as

$$y = C\mathbf{x}$$

where $C = [1 \;\; 0 \;\; 0 \ldots 0]$.

The combination of equations (3.2) and (3.3) in the form

$$\dot{\mathbf{x}} = A\mathbf{x} + Bu$$
$$y = Cx \tag{3.4}$$

are known as the state equations of the system considered. The matrix A in equation (3.2) is said to be in **companion** form.

The components of $\mathbf{x}$ are the state variables $x_1, x_2, \ldots, x_n$. They can be considered as the coordinates axes of an n-dimensional space called the **state space**. Any state of the system is represented by a point in the state space.

The choice of state variables is by no means unique; in fact, the choice is limitless. The choice made above leads to a matrix A in companion form which is not necessarily the most useful form. It is seen that $\dot{x}_n$ is a function of all the state variables. In many applications it is useful to have the state variables **decoupled**, that is, each $\dot{x}_i$ to be a function of x_i (and the input) only. This implies that the matrix A in the state equations is in a diagonal form.

Example 3.1

Obtain two forms of state equations of the system defined by

$$\dddot{y} - 2\ddot{y} + \dot{y} - 2y = u.$$

The matrix A corresponding to one of the forms should be diagonal.

Solution

(a) A in Companion Form

Using the state variables defined above, we have

$$x_1 = y, x_2 = \dot{y} \quad \text{and} \quad x_3 = \ddot{y}, \quad \text{then}$$

$$\dot{x}_1 = x_2$$

$$\dot{x}_2 = x_3$$

$$\dot{x}_3 = 2x_1 - x_2 + 2x_3 + u. \tag{3.5A}$$

Hence

$$\begin{bmatrix} \dot{x}_1 \\ \dot{x}_2 \\ \dot{x}_3 \end{bmatrix} = \begin{bmatrix} 0 & 1 & 0 \\ 0 & 0 & 1 \\ 2 & -1 & 2 \end{bmatrix} \begin{bmatrix} x_1 \\ x_2 \\ x_3 \end{bmatrix} + \begin{bmatrix} 0 \\ 0 \\ 1 \end{bmatrix} u \tag{3.5B}$$

and

$$y = [1 \quad 0 \quad 0] \begin{bmatrix} x_1 \\ x_2 \\ x_3 \end{bmatrix}.$$

(b) A in Diagonal Form

The manner of the choice of the variables made below may not be clear, but we hope that it will become so after having worked through Section 3.4. Let

$$\begin{aligned} x_1 &= \tfrac{1}{5}y + \tfrac{1}{5}\ddot{y}, \\ x_2 &= \tfrac{1}{5}(2+i)y - \tfrac{1}{2}i\dot{y} + \tfrac{1}{10}(-1+2i)\ddot{y} \\ x_3 &= \tfrac{1}{5}(2-i)y + \tfrac{1}{2}i\dot{y} - \tfrac{1}{10}(1+2i)\ddot{y} \end{aligned} \tag{3.6}$$

(where $i^2 = -1$).
Then

$$\begin{aligned} \dot{x}_1 &= \tfrac{2}{5}(\ddot{y} + y) + \tfrac{1}{5}u \quad \text{(using the fact that } \dddot{y} = 2\ddot{y} - \dot{y} + 2y + u\text{)} \\ &= 2x_1 + \tfrac{1}{5}y. \end{aligned}$$

Similarly,

$$\dot{x}_2 = ix_2 + \tfrac{1}{10}(-1+2i)u$$

and

$$\dot{x}_3 = -ix_3 - \tfrac{1}{10}(1+2i)u.$$

In vector-matrix form this is written as

$$\begin{bmatrix} \dot{x}_1 \\ \dot{x}_2 \\ \dot{x}_3 \end{bmatrix} = \begin{bmatrix} 2 & 0 & 0 \\ 0 & \mathrm{i} & 0 \\ 0 & 0 & -\mathrm{i} \end{bmatrix} \begin{bmatrix} x_1 \\ x_2 \\ x_3 \end{bmatrix} + \frac{1}{10} \begin{bmatrix} 2 \\ -1 + 2i \\ -1 - 2i \end{bmatrix} u \tag{3.7}$$

and (on addition of equations (3.6))

$$y = x_1 + x_2 + x_3,$$

that is,

$$y = [1 \quad 1 \quad 1] \begin{bmatrix} x_1 \\ x_2 \\ x_3 \end{bmatrix}.$$

For various purposes such as simulation procedures and system synthesis it is desirable to construct a block diagram representing equation (3.2). Ideally, to synthesise a dynamic system we would need components which would act as integrators and other components which would act as differentiators. In practice both types of components can be quite easily constructed using modified **summers** which are rather special amplifier circuits. Unfortunately differentiators have undesirable noise characteristics which make them useless for (analog computer) simulation purposes. There is here an analogy with integration and differentiation by numerical methods. Integration is an accurate procedure while differentiation is not. This fact will explain why block diagrams are constructed using only integrators, and never differentiators.

We shall now construct a block diagram of equation (3.5). Taking the Laplace transforms of equations (3.5A) we obtain:

$$\begin{aligned} Y = X_1 &= \tfrac{1}{s} X_2 \\ X_2 &= \tfrac{1}{s} X_3 \\ X_3 &= \tfrac{1}{s} [2X_1 - X_2 + 2X_3 + U] \end{aligned} \tag{3.8}$$

where

$$X_i = \mathcal{L}[x_i(t)] \; (i = 1, 2, 3), \quad U = \mathcal{L}[u(t)] \quad \text{and} \quad Y = \mathcal{L}[y(t)]$$

and the initial conditions are assumed to be all zero.

It will be remembered (see equation (2.5)) that $\frac{1}{s}$ represents an integration. The block diagram is shown in Fig. 3.1.

It will be seen that (for example) at the point A indicated in the figure – the outflow (X_3) does equal the inflow which is the right-hand side of the last of the equations (3.8).

We next consider – rather briefly, to see what is the form of the state equation – the case where the forcing function does involve derivative terms. We will see that this may introduce functions of the input into the output equation.

We consider the system equation

$$\overset{(n)}{y} + a_1 \overset{(n-1)}{y} + \ldots + a_{n-1} \dot{y} + a_n y = b_0 \overset{(n)}{u} + b_1 \overset{(n-1)}{u} + \ldots + b_{n-1} \dot{u} + b_n u. \tag{3.9}$$

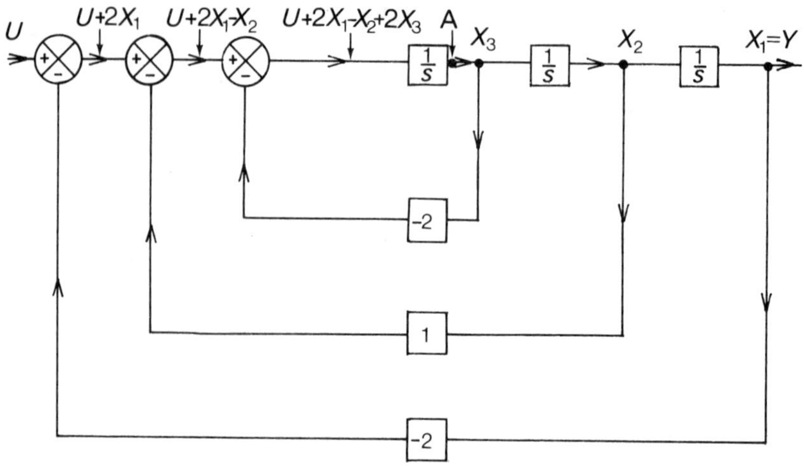

Fig. 3.1

We define

$$
\begin{aligned}
x_1 &= y - k_0 u \\
x_2 &= \dot{x}_1 - k_1 u \\
x_3 &= \dot{x}_2 - k_2 u \\
&\vdots \\
x_n &= \dot{x}_{n-1} - k_{n-1} u
\end{aligned}
\tag{3.10}
$$

where

$$
\begin{aligned}
k_0 &= b_0 \\
k_1 &= b_1 - a_1 b_0 \\
k_2 &= (b_2 - a_2 b_0) - a_1 k_1 \\
k_3 &= (b_3 - a_3 b_0) - a_2 k_1 - a_1 k_2 \\
&\vdots \\
k_n &= (b_n - a_n b_0) - a_{n-1} k_1 - a_{n-2} k_2 - \ldots - a_1 k_{n-1}.
\end{aligned}
\tag{3.11}
$$

(The algebraic manipulations involved are rather tedious, and the reader is advised not to waste time when considering this transformation, which in any case is not used much in this book and is only included here for completeness.)

The state equation of the system of equation (3.9) becomes

$$\begin{bmatrix} \dot{x}_1 \\ \dot{x}_2 \\ \vdots \\ \dot{x}_{n-1} \\ \dot{x}_n \end{bmatrix} = \begin{bmatrix} 0 & 1 & \dots & 0 \\ 0 & 0 & & 0 \\ \vdots & \vdots & & \vdots \\ 0 & 0 & & 1 \\ -a_n & -a_{n-1} & & -a_1 \end{bmatrix} \begin{bmatrix} x_1 \\ x_2 \\ \vdots \\ x_{n-1} \\ x_n \end{bmatrix} + \begin{bmatrix} k_1 \\ k_2 \\ \vdots \\ k_{n-1} \\ k_n \end{bmatrix} u \quad (3.12)$$

$$y = [1 \;\; 0 \;\; \dots \;\; 0] \begin{bmatrix} x_1 \\ x_2 \\ \vdots \\ x_n \end{bmatrix} + k_0 u,$$

that is,

$$\dot{\mathbf{x}} = A\mathbf{x} + Bu$$

$$y = C\mathbf{x} + Du \quad (3.13)$$

where A, B, C and D are indicated in equation (3.1 2).

Example 3.2

Obtain the state equation of the system defined by

$$\ddot{y} + 2\dot{y} - y = 2\ddot{u} - \dot{u} + 3u.$$

Solution

With the above notation, we have

$$a_1 = 2, \; a_2 = -1, \; b_0 = 2, \; b_1 = -1 \text{ and } b_2 = 3.$$

From equation (3.11)

$k_0 = 2$, $k_1 = -5$ and $k_2 = 15$, hence equation (3.12) becomes

$$\begin{bmatrix} \dot{x}_1 \\ \dot{x}_2 \end{bmatrix} = \begin{bmatrix} 0 & 1 \\ 1 & -2 \end{bmatrix} \begin{bmatrix} x_1 \\ x_2 \end{bmatrix} + \begin{bmatrix} -5 \\ 15 \end{bmatrix} u$$

$$y = [1 \quad 0]\begin{bmatrix} x_1 \\ x_2 \end{bmatrix} + 2u.$$

It is informative to check this.

Using equation (3.10), we obtain

$$x_1 = y - 2u \qquad \text{(a)}$$

and

$$x_2 = \dot{x}_1 + 5u. \qquad \text{(b)}$$

Equation (b) can be written as

$$x_2 = \dot{y} - 2\dot{u} + 5u.$$

On differentiating, this becomes

$$\begin{aligned} \dot{x}_2 &= \ddot{y} - 2\ddot{u} + 5\dot{u} \\ &= (2\ddot{u} - \dot{u} + 3u - 2\dot{y} + y) - 2\ddot{u} + 5\dot{u} \\ &\qquad \text{(using the original equation to obtain } \ddot{y}\text{)} \\ &= (y - 2u) - 2(\dot{y} - 2\dot{u}) + 5u \\ &= x_1 - 2\dot{x}_1 + 5u \qquad \text{(using equation (a))} \\ &= x_1 - 2(\dot{x}_1 + 5u) + 15u \\ &= x_1 - 2x_2 + 15u \qquad \text{(using equation (b)).} \end{aligned}$$

This is indeed the expression for $\dot{x}_2$ in the above vector-matrix equation.

Notice that if the order of the forcing function derivatives on the right-hand side of equation (3.9) is lower than the derivatives on the left-hand side - in the case of a *proper* system (see Section 2.3) - the state equations reduce to the form

$$\dot{\mathbf{x}} = A\mathbf{x} + Bu$$

$$y = C\mathbf{x}$$

(since $k_0 = b_0 = 0$). In general we consider only proper systems.

Up till now we have been discussing the state equations of systems having single input and single output (known as SISO) only. But there is no reason why a system should not have multiple inputs and multiple outputs (known as MIMO), in which case it is referred to as a **multivariable** system.

For example, the tank system in Fig. 1.6 could have three inflows q_1, q_2 and q_3 into tank (1) and two discharge valves offering resistances R_2 and R_3 from tank (2). It is clear that the corresponding simultaneous first order differential equations, corresponding to equations (1.7) and (1.8), are more complicated, but otherwise the mathematical problem is essentially the same.

In more general terms, the state equations of MIMO systems are of similar form to the ones for SISO systems, except that the input is now a vector **u** and the output is a vector **y**. So in its most general form, a MIMO system has state equations of the form

$$\begin{aligned} \dot{\mathbf{x}} &= A\mathbf{x} + B\mathbf{u} \\ \mathbf{y} &= C\mathbf{x} + D\mathbf{u} \end{aligned} \tag{3.14}$$

where A (the *system* matrix) is $n \times n$, B (the *input* matrix) is $n \times m$, C (the *output* matrix) is $r \times n$, and D is $r \times m$.
$\mathbf{x}(t)$ (the state vector) is $n \times 1$, $\mathbf{u}(t)$ (the *input* vector) is $m \times 1$ and $\mathbf{y}(t)$ (the *output* vector) is $r \times 1$.

As for SISO systems we usually consider the state equations in the form

$$\begin{aligned} \dot{\mathbf{x}} &= A\mathbf{x} + B\mathbf{u} \\ \mathbf{y} &= C\mathbf{x} \end{aligned}$$

3.3 USING THE TRANSFER FUNCTION TO DEFINE STATE VARIABLES

It is sometimes possible to define suitable state variables by considering the partial fraction expansion of the transfer function. For example, given the system differential equation

$$\ddot{y} + 3\dot{y} + 2y = \dot{u} + 3u$$

the corresponding transfer function is

$$G(s) = \frac{Y(s)}{U(s)} = \frac{s+3}{(s+1)(s+2)}$$

which can be written as

$$G(s) = \frac{2}{s+1} - \frac{1}{s+2}.$$

Hence $Y(s) = X_1(s) + X_2(s)$ where

$$X_1(s) = \frac{2U(s)}{s+1} \quad \text{and} \quad X_2(s) = \frac{-U(s)}{s+2}.$$

On taking the inverse Laplace transforms, we obtain

$$\dot{x}_1 = -x_1 + 2u \quad \text{and} \quad \dot{x}_2 = -2x_2 - u,$$

or

$$\begin{bmatrix} \dot{x}_1 \\ \dot{x}_2 \end{bmatrix} = \begin{bmatrix} -1 & 0 \\ 0 & -2 \end{bmatrix} \begin{bmatrix} x_1 \\ x_2 \end{bmatrix} + \begin{bmatrix} 2 \\ -1 \end{bmatrix} u$$

and

$$y = [1 \quad 1] \begin{bmatrix} x_1 \\ x_2 \end{bmatrix}.$$

We can of course make a different choice of state variables, for example

$$Y(s) = 2X_1(s) - X_2(s)$$

where

$$X_1(s) = \frac{U(s)}{s+1} \quad \text{and} \quad X_2(s) = \frac{U(s)}{s+2}.$$

then

$$\dot{x}_1 = -x_1 + u \quad \text{and} \quad \dot{x}_2 = -2x_2 + u.$$

Now the state equations are

$$\begin{bmatrix} \dot{x}_1 \\ \dot{x}_2 \end{bmatrix} = \begin{bmatrix} -1 & 0 \\ 0 & -2 \end{bmatrix} \begin{bmatrix} x_1 \\ x_2 \end{bmatrix} + \begin{bmatrix} 1 \\ 1 \end{bmatrix} u$$

$$y = [2 \quad -1] \begin{bmatrix} x_1 \\ x_2 \end{bmatrix}.$$

For both choices of the state variables, the system matrix is in a convenient diagonal form.

It would seem then that we have here a neat method for choosing suitable state variables resulting in the system matrix being in a diagonal form.

In fact for some systems there does not exist a choice of state variables leading to a diagonal form of the system matrix.

For example, consider the system equation

$$\dddot{y} + 5\ddot{y} + 8\dot{y} + 4y = \dot{u} + 3u.$$

The corresponding transfer function is

$$G(s) = \frac{Y(s)}{U(s)} = \frac{s+3}{(s+1)(s+2)^2}.$$

Even at this stage it is apparent that the squared term in the denominator of $G(s)$ is going to pose difficulties.

The partial fraction expansion of $G(s)$ (see Appendix 1) is

$$G(s) = \frac{2}{s+1} - \frac{2}{s+2} - \frac{1}{(s+2)^2}$$

so that we can choose

$$Y(s) = 2X_1(s) - 2X_2(s) - X_3(s)$$

where $\quad X_1(s) = \dfrac{U(s)}{s+1}, \; X_2(s) = \dfrac{U(s)}{s+2} \quad \text{and} \quad X_3(s) = \dfrac{U(s)}{(s+2)^2}.$

A block diagram illustrating the situation is given in Fig. 3.2.

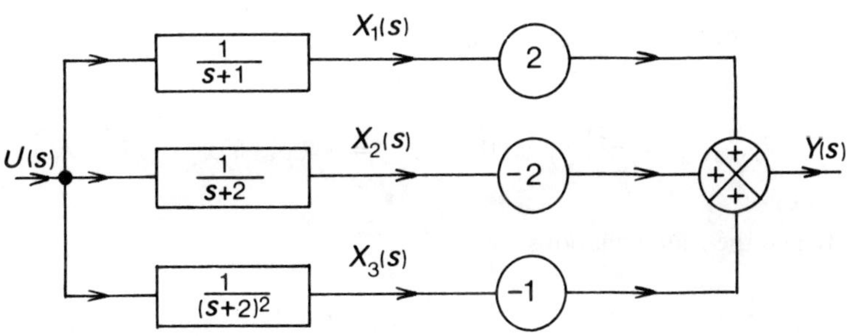

Fig. 3.2

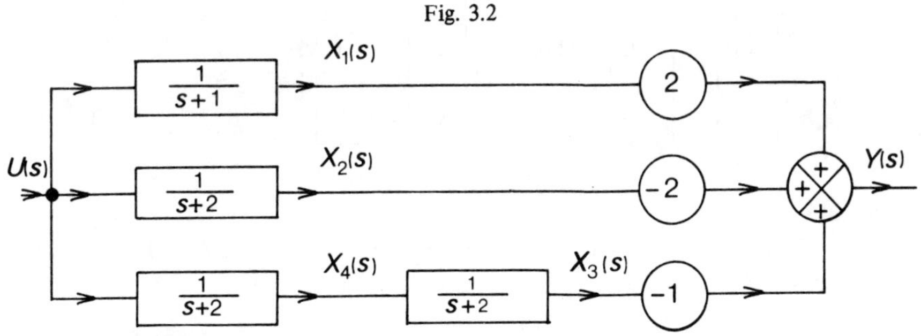

Fig. 3.3

The first two blocks in Fig. 3.2 are realisable with one integrator each, but the third block involving the squared term, needs two integrators as shown in Fig. 3.3

On inspection of Fig. 3.3 it is apparent that the signals $X_2(s)$ and $X_4(s)$ are identical, so that we can dispense with one of the integrator blocks, as shown in Fig. 3.4.

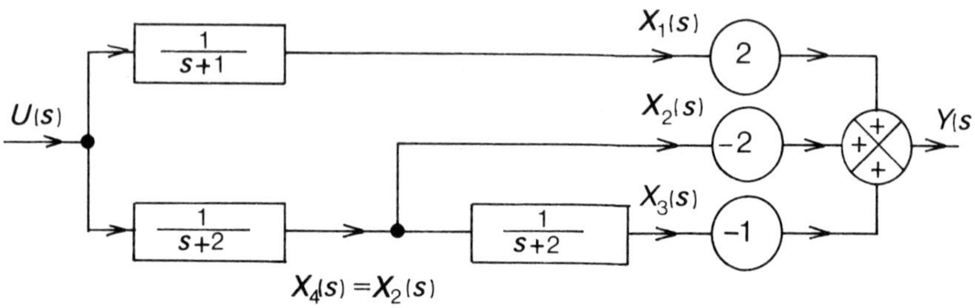

Fig. 3.4

From Fig. 3.4 we obtain

$$X_1(s) = \frac{U(s)}{s+1}, \quad \text{so that} \quad \dot{x}_1 = -x_1 + u$$

$$X_2(s) = \frac{U(s)}{s+2}, \quad \text{so that} \quad \dot{x}_2 = -2x_2 + u$$

$$X_3(s) = \frac{X_2(s)}{s+2}, \quad \text{so that} \quad \dot{x}_3 = -2x_3 + x_2 .$$

Hence the state equations are

$$\begin{bmatrix} \dot{x}_1 \\ \dot{x}_2 \\ \dot{x}_3 \end{bmatrix} = \begin{bmatrix} -1 & 0 & 0 \\ 0 & -2 & 0 \\ 0 & 1 & -2 \end{bmatrix} \begin{bmatrix} x_1 \\ x_2 \\ x_3 \end{bmatrix} + \begin{bmatrix} 1 \\ 1 \\ 0 \end{bmatrix} u$$

$$y = [2 \ -2 \ -1] \begin{bmatrix} x_1 \\ x_2 \\ x_3 \end{bmatrix} .$$

The matrix A is seen to be non-diagonal, and in fact no other choice of state variables will, in this case, result in it being in diagonal form.

Although the A matrix is not a diagonal one, it is still in a 'simple' or canonical form, in this case known as the **Jordan canonical form**. The partitioning of the matrix by the dashed lines results in there being two **Jordan blocks**, and the matrix has the form

$$A = \left[\begin{array}{c|c} J_1(-1) & 0 \\ \hline 0 & J_2(-2) \end{array}\right]$$

where $\quad J_1(-1) = [-1] \quad$ and $\quad J_2(-2) = \begin{bmatrix} -2 & 0 \\ 1 & -2 \end{bmatrix}.$

The suffix in J indicates the order of the matrix.

In general the Jordan block $J_n(\lambda)$ is a matrix of order $n \times n$ having each diagonal element equal to λ, each first infra-diagonal (that is the diagonal below the main diagonal) element equal to 1 and all other elements equal to 0. In fact there exists another Jordan form, and the reader interested in further details should consult [26].

In general, the presence of a factor of form $\dfrac{1}{(s+a)^n}$, where $n > 1$ in the transfer function $G(s)$ indicates coupling between the corresponding n state variables. In the above examples there is coupling between the variables x_2 and x_3 which implies that the time response of the system is a linear combination of the modes e^{-t} and (corresponding to the coupled variables) e^{-2t} and te^{-2t}.

Once the set of state variables is chosen, the system matrix A is defined. We can always find a transformation which will reduce A either to a diagonal form or, if that is impossible, to a Jordan block form. Such a transformation is of course equivalent to a new choice of state variables.

The Jordan canonical form arises only when the matrix A has multiple eigenvalues.

Although the results we establish in this book are valid for both diagonalisable and non-diagonalisable system matrices, the illustrative examples and indeed some of the proofs given are in terms of diagonalisable matrices only. This in no way implies that one type of system matrix is more important than the other. The reason is that we wish to keep our arguments as simple as possible, to concentrate on the principles and the concepts involved and not to cloud the issues with mathematical intricacies. We hope of course that the reader's curiosity will have been sufficiently aroused to work through proofs involving matrices with multiple eigenvalues either by himself or with the help of some of the literature mentioned in the *Bibliography*.

3.4 DIRECT SOLUTION OF THE STATE-EQUATION

By a **solution** to the state-equation

$$\dot{\mathbf{x}} = A\mathbf{x} + Bu \tag{3.15}$$

we mean finding $\mathbf{x}$ at any time t, given $u(t)$ (for all t) and the value of $\mathbf{x}$ at some specified time t_0, that is, given $\mathbf{x}(t_0) = \mathbf{x}_0$ (say). Since we are dealing with autonomous systems, there is no loss of generality in taking $t_0 = 0$. It is instructive to consider first the solution of the corresponding scalar differential equation

$$\dot{x} = ax + bu \tag{3.16}$$

given $x = x_0$ at $t = 0$.
The solution to equation (3.15) is found by an analogous method.

Multiplying the equation by the *integrating* factor e^{-at}, we obtain

$$e^{-at}(\dot{x} - ax) = e^{-at}bu$$

or
$$\frac{d}{dt}[e^{-at}x] = e^{-at}bu.$$

Integrating the result between 0 and t gives

$$e^{-at}x - x_0 = \int_0^t e^{-a\tau}bu(\tau)d\tau$$

that is,
$$x = e^{at}x_0 + \int_0^t e^{a(t-\tau)}bu(\tau)d\tau. \tag{3.17}$$

The first term on the right-hand side, dependent only on x_0, is called the **complementary function**, whereas the second term, dependent on the forcing function u, is called the **particular integral**.

Notice that the complementary function is the solution of equation (3.16) when the forcing function is zero.

Example 3.3

Solve the equation

$$\dot{x} + 3x = 4t$$

given that $x = 2$ when $t = 0$.

Solution

Since $a = -3$, $b = 4$ and $u = t$, on substituting into equation (3.17) we obtain

$$x = 2e^{-3t} + 4e^{-3t}\int_0^t e^{3\tau}\tau\,d\tau$$

$$= 2e^{-3t} + 4e^{-3t}(\tfrac{1}{3}te^{3t} - \tfrac{1}{9}e^{3t} + \tfrac{1}{9}) \text{ (on integrating by parts)}$$

$$= \tfrac{22}{9}e^{-3t} + \tfrac{4}{3}t - \tfrac{4}{9}.$$

To use this method to solve the vector matrix equation (3.15), we must first define a matrix function which is the analogue of the exponential, and we must also define the derivative and the integral of a matrix. Again we make use of analogy with scalar functions.

Definition

Since $$e^z = \sum_0^\infty \frac{1}{n!}z^n \quad \text{(all } z\text{)}$$

we define

$$e^A = \sum_0^\infty \frac{1}{n}A^n = A^0 + A + \frac{1}{2!}A^2 + \ldots$$

$$= I + A + \frac{1}{2!}A^2 + \ldots$$

(where A^0 is defined to be I) for every square matrix A.

Example 3.4

Evaluate e^{At} given $A = \begin{bmatrix} 1 & 0 & 0 \\ -1 & 0 & 0 \\ 1 & 1 & 1 \end{bmatrix}$.

Solution

We must calculate $A^2, A^3, \ldots$

$$A^2 = \begin{bmatrix} 1 & 0 & 0 \\ -1 & 0 & 0 \\ 1 & 1 & 1 \end{bmatrix} = A.$$

Hence $A^3 = A, A^4 = A \ldots$ (fortunately for us!!).

It follows that

$$e^{At} = I + At + \frac{1}{2!}A^2t^2 + \frac{1}{3!}A^3t^3 + \ldots$$

$$= I + A(t + \frac{1}{2}t^2 + \frac{1}{3!}t^3 + \ldots)$$

$$= I + A(e^t - 1)$$

$$= \begin{bmatrix} 1 & 0 & 0 \\ 0 & 1 & 0 \\ 0 & 0 & 1 \end{bmatrix} + \begin{bmatrix} e^t-1 & 0 & 0 \\ -e^t+1 & 0 & 0 \\ e^t-1 & e^t-1 & e^t-1 \end{bmatrix}$$

$$= \begin{bmatrix} e^t & 0 & 0 \\ 1-e^t & 1 & 0 \\ e^t-1 & e^t-1 & e^t \end{bmatrix}.$$

In fact, the evaluation of the matrix exponential is not quite as difficult as it may appear at first sight. We can make use of the Cayley-Hamilton theorem which assures us that every square matrix satisfies its own characteristic equation (see Appendix 2.5), indeed there may be a polynomial of lower degree (the minimal polynomial) satisfied by the matrix considered.

In the above example the minimal polynomial is

$$\lambda^2 - \lambda = 0$$

and indeed we found that

$$A^2 - A = 0 \quad \text{(the zero here is the null matrix).}$$

One direct and useful method is the following.

We know (see Appendix 2.3) that if A has distinct eigenvalues, say $\lambda_1, \lambda_2, \ldots, \lambda_n$, there exists a non-singular matrix P such that

$$P^{-1}AP = \begin{bmatrix} \lambda_1 & 0 & \dots & 0 \\ 0 & \lambda_2 & \dots & 0 \\ 0 & 0 & \dots & \lambda_n \end{bmatrix} = \text{diag}\{\lambda_1, \lambda_2, \dots, \lambda_n\} = \Lambda \text{ (say)}.$$

We then have

$$A = P\Lambda P^{-1}$$

so that

$$A^2 = (P\Lambda P^{-1})(P\Lambda P^{-1}) = P\Lambda(P^{-1}P)\Lambda P^{-1} = P\Lambda I \Lambda P^{-1} = P\Lambda^2 P^{-1}$$

and in general

$$A^r = P\Lambda^r P^{-1} \qquad (r = 1, 2, \dots).$$

If we consider a matrix polynomial, say $f(A) = A^2 - 2A + I$, we can write it as

$$\begin{aligned} f(A) &= P\Lambda^2 P^{-1} - 2P\Lambda P^{-1} + PIP^{-1} \\ &= P[\Lambda^2 - 2\Lambda + I]P^{-1} \\ &= Pf(\Lambda)P^{-1} = P \text{ diag}\{f(\lambda_1), f(\lambda_2), \dots, f(\lambda_n)\}P^{-1}. \end{aligned}$$

Since (for the above example)

$$\begin{aligned} f(\Lambda) &= \Lambda^2 - 2\Lambda + I \\ &= \begin{bmatrix} \lambda_1^2 & 0 & \dots & 0 \\ 0 & \lambda_2^2 & \dots & 0 \\ 0 & 0 & \dots & \lambda_n^2 \end{bmatrix} - 2\begin{bmatrix} \lambda_1 & 0 & \dots & 0 \\ 0 & \lambda_2 & \dots & 0 \\ 0 & 0 & \dots & \lambda_n \end{bmatrix} + \begin{bmatrix} 1 & 0 & \dots & 0 \\ 0 & 1 & \dots & 0 \\ 0 & 0 & \dots & 1 \end{bmatrix} \\ &= \begin{bmatrix} \lambda_1^2 - 2\lambda_1 + 1 & 0 & & 0 \\ 0 & \lambda_2^2 - 2\lambda_2 + 1 & \dots & 0 \\ 0 & 0 & \dots & \lambda_n^2 - 2\lambda_n + 1 \end{bmatrix} \\ &= \text{diag}\{f(\lambda_1), f(\lambda_2), \dots, f(\lambda_n)\}. \end{aligned}$$

The result holds for the general case when $f(x)$ is a polynomial of degree n.

In generalising the above, taking

$$f(A) = e^{At}, \quad \text{we obtain}$$

$$e^{At} = P \operatorname{diag}\{e^{\lambda_1 t}, e^{\lambda_2 t}, \ldots, e^{\lambda_n t}\} P^{-1}. \tag{3.18}$$

Example 3.5

Find e^{At} given $A = \begin{bmatrix} 0 & 2 \\ -1 & -3 \end{bmatrix}$.

Solution

The eigenvalues of A are $\lambda_1 = -1$ and $\lambda_2 = -2$. It can be verified that

$$P = \begin{bmatrix} 2 & 1 \\ -1 & -1 \end{bmatrix} \quad \text{and that} \quad P^{-1} = \begin{bmatrix} 1 & 1 \\ -1 & -2 \end{bmatrix}.$$

Using equation (3.18), we obtain

$$e^{At} = \begin{bmatrix} 2 & 1 \\ -1 & -1 \end{bmatrix} \begin{bmatrix} e^{-t} & 0 \\ 0 & e^{-2t} \end{bmatrix} \begin{bmatrix} 1 & 1 \\ -1 & -2 \end{bmatrix}$$

$$= \begin{bmatrix} 2e^{-t} - e^{-2t} & 2e^{-t} - 2e^{-2t} \\ e^{-2t} - e^{-t} & 2e^{-2t} - e^{-t} \end{bmatrix}.$$

From the above discussion it is clear that we can also write e^{At} in its **spectral** or **modal** form as

$$e^{At} = A_1 e^{\lambda_1 t} + A_2 e^{\lambda_2 t} + \ldots + A_n e^{\lambda_n t} \tag{3.19}$$

where $\lambda_1, \lambda_2, \ldots, \lambda_n$ are the eigenvalues of A and $A_1, A_2, \ldots, A_n$ are matrices of the same order as the matrix A.

In the above example, we can write

$$e^{At} = \begin{bmatrix} 2 & 2 \\ -1 & -1 \end{bmatrix} e^{-t} + \begin{bmatrix} -1 & -2 \\ 1 & 2 \end{bmatrix} e^{-2t}.$$

In the solution to the unforced system equation $\mathbf{x} = e^{At}\mathbf{x}_0$, the eigenvalues $\lambda_1, \ldots, \lambda_n$ are called **poles** and $e^{\lambda_1 t}, \ldots, e^{\lambda_n t}$ are called the **modes** of the system.

We now define the derivative and the integral of a matrix $A(t)$ whose elements are functions of t.

Definition

Let $\quad A(t) = [a_{ij}(t)]$, then

(1) $\quad \dfrac{\mathrm{d}}{\mathrm{d}t} A(t) = \dot{A}(t) = [\dfrac{\mathrm{d}}{\mathrm{d}t}(a_{ij})]$, and

(2) $\quad \int A(t)\mathrm{d}t = [\int a_{ij}(t)\mathrm{d}t]$,

that is, each element of the matrix is differentiated (or integrated).

For example, if

$$A = \begin{bmatrix} 6t & \sin 2t \\ t^2+2 & 3 \end{bmatrix} \text{ then}$$

$$\dot{A} = \begin{bmatrix} 6 & 2\cos 2t \\ 2t & 0 \end{bmatrix}, \text{ and } \int A\mathrm{d}t = \begin{bmatrix} 3t^2 & -\frac{1}{2}\cos 2t \\ \frac{1}{3}t^3+2t & 3t \end{bmatrix} + C$$

(where C is a constant matrix).

From the definition a number of rules follow;

If α and β are constants, and A and B are matrices,

(i) $\quad \dfrac{\mathrm{d}}{\mathrm{d}t}(\alpha A + \beta B) = \alpha\dot{A} + \beta\dot{B}$

(ii) $\quad \displaystyle\int_a^b (\alpha A + \beta B)\mathrm{d}t = \alpha\int_a^b A\mathrm{d}t + \beta\int_a^b B\mathrm{d}t$

(iii) $\quad \dfrac{\mathrm{d}}{\mathrm{d}t}(AB) = A\dot{B} + \dot{A}B$

(iv) If A is a constant matrix, then $\dfrac{\mathrm{d}}{\mathrm{d}t}(\mathrm{e}^{At}) = A\mathrm{e}^{At}$.

A word of warning – although the above rules are analogous to the rules for scalar functions, we must not be dulled into accepting for matrix functions all the rules valid for scalar functions.

For example, although

$$\frac{\mathrm{d}}{\mathrm{d}t}(x^n) = nx^{n-1}\dot{x}$$

in general, it is *not* true that

$$\frac{\mathrm{d}}{\mathrm{d}t}(A^n) = nA^{n-1}\dot{A}.$$

For example, when

$$A = \begin{bmatrix} t^2 & 2t \\ 0 & 3 \end{bmatrix}$$

then

$$\frac{\mathrm{d}}{\mathrm{d}t}[A^2] = \frac{\mathrm{d}}{\mathrm{d}t}\begin{bmatrix} t^4 & 2t^3+6t \\ 0 & 9 \end{bmatrix} = \begin{bmatrix} 4t^3 & 6t^2+6 \\ 0 & 0 \end{bmatrix}$$

and

$$2A\dot{A} = 2\begin{bmatrix} t^2 & 2t \\ 0 & 3 \end{bmatrix}\begin{bmatrix} 2t & 2 \\ 0 & 0 \end{bmatrix} = \begin{bmatrix} 4t^3 & 4t^2 \\ 0 & 0 \end{bmatrix}$$

so that $\frac{\mathrm{d}}{\mathrm{d}t}A^2 \neq 2A\dot{A}$.

We can now return to our original problem, to solve equation (3.15). We write the equation in the form;

$$\dot{\mathbf{x}} - A\mathbf{x} = Bu.$$

Pre-multiplying by the matrix e^{-At}, we obtain

$$\mathrm{e}^{-At}(\dot{\mathbf{x}} - A\mathbf{x}) = \mathrm{e}^{-At}Bu$$

or

$$\frac{\mathrm{d}}{\mathrm{d}t}[\mathrm{e}^{-At}\mathbf{x}] = \mathrm{e}^{-At}Bu.$$

On integration, this becomes

$$\mathrm{e}^{-At}\mathbf{x}(t) - \mathbf{x}(0) = \int_0^t \mathrm{e}^{-A\tau}Bu(\tau)\mathrm{d}\tau$$

so that

$$\mathbf{x}(t) = e^{At}\mathbf{x}(0) + \int_0^t e^{A(t-\tau)}Bu(\tau)d\tau \tag{3.20}$$

which is an analogous result to equation (3.17).

Example 3.6

A system is characterised by the state equation

$$\begin{bmatrix} \dot{x}_1 \\ \dot{x}_2 \end{bmatrix} = \begin{bmatrix} 0 & 2 \\ -1 & -3 \end{bmatrix}\begin{bmatrix} x_1 \\ x_2 \end{bmatrix} + \begin{bmatrix} 0 \\ 1 \end{bmatrix} u.$$

If the forcing function $u(t) = 1$ for $t \geqslant 0$ and $\mathbf{x}(0) = \begin{bmatrix} 1 \\ -1 \end{bmatrix}$ find the state $\mathbf{x}$ of the system at time t.

Solution

We have already evaluated e^{At} in Example 3.5. It follows that

$$e^{At}\mathbf{x}(0) = \begin{bmatrix} 2e^{-t} - e^{-2t} & 2e^{-t} - 2e^{-2t} \\ e^{-2t} - e^{-t} & 2e^{-2t} - e^{-t} \end{bmatrix}\begin{bmatrix} 1 \\ -1 \end{bmatrix}$$

$$= e^{-2t}\begin{bmatrix} 1 \\ -1 \end{bmatrix}.$$

(Notice that the initial condition $\mathbf{x}(0)$ is such that it eliminates the mode e^{-t} from the unforced system solution.)

The second term of equation (2.36) is

$$\int_0^t e^{A(t-\tau)}Bu(\tau)d\tau = \int_0^t e^{A(t-\tau)}\begin{bmatrix} 0 \\ 1 \end{bmatrix} d\tau$$

$$= \int_0^t \begin{bmatrix} 2e^{-(t-\tau)} - 2e^{-2(t-\tau)} \\ 2e^{-2(t-\tau)} - e^{-(t-\tau)} \end{bmatrix} d\tau$$

$$= \begin{bmatrix} 1 - 2e^{-t} + e^{-2t} \\ e^{-t} - e^{-2t} \end{bmatrix}.$$

Hence the state of the system at time t is

$$\mathbf{x}(t) = \begin{bmatrix} e^{-2t} \\ -e^{-2t} \end{bmatrix} + \begin{bmatrix} 1 - 2e^{-t} + e^{-2t} \\ e^{-t} - e^{-2t} \end{bmatrix}$$

$$= \begin{bmatrix} 1 - 2e^{-t} + 2e^{-2t} \\ e^{-t} - 2e^{-2t} \end{bmatrix}.$$

The matrix e^{At} in the solution equation (3.20) is of special interest to control engineers; they call it the **state-transition matrix** and denote it by $\Phi(t)$, that is,

$$\Phi(t) = e^{At}. \tag{3.21}$$

For the unforced system (when $u(t) = 0$) the solution (equation (3.20)) becomes

$$\mathbf{x}(t) = \Phi(t)\,\mathbf{x}(0)$$

so that $\Phi(t)$ transforms the system from its state $\mathbf{x}(0)$ at some initial time $t = 0$ to the state $\mathbf{x}(t)$ at some subsequent time t – hence the name given to the matrix.

Since

$$e^{At}\,e^{-(At)} = I, \quad \text{it follows that}$$

$$[e^{At}]^{-1} = e^{-At}.$$

Hence $\quad \Phi^{-1}(t) = e^{-At} = \Phi(-t).$

Also $\quad \Phi(t)\,\Phi(-\tau) = e^{At}e^{-A\tau} = e^{A(t-\tau)} = \Phi(t-\tau).$

With this notation equation (3.20) becomes

$$\mathbf{x}(t) = \Phi(t)\mathbf{x}(0) + \int_0^t \Phi(t-\tau)Bu(\tau)\mathrm{d}\tau. \tag{3.22}$$

3.5 SOLUTION OF THE STATE-EQUATION BY LAPLACE TRANSFORMS

Since the state equation is in a vector matrix form we must first define the Laplace transform of a vector.

Definition

Let $\mathbf{x}(t) = \begin{bmatrix} x_1 \\ x_2 \\ \vdots \\ x_n \end{bmatrix}$.

We define $\mathcal{L}[\mathbf{x}(t)] = \begin{bmatrix} \mathcal{L}[x_1(t)] \\ \mathcal{L}[x_2(t)] \\ \vdots \\ \mathcal{L}[x_n(t)] \end{bmatrix} = \begin{bmatrix} X_1(s) \\ X_2(s) \\ \vdots \\ X_n(s) \end{bmatrix} = X(s)$

From this definition we can find all the results we need. For example,

$$\mathcal{L}[\dot{\mathbf{x}}(t)] = \begin{bmatrix} \mathcal{L}[\dot{x}_1(t)] \\ \mathcal{L}[\dot{x}_2(t)] \\ \vdots \\ \mathcal{L}[\dot{x}_n(t)] \end{bmatrix} = \begin{bmatrix} sX_1(s) - x_1(0) \\ sX_2(s) - x_2(0) \\ \vdots \\ sX_n(s) - x_n(0) \end{bmatrix} \qquad \text{(by equation (2.2))}$$

$$= s\mathbf{X}(s) - \mathbf{x}(0).$$

Now we can solve the state equation (3.15). Taking the transform of

$$\dot{\mathbf{x}} = A\mathbf{x} + Bu$$

we obtain

$$s\mathbf{X}(s) - \mathbf{x}(0) = AX(s) + BU(s)$$

where $\quad U(s) = \mathcal{L}[u(t)]$

or $\quad [sI - A]\mathbf{X}(s) = \mathbf{x}(0) + BU(s).$

Unless s happens to be an eigenvalue of A, the matrix $[sI - A]$ is non-singular, so that the above equation can be solved giving

$$\mathbf{X}(s) = [sI - A]^{-1}\mathbf{x}(0) + [sI - A]^{-1}BU(s) \qquad (3.23)$$

and the solution $\mathbf{x}(t)$ is found by taking the inverse Laplace transform of equation (3.23).

Definition

$[sI - A]^{-1}$ is called the **resolvent** matrix of the system.

On comparing equations (3.22) and (3.23) we find that

$$\Phi(t) = \mathcal{L}^{-1}\{[sI - A]^{-1}\}.$$

Not only is the use of transforms a relatively simple method for evaluating the transition matrix, but indeed it allows us to calculate the state $\mathbf{x}(t)$ without having to evaluate integrals (since the tables of inverse transforms have already done this for us).

Example 3.7

Use Laplace transforms to evaluate the state $\mathbf{x}(t)$ of the system described in Example 3.6.

Solution

For this system

$$[sI - A] = \begin{bmatrix} s & -2 \\ 1 & s+3 \end{bmatrix}, \quad \text{so that}$$

$$[sI - A]^{-1} = \frac{1}{s(s+3)+2}\begin{bmatrix} s+3 & 2 \\ -1 & s \end{bmatrix} = \begin{bmatrix} \dfrac{s+3}{(s+1)(s+2)} & \dfrac{2}{(s+1)(s+2)} \\ \dfrac{-1}{(s+1)(s+2)} & \dfrac{s}{(s+1)(s+2)} \end{bmatrix}.$$

To evaluate the inverse transform, we must express each element of the matrix as a partial fraction, that is, as

$$[sI - A]^{-1} = \begin{bmatrix} \dfrac{2}{s+1} - \dfrac{1}{s+2} & \dfrac{2}{s+1} - \dfrac{2}{s+2} \\ \dfrac{-1}{s+1} + \dfrac{1}{s+2} & \dfrac{-1}{s+1} + \dfrac{2}{s+2} \end{bmatrix}$$

so that $$\mathcal{L}^{-1}\{[sI - A]^{-1}\} = \begin{bmatrix} 2e^{-t} - e^{-2t} & 2e^{-t} - 2e^{-2t} \\ e^{-2t} - e^{-t} & 2e^{-2t} - e^{-t} \end{bmatrix} = \Phi(t).$$

Hence the complementary function is as in Example 3.6. For the particular integral, we note that since

$$\mathcal{L}[u(t)] = \frac{1}{s}, \quad \text{then}$$

$$[sI - A]^{-1}BU(s) = \frac{1}{(s+1)(s+2)}\frac{1}{s}\begin{bmatrix} s+3 & 2 \\ -1 & s \end{bmatrix}\begin{bmatrix} 0 \\ 1 \end{bmatrix}$$

$$= \begin{bmatrix} \dfrac{2}{s(s+1)(s+2)} \\ \dfrac{1}{(s+1)(s+2)} \end{bmatrix} = \begin{bmatrix} \dfrac{1}{s} - \dfrac{2}{s+1} + \dfrac{1}{s+2} \\ \dfrac{1}{s+1} - \dfrac{1}{s+2} \end{bmatrix}.$$

On taking the inverse Laplace transform, we obtain

$$\begin{bmatrix} 1 - 2e^{-t} + e^{-2t} \\ e^{-t} - e^{-2t} \end{bmatrix}$$

which is the particular integral part of the solution obtained in Example 3.6.

3.6 THE TRANSFORMATION FROM THE COMPANION TO THE DIAGONAL STATE FORM

It was shown in Section 3.2 that the choice of the state vector is not unique.

We now assume that with one choice of the state vector the state equations are

$$\dot{\mathbf{x}} = A\mathbf{x} + B\mathbf{u} \tag{3.24}$$

$$\mathbf{y} = C\mathbf{x}$$

where A is a matrix of order $n \times n$ and B and C are matrices of appropriate order (see equation (3.14)).

Consider any non-singular matrix T of order $n \times n$. Let

$$\mathbf{x} = T\mathbf{z} \tag{3.25}$$

then $\mathbf{z}$ is also a state vector and equation (3.24) can be written as

$$T\dot{\mathbf{z}} = AT\mathbf{z} + B\mathbf{u}$$

$$\mathbf{y} = CT\mathbf{z}$$

or as

$$\dot{\mathbf{z}} = A_1\mathbf{z} + B_1\mathbf{u} \tag{3.26}$$

$$\mathbf{y} = C_1\mathbf{z}$$

where $A_1 = T^{-1}AT$, $B_1 = T^{-1}B$ and $C_1 = CT$.

The transformation defined by equation (3.25) is called a **state-transformation**, and the matrices A and A_1 are **similar** (see Appendix 2.3). We are particularly interested in the transformation when A_1 is diagonal (usually denoted by Λ) and A is in the companion form (see equation (3.2)). It is assumed that the matrix A has distinct eigenvalues $\lambda_1, \lambda_2, \ldots, \lambda_n$.

Corresponding to the eigenvalue λ_i there is the eigenvector $\mathbf{x}_i$ such that

$$A\mathbf{x}_i = \lambda_i\mathbf{x}_i \quad (i = 1, 2, \ldots, n). \tag{3.27}$$

We define the matrix V whose columns are the eigenvectors $\mathbf{x}_1, \mathbf{x}_2, \ldots, \mathbf{x}_n$, that is

$$V = [\mathbf{x}_1, \mathbf{x}_2, \ldots, \mathbf{x}_n].$$

V is called the **modal** matrix, it is non-singular (see [26]) and can be used as the transformation matrix T above. We can write the n equations defined by equation (3.27) as

$$AV = \Lambda V \tag{3.28}$$

where

$$\Lambda = \begin{bmatrix} \lambda_1 & 0 & \ldots & 0 \\ 0 & \lambda_2 & \ldots & 0 \\ 0 & 0 & \ldots & \lambda_n \end{bmatrix} = \text{diag}\,\{\lambda_1, \lambda_2, \ldots, \lambda_n\}.$$

From equation (3.28) we obtain

$$\Lambda = V^{-1}AV. \tag{3.29}$$

The matrix A has the form (companion form)

$$A = \begin{bmatrix} 0 & 1 & \dots & 0 \\ 0 & 0 & \dots & 0 \\ \vdots & \vdots & & \vdots \\ 0 & 0 & & 1 \\ -a_0 & -a_1 & \dots & -a_{n-1} \end{bmatrix}$$

so that its characteristic equation is (see Appendix 2.3)

$$|\lambda I - A| = \lambda^n + a_{n-1}\lambda^{n-1} + \dots + a_1\lambda + a_0 = 0.$$

By solving this equation we obtain the eigenvalues $\lambda_1, \lambda_2, \dots, \lambda_n$. The corresponding eigenvectors have an interesting form. Consider one of the eigenvalues λ and the corresponding eigenvector

$$\mathbf{x} = \begin{bmatrix} \alpha_1 \\ \alpha_2 \\ \vdots \\ \alpha_n \end{bmatrix}.$$

Then the equation $A\mathbf{x} = \lambda\mathbf{x}$, corresponds to the system of equations

$$\begin{aligned} \alpha_2 &= \lambda\alpha_1 \\ \alpha_3 &= \lambda\alpha_2 \\ &\vdots \\ \alpha_n &= \lambda\alpha_{n-1}. \end{aligned}$$

Setting $\alpha_1 = 1$, we obtain $\mathbf{x} = \begin{bmatrix} 1 \\ \lambda \\ \lambda^2 \\ \vdots \\ \lambda^{n-1} \end{bmatrix}.$

Hence the modal matrix in this case takes the form

$$V = \begin{bmatrix} 1 & 1 & \dots & 1 \\ \lambda_1 & \lambda_2 & & \lambda_n \\ \vdots & \vdots & & \vdots \\ \lambda_1^{n-1} & \lambda_2^{n-1} & & \lambda_n^{n-1} \end{bmatrix} \tag{3.30}$$

In this form the matrix V is called a **Vandermonde** matrix; it is non-singular since it has been assumed that $\lambda_1, \lambda_2, \ldots, \lambda_n$ are distinct. Special methods exist for inverting a Vandermonde matrix – the interested reader should consult [26]. Of course, any method for matrix inversion can be used to evaluate V^{-1}. We now consider the problem, first discussed in Example 3.1, to obtain the transformation from the companion form to the diagonal form.

Example 3.8

Having chosen the state vector (see Example 3.1) so that

$$\dddot{y} - 2\ddot{y} + \dot{y} - 2y = u$$

is written as the state equation (in companion form)

$$\begin{bmatrix} \dot{x}_1 \\ \dot{x}_2 \\ \dot{x}_3 \end{bmatrix} = \begin{bmatrix} 0 & 1 & 0 \\ 0 & 0 & 1 \\ 2 & -1 & 2 \end{bmatrix} \begin{bmatrix} x_1 \\ x_2 \\ x_3 \end{bmatrix} + \begin{bmatrix} 0 \\ 0 \\ 1 \end{bmatrix} u$$

$$y = [1 \quad 0 \quad 0] \begin{bmatrix} x_1 \\ x_2 \\ x_3 \end{bmatrix},$$

find the state transformation which will transform this into a state equation with A in diagonal form.

Solution

The characteristic equation of A is

$$|\lambda I - A| = \lambda^3 - 2\lambda^2 + \lambda - 2 = 0$$

that is, $(\lambda - 2)(\lambda^2 + 1) = 0$

hence $\lambda_1 = 2, \lambda_2 = i$ and $\lambda_3 = -i$.

From equation (3.30) the modal matrix is

$$V = \begin{bmatrix} 1 & 1 & 1 \\ 2 & i & -i \\ 4 & -1 & -1 \end{bmatrix}$$

and its inverse can be shown to be

$$V^{-1} = \frac{1}{20}\begin{bmatrix} 4 & 0 & 4 \\ 8+4i & -10i & -2+4i \\ 8-4i & 10i & -2-4i \end{bmatrix}.$$

The transformation is defined by equation (3.25), that is,

$$\mathbf{x} = V\mathbf{z} \quad \text{or} \quad \mathbf{z} = V^{-1}\mathbf{x}.$$

The original choice for **x** (see Example 3.1) was

$$\mathbf{x} = [y,\ \dot{y},\ \ddot{y}]^{-}$$

so the transformation $V^{-1}\mathbf{x}$ gives the new choice of the state vector

$$\begin{bmatrix} z_1 \\ z_2 \\ z_3 \end{bmatrix} = \frac{1}{20}\begin{bmatrix} 4 & 0 & 4 \\ 8+4i & -10i & -2+4i \\ 8-4i & 10i & -2-4i \end{bmatrix}\begin{bmatrix} y \\ \dot{y} \\ \ddot{y} \end{bmatrix}$$

which, written out in its components, is

$$z_1 = \tfrac{1}{5}y + \tfrac{1}{5}\ddot{y}$$

$$z_2 = \tfrac{1}{5}(2+i)y - \tfrac{1}{2}i\dot{y} + \tfrac{1}{10}(-1+2i)\ddot{y}$$

$$z_3 = \tfrac{1}{5}(2-i)y + \tfrac{1}{2}i\dot{y} - \tfrac{1}{10}(1+2i)\ddot{y}.$$

So we obtain the state vector already defined in equation (3.6). Notice that z_3 is the complex conjugate of z_2.

The state equations are now in the form of equation (3.26), that is,

$$\dot{\mathbf{z}} = A_1\mathbf{z} + B_1 u$$

$$y = C_1\mathbf{z}$$

where

$$A_1 = V^{-1}AV = \text{diag}\,\{2, i, -i\}$$

$$B_1 = V^{-1}B = \tfrac{1}{10}\begin{bmatrix} 2 \\ -1+2i \\ -1-2i \end{bmatrix}$$

$$C_1 = CV = [1 \quad 1 \quad 1]$$

(see equation (3.7)).

We can also obtain a transformation transforming a system having a state matrix A into one with the state matrix in a companion form. Because this transformation is only possible when the system is controllable, we defer the discussion until Chapter 6.

3.7 THE TRANSFER FUNCTION FROM THE STATE EQUATION

We have already defined the transfer function of a system as the ratio of the Laplace transform of the input to the Laplace transform of the output when the initial conditions are all zero.

Taking the Laplace transforms of equation (3.24), we obtain

$$sX(s) = AX(s) + BU(s)$$

$$Y(s) = CX(s).$$

From the first of these equations, we obtain;

$$X(s) = [sI - A]^{-1}BU(s)$$

(compare this with equation (3.23) with $\mathbf{x}(0) = \mathbf{0}$).
So that, from the second equation,

$$G(s) = \frac{Y(s)}{U(s)} = C[sI - A]^{-1}B \tag{3.31}$$

Example 3.9

Calculate the transfer function for the system whose state equations are

$$\dot{\mathbf{x}} = \begin{bmatrix} 1 & -2 \\ 3 & -4 \end{bmatrix} \mathbf{x} + \begin{bmatrix} 1 \\ 2 \end{bmatrix} u$$

$$y = [1 \quad -2]\mathbf{x}.$$

Solution

$$G(s) = [1 \quad -2] \begin{bmatrix} s-1 & 2 \\ -3 & s+4 \end{bmatrix}^{-1} \begin{bmatrix} 1 \\ 2 \end{bmatrix}$$

$$= [1 \quad -2] \begin{bmatrix} \dfrac{s+4}{(s+1)(s+2)} & \dfrac{-2}{(s+1)(s+2)} \\ \dfrac{3}{(s+1)(s+2)} & \dfrac{s-2}{(s+1)(s+2)} \end{bmatrix} \begin{bmatrix} 1 \\ 2 \end{bmatrix}.$$

$$= -\frac{(3s+2)}{(s+1)(s+2)}.$$

Example 3.10

Given that the system

$$\dot{\mathbf{x}} = A\mathbf{x} + B\mathbf{u}$$

$$\mathbf{y} = C\mathbf{x}$$

has transfer function $G(s)$, find the transrer function of the system

$$\dot{\mathbf{z}} = T^{-1}AT\mathbf{z} + T^{-1}B\mathbf{u}$$

$$\mathbf{y} = CT\mathbf{z}.$$

Solution

If the transfer function is $G_1(s)$

$$G_1(s) = CT[sI - T^{-1}AT]^{-1}T^{-1}B$$

$$= CT\{T^{-1}[sI - A]T\}^{-1}T^{-1}B$$

$$= CT\{T^{-1}[sI - A]^{-1}T\}T^{-1}B$$

(using the relation $[ABC]^{-1} = C^{-1}B^{-1}A^{-1}$)

$$= C[sI - A]^{-1}B$$

$$= G(s).$$

so that $\quad G_1(s) = G(s).$

PROBLEMS

(1) Express the following differential equations in vector-matrix form

(a) $\ddot{y} + 2\dot{y} - 3y = u$

(b) $\ddot{y} - 3\dot{y} + y = 2\dot{u} + u.$

(2) Express the differential equation

$$\ddot{y} + 2\dot{y} - 3y = 5u$$

in companion form as in equation (3.2).
(Note the difference between this equation and the one of Problem 1(a).)

(3) (i) The behaviour of two systems are respectively governed by the following differential equations

(a) $\dddot{y} + 2\ddot{y} - \dot{y} - 2y = 3u$

(b) $\ddot{y} + 3\dot{y} + 2y = 2\dot{u} + u.$

In each case find the transfer function and write it as a partial fraction expansion. Hence suggest appropriate transformations to express the differential equations in vector-matrix forms.

(ii) Draw the block diagram representation of the state equations in (1).

(4) The two tank system in Fig. 3.5 has two inputs q_1 and q_2 and two outputs through the valves R_1 and R_2.

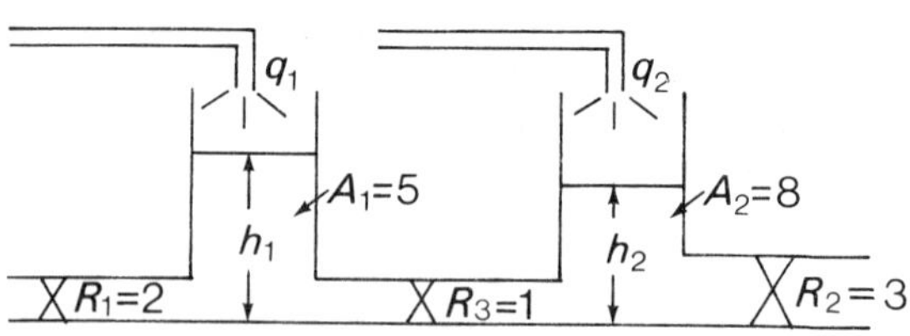

Fig. 3.5

Find the state space representation of this multivariable system in terms of the state variables h_1 and h_2.

(5) Find the state transition matrix for

(i) $$A = \begin{bmatrix} \lambda_1 & 0 \\ 0 & \lambda_2 \end{bmatrix}$$

(ii) $$A = \begin{bmatrix} 1 & 1 \\ 0 & 2 \end{bmatrix}.$$

(6) Find **x**, given

$$\begin{bmatrix} \dot{x}_1 \\ \dot{x}_2 \end{bmatrix} = \begin{bmatrix} -1 & 0 \\ 2 & -2 \end{bmatrix} \begin{bmatrix} x_1 \\ x_2 \end{bmatrix} + \begin{bmatrix} 0 \\ 1 \end{bmatrix} \mu$$

with $\mathbf{x}(0) = \begin{bmatrix} -1 \\ 2 \end{bmatrix}$ and $u(t) = 1$ for $t \geqslant 0$.

(7) A system has the transfer function

(i) $$G(s) = \frac{1}{s^2 + 5s + 2}$$

(ii) $$G(s) = \frac{1}{s^3 + 2s^2 + 5s + 3}.$$

Write the state-space equations for the system so that the state matrix is in a companion form.

(8) Determine state variables for the system defined by

$$\begin{bmatrix} \dot{x}_1 \\ \dot{x}_2 \end{bmatrix} = \begin{bmatrix} 0 & 1 \\ -2 & -3 \end{bmatrix} \begin{bmatrix} x_1 \\ x_2 \end{bmatrix} + \begin{bmatrix} 0 \\ 1 \end{bmatrix} u$$

$$y = [1 \quad 0] \begin{bmatrix} x_1 \\ x_2 \end{bmatrix}$$

so that the state matrix in the resulting equation is in a diagonal form.

(9) Find the system transfer matrix for

(i) the system in problem (8) above,

(ii) the system

$$\dot{\mathbf{x}} = \begin{bmatrix} -1 & 0 & 0 \\ 0 & -2 & 0 \\ 0 & 0 & -3 \end{bmatrix} \mathbf{x} + \begin{bmatrix} 0 & 1 \\ 1 & 1 \\ -1 & 0 \end{bmatrix} \mathbf{u}$$

$$y = \begin{bmatrix} 0 & 1 & 1 \\ 1 & 1 & 1 \end{bmatrix} \mathbf{x}.$$

(10) A system is described by the differential equation

$$\ddddot{y} + 6\dddot{y} + 13\ddot{y} + 12\dot{y} + 4y = \dot{u} + 3u.$$

(a) Obtain the transfer function $T(s)$ for this system.

(b) By a partial fraction expansion of $T(s)$ obtain the state variable representation of the septem in which the state matrix is in a Jordan canonical form.

Obtain an appropriate block diagram for this representation.

CHAPTER 4

Transient and steady state response analysis

4.1 INTRODUCTION

Many applications of control theory are to **servomechanisms** (or **servos** for short) which are systems using the feedback principle designed so that the output will follow the input. For example, an anti-aircraft gun must follow accurately and compensate for the rapidly varying signals received from an incoming plane or missile.

In general, of course, the input signals into a system are not known, and even if they were, it would be difficult to express them analytically in a form suitable for system analysis.

Because of this it is a frequent practice to apply a reference input into the system and to study its response in the time domain. The type of reference input applied does depend on the purpose of the system - for example, if the system may be subjected to sudden disturbances, a step reference input may be best; on the other hand, if the system is used to follow a gradually changing situation, a ramp reference input may be better.

The time response of a system may be considered in two parts. One part of the response, known as the **transient response** (corresponding to the complementary function), which reduces to zero as time increases, the other part, known as **steady-state** (corresponding to the particular integral), is the response of the system when time tends to infinity.

As already pointed out in Chapter 1, because of friction, inertia etc., the output of a system cannot follow an instantaneous change in the input (such as a reference step signal), so that the system response will always involve a transient part. To a control engineer, the time required for a system to reach a new steady state is of great importance, but there are various other factors he must consider. Apart from stability, which is, of course, the most important factor of all, he may have to consider factors such as **steady-state error** which is the difference between the steady-state response and the input.

We begin by considering first and second order systems, and it will then be shown that the time responses of higher order systems are a collection of functions of the same type as those for first and second order ones.

4.2 RESPONSE OF FIRST ORDER SYSTEMS

We have already noted – (see equation (2.13)) that the Laplace transform $Y(s)$ of the output of a linear system can be written as

$$Y(s) = G(s)U(s) \tag{4.1}$$

where $G(s)$ is its transfer function and $U(s)$ the Laplace transform of the input.

A first order system has an associated differential equation of the form

$$aDy + y = u, \qquad \text{hence (from Section 2.4) with } y(0) = 0$$

$$Y(s) = \frac{1}{as+1}U(s).$$

For transient response analysis it is customary to use a reference unit step function $u(t)$ for which

$$U(s) = 1/s.$$

It then follows that

$$Y(s) = \frac{1}{(as+1)s} = \frac{1/s}{(s+1/a)s} = \frac{1}{s} - \frac{1}{s+1/a}. \tag{4.2}$$

On taking the inverse Laplace of equation (4.2), we obtain;

$$y(t) = 1 - e^{-t/a} \qquad (t \geqslant 0). \tag{4.3}$$

Both the input and the output to the system are shown in Fig. 4.1. The response has an exponential form. The constant a is called the **time constant** of the system.

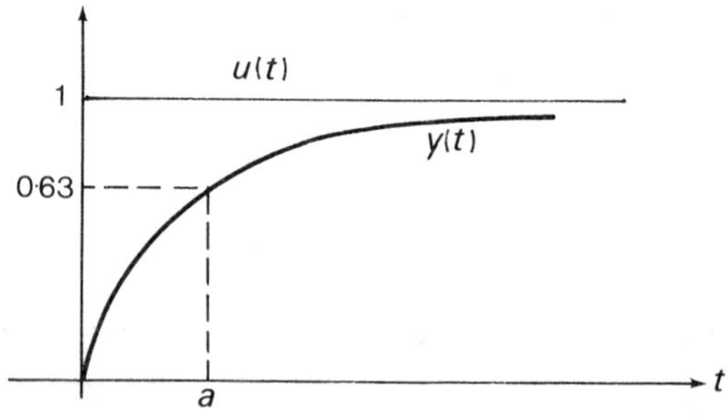

Fig. 4.1

Notice that when $t = a$, then $y(t) = y(a) = 1 - e^{-1} = 0.63$.

The response is in two parts, the transient part $e^{-t/a}$, which approaches 0 as $t \to \infty$ and the steady-state part 1, which means that eventually (when $t = \infty$) the output is 1 (that is, equal to the input).

If the derivatives of the input are involved in the differential equation of the system, that is, if

$$aDy + y = bDu + u$$

then

$$Y(s) = \frac{(bs+1)}{(as+1)} U(s) = K \frac{(s+z)}{(s+p)} U(s) \tag{4.4}$$

where $K = \dfrac{b}{a}$, $z = \dfrac{1}{b}$ and $p = \dfrac{1}{a}$.

When $U(s) = 1/s$, equation (4.4) can be written as

$$\frac{K_1}{s} - \frac{K_2}{s+p}$$

where $K_1 = K\dfrac{z}{p}$ and $K_2 = K\dfrac{z-p}{p}$.

Hence,

$$y(t) = K_1 - K_2 e^{-pt}. \tag{4.5}$$

With the the assumption that $z > p > 0$, this response is shown in Fig. 4.2.

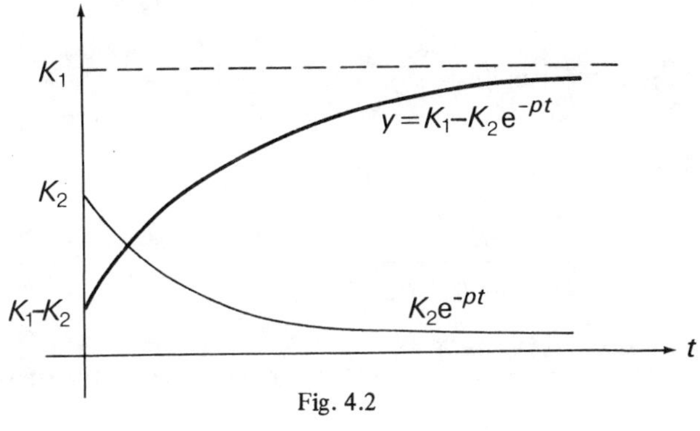

Fig. 4.2

We note that the responses to the two systems (Figs. 4.1, 4.2) have the same form, except for the constant terms K_1 and K_2. It appears that the role of the numerator of the transfer function is to determine these constants, that is, the size of $y(t)$, but its form is determined by the demoninator.

4.3 RESPONSE OF SECOND ORDER SYSTEMS

An example of a second order system is a spring–dashpot arrangement shown in Fig. 4.3. Applying Newton's law, we find

$$M\frac{d^2y}{dt^2} = -\mu\frac{dy}{dt} - ky + u(t)$$

where k = spring constant and μ = damping coefficient, y is the distance of the system from its position of equilibrium, and it is assumed that $y(0) = y'(0) = 0$.

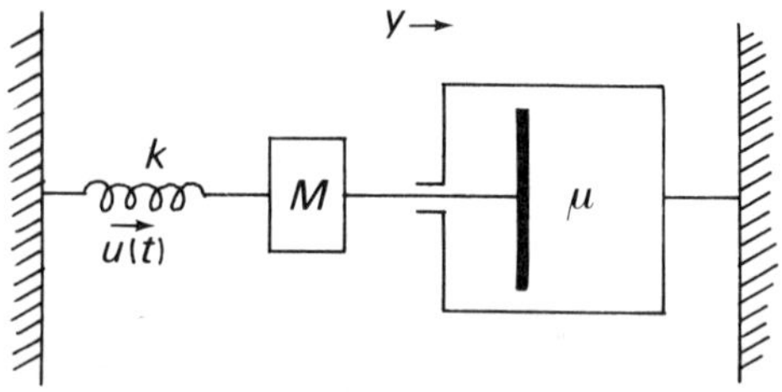

Fig. 4.3

Hence $\quad u(t) = M\dfrac{d^2y}{dt^2} + \mu\dfrac{dy}{dt} + ky.$

On taking Laplace transforms, we obtain

$$Y(s) = \frac{1}{Ms^2 + \mu s + k}U(s) = \frac{K}{s^2 + a_1 s + a_2}U(s)$$

where $K = \dfrac{1}{M}$, $a_1 = \dfrac{\mu}{M}$, and $a_2 = \dfrac{k}{M}$.

Applying a unit step input, we obtain

$$Y(s) = \frac{K}{s(s+p_1)(s+p_2)} \tag{4.6}$$

where $p_1, p_2 = \dfrac{a_1 \pm \sqrt{(a_1^2 - 4a_2)}}{2}$. p_1 and p_2 are the poles of the transfer function

$$G(s) = \frac{K}{s^2 + a_1 s + a_2},$$

that is, the zeros of the denominator of $G(s)$.

There are three cases to be considered:

Case 1. $a_1^2 > 4a_2$ (overdamped system)
In this case p_1 and p_2 are both real and unequal. Equation (4.6) can be written as

$$Y(s) = \frac{K_1}{s} + \frac{K_2}{s+p_1} + \frac{K_3}{s+p_2} \tag{4.7}$$

where $K_1 = \dfrac{K}{p_1 p_2} \left(= \dfrac{K}{a_2}\right)$, $K_2 = \dfrac{K}{p_1(p_1 - p_2)}$ and $K_3 = \dfrac{K}{p_2(p_2 - p_1)}$

(notice that $K_1 + K_2 + K_3 = 0$).

On taking the inverse Laplace transform of equation (4.7), we obtain

$$y(t) = K_1 + K_2 e^{-p_1 t} + K_3 e^{-p_2 t}. \tag{4.8}$$

The transient part of the solution is seen to be $K_2 e^{-p_1 t} + K_3 e^{-p_2 t}$.

Case 2. $a_1^2 = 4a_2$ (critically damped system)
In this case, the poles p_1 and p_2 are real and equal:

$$p_1 = p_2 = \frac{a_1}{2} = p \quad \text{(say)}$$

and

$$Y(s) = \frac{K}{s(s+p)^2} = \frac{K_1}{s} + \frac{K_2}{s+p} + \frac{K_3}{(s+p)^2}. \tag{4.9}$$

Hence $\quad y(t) = K_1 + K_2 e^{-pt} + K_3 t e^{-pt}$, where

$$K_1 = \frac{K}{p^2}, \quad K_2 = -\frac{K}{p^2} \quad \text{and} \quad K_3 = -\frac{K}{p}$$

so that

$$y(t) = \frac{K}{p^2}[1 - e^{-pt} - pt e^{-pt}]. \tag{4.10}$$

Case 3. $4a_2 > a_1^2$ (underdamped system)
In this case, the poles p_1 and p_2 are complex conjugate having the form

$$p_1, p_2 = \alpha \pm i\beta,$$

where

$$\alpha = \frac{a_1}{2} \quad \text{and} \quad \beta = \tfrac{1}{2}\sqrt{(4a_2 - a_1^2)}.$$

Hence

$$Y(s) = \frac{K_1}{s} + \frac{K_2}{s+p_1} + \frac{K_3}{s+p_2},$$

where

$$K_1 = \frac{K}{\alpha^2+\beta^2}, \quad K_2 = \frac{K(-\beta - i\alpha)}{2\beta(\alpha^2+\beta^2)} \quad \text{and} \quad K_3 = \frac{K(-\beta + i\alpha)}{2\beta(\alpha^2+\beta^2)}.$$

(Notice that K_2 and K_3 are complex conjugates.)

It follows that

$$y(t) = K_1 + K_2 e^{-(\alpha+i\beta)t} + K_3 e^{-(\alpha-i\beta)t}$$

$$= K_1 + e^{-\alpha t}[(K_2 + K_3)\cos\beta t + (K_3 - K_2)i \sin\beta t]$$

(using the relation $e^{i\beta t} = \cos\beta t + i\sin\beta t$)

$$= \frac{K}{\alpha^2+\beta^2} + \frac{K}{\alpha^2+\beta^2} e^{-\alpha t}\left[-\cos\beta t - \frac{\alpha}{\beta}\sin\beta t\right]. \tag{4.11}$$

$$= \frac{K}{\alpha^2+\beta^2} - \frac{K}{\alpha^2+\beta^2}\left[1 + \frac{\alpha^2}{\beta^2}\right]^{\frac{1}{2}} e^{-\alpha t}\sin(\beta t + \epsilon) \tag{4.12}$$

where

$$\tan\epsilon = \frac{\beta}{\alpha}.$$

Notice that when $t = 0$, $y(t) = 0$ (see equation (4.11).

The three cases discussed above are plotted in Fig. 4.4.

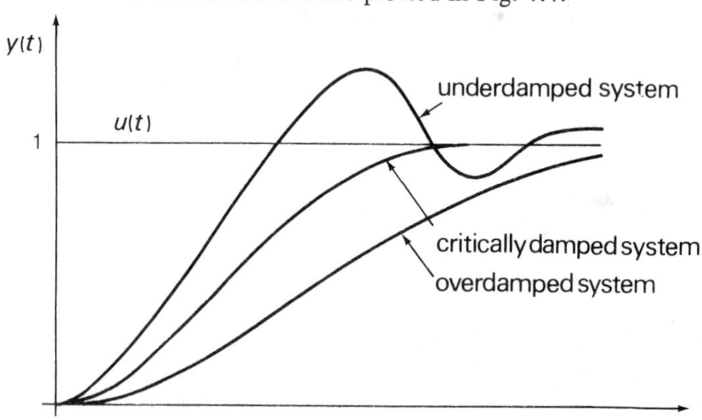

Fig. 4.4

From Fig. 4.4, we see the importance of damping (note that $a_1 = \mu/M$, μ being the damping factor). We would expect that when the damping is 0 (that is, $a_1 = 0$) the system should oscillate indefinitely. Indeed when $a_1 = 0$, then

$$\alpha = 0, \quad \text{and} \quad \beta = \sqrt{a_2}$$

and since $\sin \epsilon = 1$ and $\cos \epsilon = 0$, then $\epsilon = \pi/2$, equation (4.12) becomes

$$y(t) = \frac{K}{a_2}\left[1 - \sin\left(\sqrt{a_2}\,t + \frac{\pi}{2}\right)\right] = \frac{K}{a_2}[1 - \cos\sqrt{a_2}\,t].$$

This response of the undamped system is shown in Fig. 4.5.

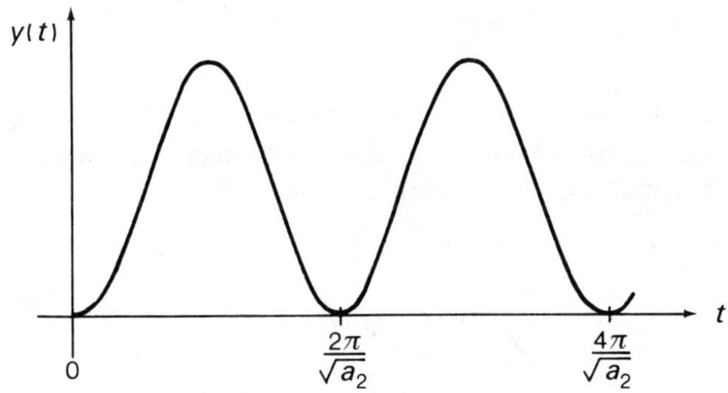

Fig. 4.5

Notice that this is the response of the system when the roots of $s^2 + a_1 s + a_2 = 0$ are pure imaginary, that is, when the poles of the transfer functions are pure imaginary.

There are two important constants associated with each second order system.

The **undamped natural frequency** ω_n of the system is the frequency of the response shown in Fig. 4.5, that is,

$$\omega_n = \sqrt{a_2}$$

The **damping ratio** ζ of the system is the ratio of the actual damping $\mu(= a_1 M)$ to the value of the damping μ_c, which results in the system being critically damped (that is, when $a_1 = 2\sqrt{a_2}$). Hence

$$\zeta = \frac{\mu}{\mu_c} = \frac{a_1}{2\sqrt{a_2}}.$$

We can write equation (4.12) in terms of these constants. We note that

$$a_1 = 2\omega_n\zeta \quad \text{and} \quad a_2 = \omega_n^2.$$

Hence
$$(1 + \alpha^2/\beta^2)^{\frac{1}{2}} = \left[1 + \frac{a_1^2}{4a_2 - a_1^2}\right]^{\frac{1}{2}} = \frac{2\sqrt{a_2}}{\sqrt{(4a_2 - a_1^2)}} = \frac{1}{\sqrt{(1-\zeta^2)}}.$$

Equation (4.12) becomes

$$y(t) = \frac{K}{\omega_n^2}\left[1 - \frac{1}{\sqrt{(1-\zeta^2)}}e^{-\omega_n\zeta t}\sin(\omega t + \epsilon)\right] \tag{4.13}$$

where
$$\omega = \omega_n\sqrt{(1-\zeta^2)} \quad \text{and} \quad \tan\epsilon = \frac{\sqrt{(1-\zeta^2)}}{\zeta}.$$

It is conventional to choose $K/a_2 = 1$ and then plot graphs of the 'normalised' response $y(t)$ against ωt for various values of the damping ratio ζ.

Three typical graphs are shown in Fig. 4.6.

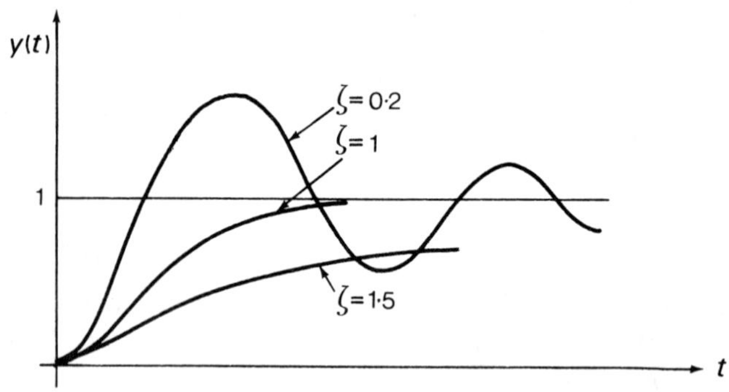

Fig. 4.6

It is obvious from Fig. 4.4 that if the speed of response is an important criterion in the design of the system, the engineer will choose the system to function in the underdamped state, but he will then have to consider various other factors or parameters, some of which are shown in Fig. 4.7.

(1) **Overshoot** defined as

$$\frac{\text{maximum overshoot} \times 100}{\text{final desired value}}.$$

(2) **Time delay** t_d, the time required for the system response to reach 50% of its final value.

(3) **Rise time**, the time required for the system response to rise from 10% to 90% of its final value.

(4) **Settling time**, the time required for the eventual settling down of the system response to be within (normally) 5% of its final value.

(5) **Steady state error** e_{ss}, the difference between the steady state response and the input.

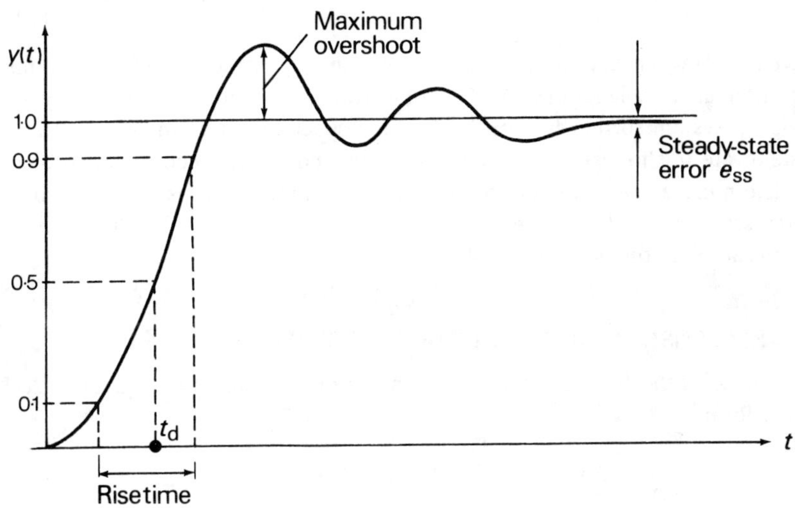

Fig. 4.7

In fact, one can often improve one of the parameters but at the expense of the others. For example, the overshoot can be decreased at the expense of the time delay.

In general, the quality of a system may be judged by various standards. Since the purpose of a servomechanism is to make the output follow the input signal, we may define expressions which will describe dynamic accuracy in the transient state. Such expressions are based on the **system error**, which in its simplest form is the difference between the system input and the output and is denoted by $e(t)$, that is,

$$e(t) = y(t) - u(t)$$

where $y(t)$ is the actual output and $u(t)$ is the desired output; ($u(t)$ is the input).

The expression called the **performance index** can take on various forms; typical among them are:

$$\int_0^\infty e^2(t)\mathrm{d}t, \quad \text{integral of error squared (IES)}$$

$$\int_0^\infty |e(t)|\mathrm{d}t, \quad \text{integral of absolute error (IAE)}$$

and

$$\int_0^\infty t|e(t)|\mathrm{d}t, \quad \text{integral of time multiplied absolute error criterion (ITAE).}$$

Having chosen an appropriate performance index, the system which minimises the integral is called **optimal**. The objective of modern control theory is to design a system so that it is optimal with respect to a performance index. The systems are not necessarily servomechanisms, and in fact, a common problem is to determine a signal vector so that the state vector $\mathbf{x}(t)$ is brought from an initial state $\mathbf{x}_0$ to a final state $\mathbf{x}_\mathrm{f}$ in the shortest possible time. We discuss various optimisation methods in Part II of the book.

4.4 RESPONSE OF HIGHER ORDER SYSTEMS

We can write the transfer function of an nth order system (see equation (2.12)) in the form

$$G(s) = \frac{K(s^m + b_1 s^{m-1} + \ldots + b_m)}{s^n + a_1 s^{n-1} + \ldots + a_n}. \tag{4.14}$$

Systems having a transfer function of high order can arise when several components, each involving a 'low' order transfer function, make up the system as illustrated by the following simple example.

Example 4.1

With reference to Fig. 2.11 (see page 43), calculate the closed loop transfer function ($G(s)$) given the transfer functions

$$A(s) = \frac{1}{s+3} \quad \text{and} \quad B(s) = \frac{2}{s}.$$

Solution

Using equation (2.16), that is, $G(s) = \dfrac{A(s)}{1 + A(s)B(s)}$, we obtain

$$G(s) = \frac{s}{s^2 + 3s + 2} = \frac{s}{(s+1)(s+2)}.$$

The response of the system having the transfer function (4.14) to a unit step input can be written in the form

$$Y(s) = \frac{K(s+z_1)(s+z_2)\ldots(s+z_m)}{s(s+p_1)(s+p_2)\ldots(s+p_n)} \tag{4.15}$$

where $z_1, z_2, \ldots, z_m$ and $p_1, p_2, \ldots, p_n$ are respectively the zeros of the numerator and the denominator polynomials of equation (4.14).

The zeros are either real or complex numbers; in the case of a complex zero there must be another zero which is its conjugate, since the a's and b's in equation (4.14) are all real.

We first assume that $n \geqslant m$ in equation (4.14); we then have two cases to consider:

Case 1

Here $p_1, p_2, \ldots, p_n$ are all distinct numbers. The partial fraction expansion of equation (4.15) has the form

$$Y(s) = \frac{K_1}{s} + \frac{K_2}{s+p_1} + \ldots + \frac{K_{n+1}}{s+p_n}. \tag{4.16}$$

$K_1, K_2, \ldots, K_{n+1}$ are called the **residues** of the expansions. The response has the form

$$y(t) = K_1 + K_2 \mathrm{e}^{-p_1 t} + \ldots + K_{n+1} \mathrm{e}^{-p_n t}.$$

As we have already noticed in the previous section, if p_1 is complex and p_2 is its conjugate then the corresponding residues (K_2 and K_3) are complex conjugate. The two terms can be combined as illustrated in equation (4.12).

Case 2

Here at least one of the roots, say p_1, is of multiplicity r, that is,

$$Y(s) = \frac{K(s+z_1)(s+z_2)\ldots(s+z_m)}{s(s+p_1)^r(s+p_2)\ldots(s+p_{n-r+1})}. \tag{4.17}$$

The partial fraction expansion has the form

$$Y(s) = \frac{K_1}{s} + \frac{K_{21}}{s+p_1} + \frac{K_{22}}{(s+p_1)^2} + \ldots + \frac{K_{2r}}{(s+p_1)^r} + \ldots + \frac{K_{n-r+2}}{s+p_{n-r+1}}. \tag{4.18}$$

Since $\mathcal{L}^{-1}\left\{\dfrac{K}{(s+p)^j}\right\} = \dfrac{K}{(j-1)!} t^{j-1} \mathrm{e}^{-pt} \quad (j = 1, 2, \ldots, r)$

the response has the form

$$y(t) = K_1 + K_{21}\mathrm{e}^{-p_1 t} + K_{22}\, t\, \mathrm{e}^{-p_1 t} + \ldots + \frac{K_{2r}}{(r-1)!} t^{r-1} \mathrm{e}^{-p_1 t}$$

$$+ K_3 \mathrm{e}^{-p_2 t} + \ldots + K_{n-r+2}\, \mathrm{e}^{-p_{n-r+1} t}. \tag{4.19}$$

We now consider that $m > n$ in equation (4.14), which is the case (see section 2.4) when the system is improper; that is, it can happen when we consider idealised and physically non-realisable systems, such as resistanceless circuits. We then divide the numerator by the denominator until we obtain a proper fraction so that when applying a unit step input, we can write $Y(s)$ as

$$Y(s) = K[c_1 + c_2 s + \ldots + c s^{n-m-1}] + \frac{K_1[s^n + d_1 s^{n-1} + \ldots + d_n]}{s(s^n + a_1 s^{n-1} + \ldots + a_n)} \tag{4.20}$$

where the c_s, d_s, K and K_1 are all constants.

The second term of the right-hand side is treated as in Case 1 or Case 2 above; the inverse Laplace transform of the first term involves the impulse function (see equation (2.7)) and various derivatives of it.

Example 4.2

Find the response of the system having a transfer function

$$G(s) = \frac{5(s^2 + 5s + 6)}{s^3 + 6s^2 + 10s + 8}$$

to a unit step input.

Solution

In this case,

$$Y(s) = \frac{5(s^2 + 5s + 6)}{s(s^3 + 6s^2 + 10s + 8)} = \frac{5(s+2)(s+3)}{s(s+4)(s^2 + 2s + 2)}.$$

The term $s^2 + 2s + 2 = (s + 1 + i)\,(s + 1 - i)$, hence $Y(s)$ can be written as the partial fraction expansion as

$$Y(s) = \frac{K_1}{s} + \frac{K_2}{s+4} + \frac{K_3}{s+1+i} + \frac{K_4}{s+1-i}$$

where K_3 is the conjugate of K_4. In fact,

$$K_1 = \frac{15}{4}, K_2 = -\frac{1}{4}, K_3 = \frac{1}{4}(-7+i) \quad \text{and} \quad K_4 = \frac{1}{4}(-7-i).$$

Hence

$$y(t) = \frac{15}{4} - \frac{1}{4}e^{-4t} + \frac{1}{4}e^{-t}[-14\cos t + 2\sin t]$$

$$= \frac{15}{4} - \frac{1}{4}e^{-4t} + \frac{\sqrt{200}}{4}e^{-t}\sin(t + 352).$$

4.5 STEADY STATE ERROR

It was stated in the previous section that one of the important considerations in system design is the steady state error e_{ss} shown in Fig. 4.7. We would expect that e_{ss} would depend on both the transfer function of the system and on the reference input. In this chapter we have used the notation $u(t)$ to denote the input to a system - generally to denote a (unit) step input. We shall now change this notation to conform with the general usage and the fact that we consider different types of reference inputs. We shall use $r(t)$ to denote the (reference) input and $c(t)$ the output. A unity feedback system with an open loop transfer function $A(s)$ will then be represented (as in Fig. 2.12) by Fig. 4.8.

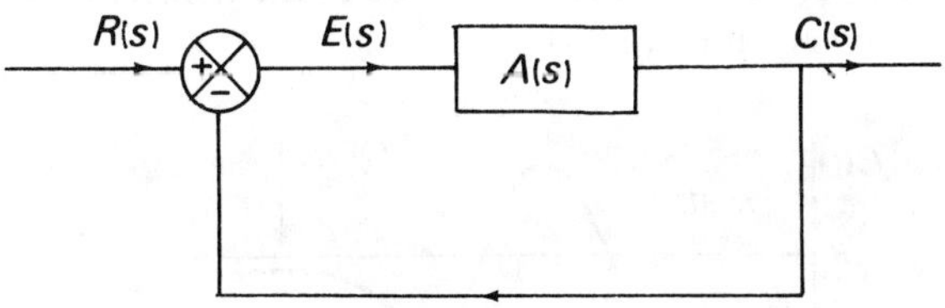

Fig. 4.8
$E(s)$, $R(s)$ and $C(s)$ are the Laplace transforms of $e(t)$, $r(t)$ and $c(t)$ respectively.

We define the **error function†**

$$e(t) = r(t) - c(t). \tag{4.21}$$

Hence

$$e_{ss} = \lim_{t\to\infty} e(t).$$

†This is assuming that $r(t)$ and $c(t)$ are of the same dimensions, that is, are both voltages or speeds, etc., otherwise a non-unity feedback system must be used.

Since $E(s) = R(s) - A(s)E(s)$, it follows that

$$E(s) = \frac{R(s)}{1 + A(s)}$$

and by the final value theorem (equation (2.10))

$$e_{ss} = \lim_{s\to 0} sE(s)$$

$$= \lim_{s\to 0} \frac{sR(s)}{1 + A(s)}. \qquad (4.22)$$

We now define three **error coefficients** which indicate the steady state error when the system is subjected to three different standard reference inputs $r(t)$.

(1) *Step input,* $r(t) = ku(t)$ (k is a constant) (Fig. 4.9).

$$e_{ss} = \lim_{s\to 0} \frac{sk/s}{1 + A(s)} = \frac{k}{1 + \lim_{s\to 0} A(s)}.$$

$\lim_{s\to 0} A(s) = K_p$, called the **position error constant**, then

$$e_{ss} = \frac{k}{1 + K_p} \quad \text{or} \quad K_p = \frac{k - e_{ss}}{e_{ss}} \qquad (4.23)$$

Hence if k and e_{ss} are specified, the value of the transfer function at $s = 0$ is determined by equation (4.23).

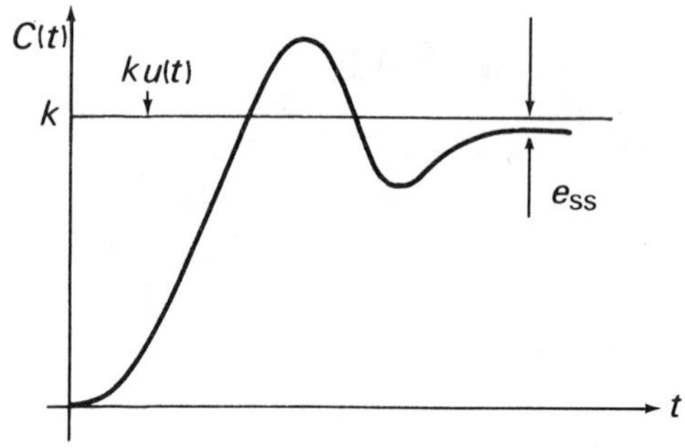

Fig. 4.9

(2) *Ramp input,* $r(t) = ktu(t)$ (Fig. 4.10).

In this case, $R(s) = \frac{k}{s^2}$, so that $e_{ss} = \frac{k}{K_v}$ or $K_v = \frac{k}{e_{ss}}$ where $K_v = \lim_{s\to 0} sA(s)$, is called the **velocity error constant**.

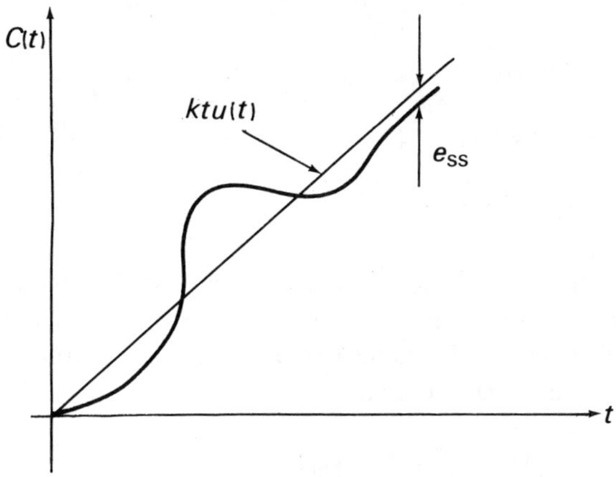

Fig. 4.10

(3) *Parabolic input*, $r(t) = \dfrac{1}{2}kt^2u(t)$ (Fig. 4.11).

$R(s) = \dfrac{k}{s^3}$, so that $e_{ss} = \dfrac{k}{K_a}$ or $K_a = \dfrac{k}{e_{ss}}$ where $K_a = \lim_{s\to 0} s^2A(s)$, called the **acceleration error constant**.

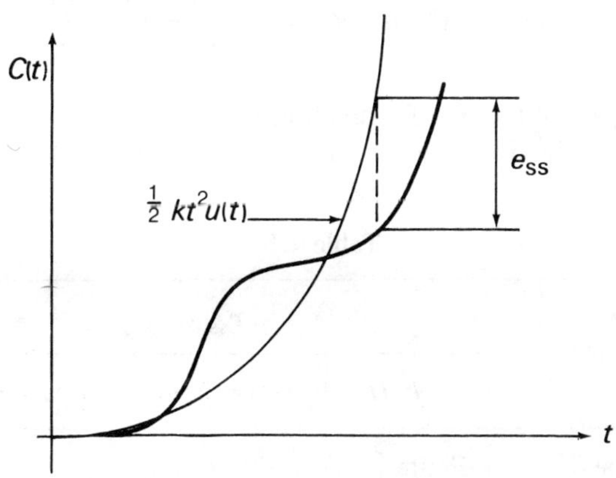

Fig. 4.11

Example 4.3

Find the (static) error coefficients for the system having the open loop transfer function

$$A(s) = \frac{8}{s(4s+2)}.$$

Solution

$$K_p = \lim_{s \to 0} A(s) = \infty$$

$$K_v = \lim_{s \to 0} sA(s) = 4$$

$$K_a = \lim_{s \to 0} s^2 A(s) = 0.$$

From the definitions of the error coefficients, it is seen that e_{ss} depends on the number of poles (or integrations) at $s = 0$ of the transfer function. This leads to the following classification. A transfer function is said to be of type N if it has N poles at the origin. Thus if

$$A(s) = \frac{K(s - z_1) \ldots (s - z_m)}{s^j (s - p_1) \ldots (s - p_n)} \tag{4.24}$$

where

$j = 0$, $A(s)$ is of type 0,

$j = 1$, $A(s)$ is of type 1,

$j = 2$, $A(s)$ is of type 2.

At $s = 0$, $$A(s) = \lim_{s \to 0} \frac{K_1}{s^j} \quad \text{where} \quad K_1 = \frac{K(-z_1) \ldots (-z_m)}{(-p_1) \ldots (-p_n)}. \tag{4.25}$$

(K_1 is called the **gain** of the transfer function.)

Hence the steady state error e_{ss} depends on j and $r(t)$ as summarised in Table 4.1.

Table 4.1

j	System	e_{ss}		
		$r(t) = ku(t)$	$r(t) = ktu(t)$	$r(t) = \frac{1}{2}t^2u(t)$
0	Type 0	Finite	∞	∞
1	Type 1	0	Finite	∞
2	Type 2	0	0	Finite

It would seem that for a system to have a rapid response, capable of following various inputs with zero steady state error, the engineer must ensure that the open loop transfer function has a pole of high order at the origin. Unfortunately, such systems are prone to instability so that it is unusual to design a system having a pole at the origin of order greater than two (that is, $j > 2$).

4.6 FEEDBACK CONTROL

In this section we shall discuss certain aspects of the effect of feedback on the system response. This is in fact the very basis of classical control theory, and much literature is devoted to it. We shall only consider certain general principles and not detailed system design considerations.

We have already shown (see Fig. 2.12 and equation (2.16)) that the *closed loop* transfer function $G(s)$ of the negative feedback system (Fig. 4.12) is related to the *feed-forward* transfer function $A(s)$ and *feed-back* transfer function $B(s)$ by

$$G(s) = \frac{A(s)}{1 + A(s)B(s)}. \tag{4.26}$$

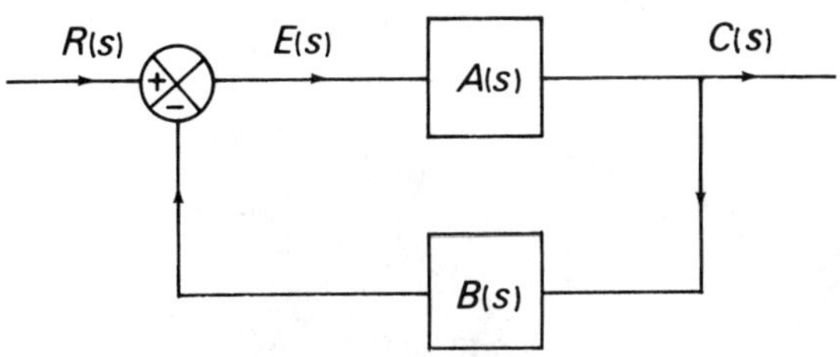

Fig. 4.12

We consider a simple example of a first order system for which

$$A(s) = \frac{K}{as+1}$$

and $$B(s) = c \quad \text{(a constant)}$$

so that

$$G(s) = \frac{K}{as + Kc + 1} = \frac{K/a}{s + \dfrac{Kc+1}{a}}.$$

We now take the inverse Laplace transform, that is, obtain the impulse response (see equation (2.14)) of the system, where

(1) $c = 0$ (that is, response of the open loop system)

$$g(t) = \frac{K}{a}\,e^{-t/a}, \quad \text{and}$$

(2) $c \neq 0$

$$g(t) = \frac{K}{a}\,e^{-\frac{Kc+1}{a}t} = \frac{K}{a}\,e^{-\frac{t}{\alpha}}$$

where $\alpha = \dfrac{a}{Kc+1}$.

a and α are respectively the time-constants of the open loop and the closed loop systems.

a is always positive (since we are considering a physical system), but since c can be made positive or negative, α may be either positive or negative.

Fig. 4.13 shows how the time responses vary with different values of Kc.

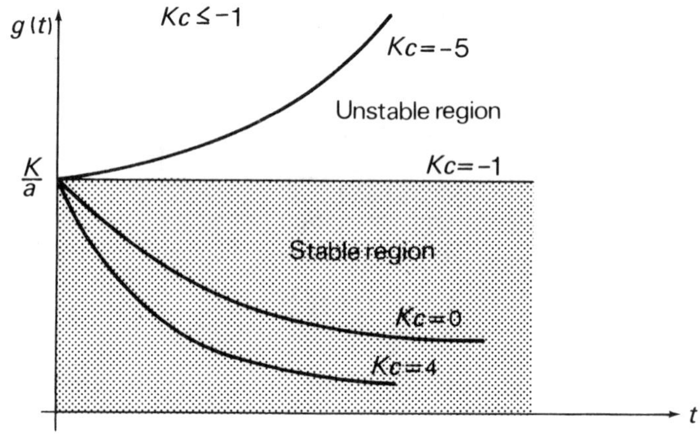

Fig. 4.13

If the impulse response does not decay to zero as t increases, the system is **unstable**. From Fig. 4.13 it is seen that the instability region is defined by $Kc \leqslant -1$.

In many applications the control system consists basically of a plant having the transfer function $A(s)$ and a *controller* having a transfer function $B(s)$, as in Fig. 4.14.

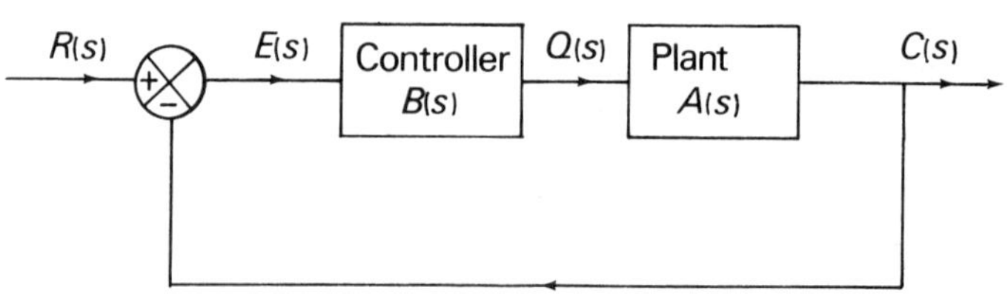

Fig. 4.14

The transfer function of the closed loop system can be found by application of equation (4.26) with feed-forward transfer function $A(s)B(s)$ (see equation (2.15)) and a feed-back transfer function 1; then

$$G(s) = \frac{A(s)B(s)}{1 + A(s)B(s)}. \tag{4.27}$$

The controllers can be of various types.

(1) *The on–off controller*
The action of such a controller is very simple. With reference to Fig. 4.14, the control depends on the actuating signal $e(t)$. Thus if

$$e(t) > 0 \quad \text{then} \quad q(t) = Q_1$$

$$\text{and if} \quad e(t) < 0 \quad \text{then} \quad q(t) = Q_2$$

where $q(t)$ is the output signal from the controller and Q_1 and Q_2 are some constants.

The on–off controller is obviously a nonlinear device and it cannot be described by a transfer function. An analysis of such systems can be accomplished by 'describing functions' which we do not consider in this book.

(2) *Proportional Controllers*
For this control action

$$q(t) = K_p e(t)$$

where K_p is a constant, called the **gain** of the controller. Thus the transfer function for this controller is

$$B(s) = \frac{Q(s)}{E(s)} = K_p. \tag{4.28}$$

(3) *Integral Controller*
In this case

$$q(t) = K \int_0^t e(t)\,dt, \quad \text{hence}$$

$$B(s) = K/s. \tag{4.29}$$

(4) *Derivative Controller*
In this case

$$q(t) = K\frac{\mathrm{d}e}{\mathrm{d}t}, \quad \text{hence}$$

$$B(s) = Ks.$$

This type of controller is never used on its own, since its action is only effective while $e(t)$ is changing, that is, during the transient state period, and has no effect in the steady state.

(5) *Proportional-plus-derivative (PD) Controller*
In this case

$$q(t) = K_\mathrm{p}e(t) + K_1\frac{\mathrm{d}e}{\mathrm{d}t}, \quad \text{hence}$$

$$B(s) = K_\mathrm{p}(1 + \frac{K_1}{K_\mathrm{p}}s)$$

$$= K_\mathrm{p}(1 + Ks)$$

where $K = \dfrac{K_1}{K_\mathrm{p}}$. (4.30)

(6) *Proportional-plus-integral (PI) Controller*
In this case

$$q(t) = K_\mathrm{p}e(t) + K_1\int_0^t e(t)\mathrm{d}t, \quad \text{hence}$$

$$B(s) = K_\mathrm{p}(1 + \frac{K}{s}) \quad \text{where} \quad K = \frac{K_1}{K_\mathrm{p}}. \tag{4.31}$$

(7) *Proportional-plus-derivative-plus-integral (PDI) Controller*
In this case

$$q(t) = K_\mathrm{p}e(t) + K_1\frac{\mathrm{d}e}{\mathrm{d}t} + K_2\int_0^t e(t)\mathrm{d}t, \quad \text{hence}$$

$$B(s) = K_\mathrm{p}(1 + k_1s + k_2/s)$$

where $k_1 = \dfrac{K_1}{K_\mathrm{p}}$ and $k_2 = \dfrac{K_2}{K_\mathrm{p}}$.

Controllers of the various types described above can be constructed for electrical, mechanical, and other systems. This was pointed out for electrical systems when discussing simulation in section 3.2. We shall not discuss the details of how such controllers are designed. For our purposes it is sufficient to appreciate that such controllers are at our disposal.

Example 4.4

Design a controller for a plant having the transfer function $A(s) = 1/(s+2))$ so that the resulting closed loop system has a zero steady state error to a reference ramp input.

Solution

From the Table 4.1 (on p. 98) we note that for zero steady state error to a ramp input, the system must be of type 2. Hence if we choose an integral controller with

$$B(s) = K/s$$

then the transfer function of the closed loop system including the plant and the controller (see Fig. 2.11) is

$$\frac{A(s)B(s)}{1+A(s)B(s)} = \frac{K}{s^3+2s^2+K}.$$

The response of this control system depends on the roots of the denominator polynomial

$$s^3 + 2s^2 + K = 0.$$

It will be shown in the next chapter (example 5.3) that for all values of K at least one of the roots has a positive real part, which implies that the system is unstable. We cannot use an integral controller in this case. But if we use a PI (proportional-plus-integral) controller, say

$$B(s) = K_p(1 + \frac{K}{s}),$$

the system is of type 2 and response of the system depends on the roots of

$$s^3 + 2s^2 + K_p s + KK_p = 0.$$

In this case, values of K and K_p can be found so that the system is stable (see example 5.3).

PROBLEMS

(1) A unity feedback system has the open-loop transfer function

$$G(s) = \frac{2}{s^3 + 5s^2 + 6s}.$$

(i) Calculate the positional, velocity and acceleration error constants.
(ii) Determine the steady-state error where the input for $t \geqslant 0$ is
(a) $r(t) = 2$, (b) $r(t) = 2t$, (c) $r(t) = \frac{3}{2}t^2$ and (d) $r(t) = 3 - 2t$.

(2) Find the time constants of the system
(i) whose behaviour is defined by the differential equation

$$\text{(a)}\quad 3\frac{dy}{dt} + 2y = 2$$

$$\text{(b)}\quad \frac{1}{2}\frac{dy}{dt} + 3y = 3$$

given that at $t = 0$, $y = 0$.
(ii) whose (open-loop) transfer function is

$$\text{(a)}\quad G(s) = \frac{1}{1 + 3s}$$

$$\text{(b)}\quad G(s) = \frac{1}{1 + 2s}.$$

(iii) Unity feedback systems have open-loop transfer functions as in (ii) above. Determine the system time-constants. What is the effect on the response of the system on the addition of a closed loop?

(3) Investigate the response to a unit step input of the system having the transfer function

$$G(s) = \frac{4}{s^2 + 4s + 5}.$$

(4) Obtain the response to a unit step input of the system shown in Fig. 4.15 for $K = 1, 5.25$, and 10.

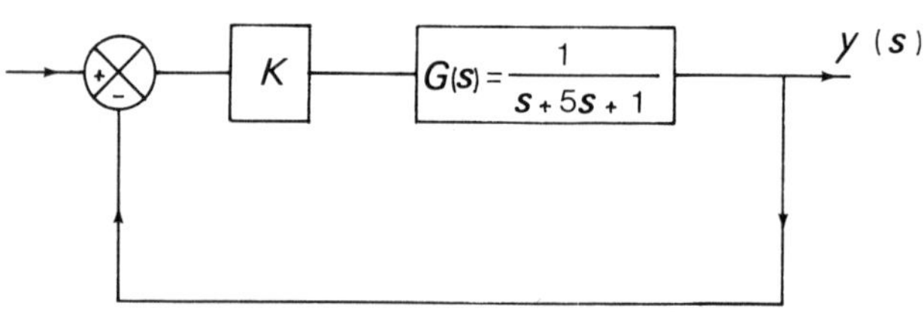

Fig. 4.15

(5) The output of a system to a unit step input is

$$Y(s) = \frac{a_2}{s(s^2 + a_1 s + a_2)}.$$

(a) Write this output (in its normalised form) as a function of the system natural frequency ω_n and the damping ratio τ.

(b) The transfer function of the system is written as

$$G(s) = \frac{K}{(s + p_1)(s + p_2)}.$$

Find the values of p_1 and p_2 in terms of the above two system constants.

(6) A second order underdamped system has a response $y(t)$ to a unit step input. Obtain the time of occurrence and the height of the maximum overshoot in terms of the system natural frequency ω_n and the damping ratio τ.

CHAPTER 5

Stability

5.1 INTRODUCTION

It has already been mentioned that stability is possibly the most important consideration when designing a control system. The problems involved are not only important but extremely complicated - indeed much present day research in control is concerned with stability. That is not to say that methods for determining stability do not exist - on the contrary many methods do exist. Unfortunately they deal only with certain aspects of stability. For nonlinear systems and even for linear systems it is not always possible to determine analytically for what ranges of the relevant parameters the system remains stable.

The difficulties begin with the definitions of stability; we all know roughly what is involved, but it is very difficult to cover all possible cases. Perhaps a suitable analogy is the definition of probability - we all know roughly what is meant, but its definition is very 'theoretical' and is difficult to interpret in practical situations. In this chapter we shall be content to explore only some of the many aspects of the stability problem for ***linear*** systems.

5.2 THE CONCEPT OF STABILITY

Within the context of this book it would be intuitively reasonable to define a linear system to be stable if its output is bound for every bounded input.

This definition states that stability is dependent on the system itself. The properties of a system are characterised by its weighting function $g(t)$, that is, its response to a unit impulse (see section 2.4) or equivalently by the Laplace transform of the weighting function - the transfer function.

Some methods of testing for system stability use the transfer function as the starting point; others make use of the state-space equations. It must be remembered that the state-space equations are themselves a representation of the differential equations characterising the system. The transfer function of the system is (at least) a transformation of these equations when the initial conditions are assumed zero. The transfer functions can also be obtained by the practical 'frequency-response' method, in which case it may happen that some of the system information may have been lost, owing to the cancellation of poles and zeros of the transfer function - a problem we further investigate when discussing system controllability and observability. It follows that a stability analysis based on the transfer function of a system must be used with care, especially if the system is uncontrollable and/or unobservable (see Chapter 6).

In terms of its weighting function we say that the system is stable if the function respose remains bounded as the time t tends to infinity. For most control engineering purposes and our own in this chapter, an even stronger concept of stability is required:

> A system is said to be **asymptotically stable** if its weighting function response decays to zero as t tends to infinity.

When we considered, in Chapter 4, an undamped second order system (Fig. 4.5) we noted that the response was indefinitely oscillatory. This is an example of a system which is stable but not asymptotically stable. This situation is in fact known as **marginally stable**. Otherwise, the system is said to be unstable.

We have noted in the previous chapter that the transfer function $G(s)$ (or the Laplace transform of the system response to a unit impulse) has the form (see equation (4.14))

$$G(s) = \frac{K(s^m + b_1 s^{m-1} + \ldots + b_m)}{s^n + a_1 s^{n-1} + \ldots + a_n} \tag{5.1}$$

and the system response is determined, except for the residues, by the poles of $G(s)$, that is by the solutions of

$$s^n + a_1 s^{n-1} + \ldots + a_n = 0. \tag{5.2}$$

Equation (5.2) is called the **characteristic equation** of the system. It follows that the system is asymptotically stable if and only if the zeros of the characteristic equation (that is, the finite poles of the transfer function) $p_1, p_2, \ldots, p_n$ are negative (for the real zeros) or have negative real parts (for the complex zeros).

The case of purely imaginary zeros has already been mentioned briefly. Like complex roots, all purely imaginary roots occur in conjugate pairs. If the multiplicity of these roots is $r = 1$, the corresponding response is oscillatory and of constant amplitude as indicated by the second term of equation (4.12) (with $\alpha = 0$).

On the other hand, if the multiplicity of the imaginary roots is $r > 1$, the corresponding response terms are oscillatory but involve factors of $t, t^2, \ldots, t^{n-1}$ as indicated by equation (4.19), so that the oscillations are of increasing amplitude.

Hence for $r = 1$ the system is marginally stable, but for $r > 1$ the system is unstable.

Type of root	s-plane graph $s=\sigma+i\omega$	Response graph	Remarks
Real and negative	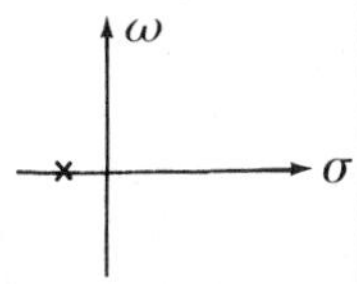		Asymptotically stable
Real and positive	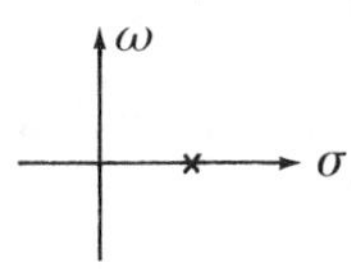		Unstable
Zero	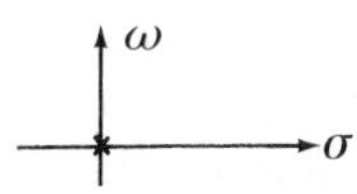		Marginally stable
Conjugate complex with negative real part	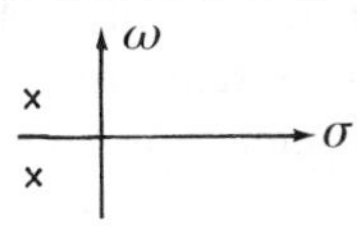	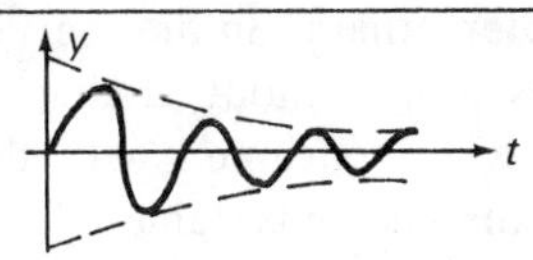	Asymptotically stable
Conjugate imaginary (multiplicity $r=1$)	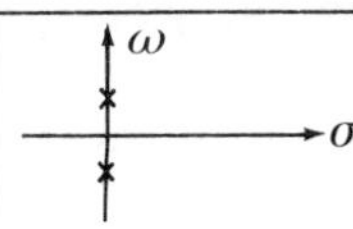	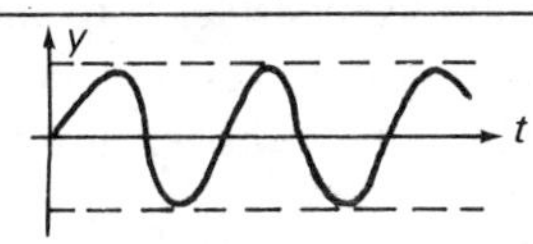	Marginally stable
Conjugate imaginary (multiplicity $r=2$)	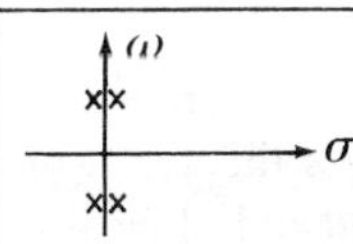	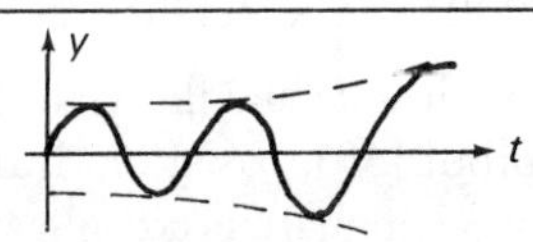	Unstable
Conjugate with positive real part	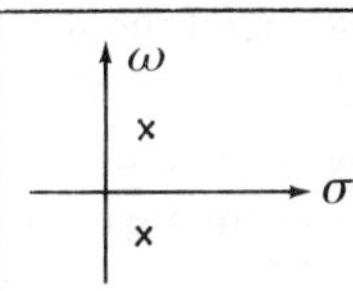	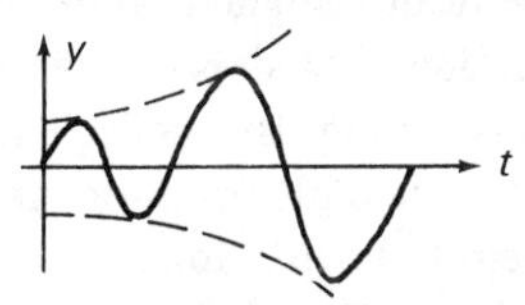	Unstable
Roots of multiplicity $r=2$ at the origin	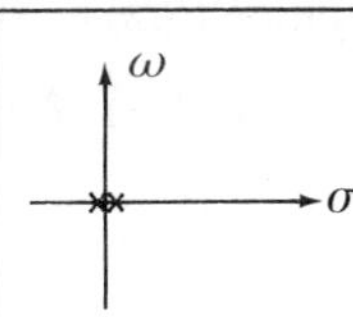	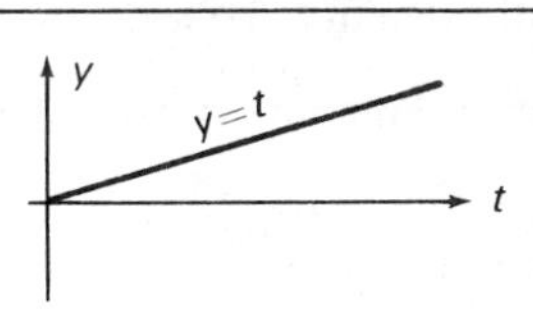	Unstable

Fig. 5.1

Similarly, if there is a root of multiplicity $r > 1$ at the origin, then since

$$\mathcal{L}^{-1}\left\{\frac{1}{s^r}\right\} = \frac{1}{(r-1)!}\,t^{r-1}$$

the system is unstable.

We can illustrate graphically the various cases discussed above by plotting the position of the roots of the characteristic equation on the s-plane and the corresponding response $y(t)$ to a unit impulse on a graph next to it as shown in Fig. 5.1.

It would seem that to determine whether a system is stable all we need to do is to evaluate and examine the roots of the characteristic equation (5.2). (Note that although in some cases the coefficients $a_1, a_2, \ldots, a_n$ may be complex, we shall assume that they are real.)

Unfortunately, even with the aid of modern computers there are difficulties in evaluating the roots of polynomials (because of rapid loss in accuracy). Apart from this, we may also be concerned with obtaining the range of the variations of certain parameters (that is, in the coefficients a_i) which will allow for changes in the system response without driving the system into an unstable state. In example 4.4 we were concerned with the choice of values of two constants K and K_{p} so that the system is stable.

The first method we consider is to answer the question of absolute stability - whether the system is stable or not. In the case of the system being unstable, the criterion will also indicate how many of the roots have positive real parts.

5.3 ROUTH STABILITY CRITERION

In this section we shall state and illustrate Routh's criterion but make no attempt to prove it. The proof [50], based on Sturm's theorem which is concerned with sequences of real polynomials, is complicated and of great algebraic interest, but not, except for the result, of great interest in control theory.

The criterion does not depend on finding the roots of the characteristic equation but on inspecting the coefficients $a_1, a_2, \ldots, a_n$ and determining under what conditions it is possible for a root to be in the right half of the s-plane and hence for the system to be unstable.

For **asymptotic stability**, the roots of equation (5.2) must be either

(1) real and negative, implying factors of equation (5.2) of the form

$$(s + \alpha) \quad (\alpha > 0), \qquad \text{or}$$

(2) complex with negative real parts, involving factors of the form (since the coefficients a_i are real)

$$(s + \alpha + i\beta)(s + \alpha - i\beta) \quad (\alpha > 0)$$

$$= s^2 + 2\alpha s + \alpha^2 + \beta^2.$$

The characteristic polynomial

$$P(s) = s^n + a_1 s^{n-1} + \ldots + a_n \tag{5.3}$$

is the product of these two types of factors, all having positive coefficients, hence

(a) all the coefficients $a_1, a_2, \ldots, a_n$ must be present and

(b) they must all be positive. — it -ve explain how many etc...

(c) cannot be 0 in 1st column

If (a) and (b) are satisfied, Routh's criterion requires the construction of the following array of coefficients and values dependent on the coefficients:

(d) Coefficients cannot be of different signs — either all +ve or all -ve. No mixtures!

row 1	1	a_2	a_4	a_6
row 2	a_1	a_3	a_5	a_7
row 3	b_1	b_2	b_3	b_4
row 4	c_1	c_2	c_3	c_4
row 5	d_1	d_2	d_3	d_4
$\vdots$		e_2		
row $(n+1)$		$\ldots$	$\ldots$	$\ldots$

In row 1 we insert the coefficients of $s^n, s^{n-2}, \ldots,$ in $P(s)$. When the coefficients 'run-out' we insert zeros, hence if $n = 5$, then $a_6 = a_7 = \ldots = 0$.

In row 2 we insert the coefficients of $s^{n-1}, s^{n-3}, \ldots,$ in $P(s)$. To generate the coefficients $b_1, b_2, \ldots, c_1, c_2$, etc. the following rules are used:

$$b_1 = -\frac{1}{a_1}\begin{vmatrix} 1 & a_2 \\ a_1 & a_3 \end{vmatrix}, \quad b_2 = -\frac{1}{a_1}\begin{vmatrix} 1 & a_4 \\ a_1 & a_5 \end{vmatrix}, \quad b_3 = -\frac{1}{a_1}\begin{vmatrix} 1 & a_6 \\ a_1 & a_7 \end{vmatrix}, \ldots$$

$$c_1 = -\frac{1}{b_1}\begin{vmatrix} a_1 & a_3 \\ b_1 & b_2 \end{vmatrix}, \quad c_2 = -\frac{1}{b_1}\begin{vmatrix} a_1 & a_5 \\ b_1 & b_3 \end{vmatrix}, \quad c_3 = -\frac{1}{b_1}\begin{vmatrix} a_1 & a_7 \\ b_1 & b_4 \end{vmatrix}, \ldots$$

$$d_1 = -\frac{1}{c_1}\begin{vmatrix} b_1 & b_2 \\ c_1 & c_2 \end{vmatrix}, \quad d_2 = -\frac{1}{c_1}\begin{vmatrix} b_1 & b_3 \\ c_1 & c_3 \end{vmatrix}, \quad d_3 = -\frac{1}{c_1}\begin{vmatrix} b_1 & b_4 \\ c_1 & c_4 \end{vmatrix}, \ldots$$

$\ldots \quad \ldots$

Having writen down the first two rows, the subsequent rows are easily constructed, using the simple rule defined by the above relations.

Example 5.1
Construct the Routh array for the polynomial

$$P(s) = s^4 + 2s^3 + s^2 + 6s + 4.$$

Solution

Row 1	1	1	4	0	...
Row 2	2	6	0	0	...
Row 3	−2	4	0	0	...
Row 4	10	0	0	0	...
Row 5	4	0	0	0	...

Routh's Stability Criterion

(a) The roots of the characteristic equations all have negative real parts, if and only if the elements of the first column of the Routh array are positive.

(b) The number of roots with positive real parts is equal to the number of changes in sign of the elements of the first column.

Note 1 In example 5.1, we were fortunate (from the numeric simplicity point of view) not to have any fractions. If a fraction is encountered we can multiply (or divide) the entire row by an appropriate positive constant without affecting the conclusion.

Note 2 If an element in the first column is zero, Routh's criterion is violated and the system is not asymptotically stable. There exist methods for completing the array in these circumstances, but we shall not discuss them here.

Note 3 The element a_n (the constant coefficient in $P(s)$) is propagated through the Routh array by the rule

'one to the left and two down'

starting from its original position, which is the last non-zero element in the first or the second row, depending on whether the $P(s)$ is an odd or even powered polynomial. Thus for

$$P(s) = s^5 + a_1 s^4 + a_2 s^3 + a_3 s^2 + a_4 s + a_5$$

the Routh array is

$$\begin{array}{ccc}
1 & a_2 & a_4 \\
a_1 & a_3 & \boxed{a_5} \\
b_1 & b_2 & 0 \\
c_1 & \boxed{a_5} & 0 \\
d_1 & 0 & 0 \\
\boxed{a_5} & 0 & 0\,.
\end{array}$$

To show in the fourth row that $c_2 = a_5$, notice that

$$c_2 = -\frac{1}{b_1}\begin{vmatrix} a_1 & a_5 \\ b_1 & 0 \end{vmatrix} = \frac{a_5\,b_1}{b_1} = a_5\,.$$

Similarly in the sixth row

$$e_1 = -\frac{1}{d_1}\begin{vmatrix} c_1 & a_5 \\ d_1 & 0 \end{vmatrix} = a_5\,.$$

Example 5.2

What does Routh's criterion indicate about the roots of

$$P(s) = s^4 + s^3 + 2s^2 + 3s + 3?$$

Solution

In this case the Routh's array has the following form:

$$\begin{array}{rrr}
1 & 2 & 3 \\
1 & 3 & 0 \\
-1 & 3 & 0 \\
6 & 0 & 0 \\
3 & 0 & 0\,.
\end{array}$$

Since there are two sign variations in the first column, two of the zeros have positive real parts. (In fact, the roots of $P(s) = 0$ are approximately $s = 0.33 \pm 1.44i$ and $s = -.83 \pm 0.48i$.)

Example 5.3

Withe reference to example 4.4, show that, using an integral controller, the system is unstable. Suggest a choice of constants so that the system using a PI controller is stable.

Solution

We have shown (example 4.4) that when an integral controller is used the characteristic equation is

$$s^3 + 2s^2 + K = 0.$$

Since the coefficient of s is zero, the system is unstable for all K.

Using a PI controller, the characteristic equation is

$$s^3 + 2s^2 + K_p s + KK_p = 0.$$

Choose $K_p = 1$, so that

$$s^3 + 2s^2 + s + K = 0$$

and the Routh array in this case is

$$\begin{array}{cc} 1 & 1 \\ 2 & K \\ \dfrac{2-K}{2} & 0 \\ K & 0. \end{array}$$

Hence for stability we must have:

$$K > 0 \quad \text{and} \quad \frac{2-K}{2} > 0, \quad \text{that is,} \quad K < 2.$$

Hence the choice $K_p = 1$ and $K < 2$ leads to an asymptotically stable system. For example, when $K = 1.5$, the roots are (approximately)

$$s = -1.76 \quad \text{and} \quad s = -0.12 \pm 0.74i$$

so that the system is stable. But when $K = 2.5$, the roots are

$$s = -2.09 \quad \text{and} \quad s = 0.5 \pm 1.09i$$

so that the system is unstable.

The Routh criterion can be further exploited to give more information about the system stability. Although we are most interested in the question of the system absolute stability, we often need some indication of the system relative stability. In other words, although we require all the roots of the characteristic equations to have negative real parts, sometimes we need to know whether these roots have real parts which are less than $(-\sigma)$ where σ is some positive real number. For example, we may wish all the system modes to decay at a rate which is at least as fast as $e^{-\sigma t}$. If this is the case, the system is said to have a **stability margin** σ.

The characteristic equation

$$P(s) = 0$$

has roots $\lambda_1, \lambda_2, \ldots, \lambda_n$ so that the equation

$$P(s - \sigma) = 0$$

has roots $\lambda_1 + \sigma, \lambda_2 + \sigma, \ldots, \lambda_n + \sigma$.

Now applying the Routh criterion to $P(s - \sigma) = 0$ will determine whether

$$\lambda_i + \sigma \qquad (i = 1, 2, \ldots, n)$$

will all have negative real parts, that is whether

$$\text{Re}\,\lambda_i < -\sigma \qquad (i = 1, 2, \ldots, n).$$

Example 5.4

A system has an associated characteristic polynomial

$$P(s) = s^3 + 9s^2 + 26s + 24.$$

Determine whether the system stability margin is

(a) 1

(b) 2.

Solution

(a) $P(s-1) = (s-1)^3 + 9(s-1)^2 + 26(s-1) + 24$

$$= s^3 + 6s^2 + 11s + 6.$$

The Routh array for $P(s-1)$ is

$$\begin{array}{cc} 1 & 11 \\ 6 & 6 \\ 10 & 0 \\ 6 & 0. \end{array}$$

All the roots of $P(s-1)=0$ have negative real parts, hence the system stability margin is 1.

(b) $P(s-2) = s^3 + 3s^2 + 28s.$

Since the constant coefficient of $P(s-2)$ is zero, at least one of the roots of $P(s-2)=0$ has a non-negative real part, hence the system stability margin is less than 2.

Note: In this example, $P(s)=0$ has roots $-2, -3$, and -4.

5.4 INTRODUCTION TO LIAPUNOV'S METHOD

Liapunov's method is the most general method we know for system stability analysis. It is applicable to both linear and non-linear systems of any order. For linear systems it provides both the necessary and sufficient conditions, whereas for nonlinear systems it provides only the sufficient conditions for asymptotic stability. It will be shown that it is possible to determine the system's stability without the necessity of solving the state equations and determining the roots of the system characteristic equation. The method depends on defining a Liapunov function, closely related to the system energy function. For linear systems, such functions are easily determined; unfortunately for nonlinear systems it can be extremely difficult to do so.

We only consider linear systems, although the simple illustrative examples used can be solved by other elementary methods. The object is to present a method, quite different from all other methods, whose true worth can only be fully appreciated for problems which are beyond the scope of this book. We nevertheless hope that this outline will stimulate the reader to consult more detailed works on this topic (some of which are quoted in the *Bibliography*) especially when applied to nonlinear systems.

Although Liapunov published his treatise on the stability of motion in 1892, it is only recently, in the 1960s, that it has gained the enormous popularity that it deserves.

Actually Liapunov suggested two methods. It is the second method, known as the **'direct' method**, which we now consider.

Liapunov considered the solution $\mathbf{x}(t)$ of the state equations to trace out a **trajectory** (also known as an **orbit** or a **path**) which, for an asymptotically stable system, terminates at the origin of the n-dimensional state-space – known as the **phase plane** when $n = 2$.

As a very simple illustrative example of a trajectory, we consider the motion of an undamped pendulum (see Fig. 5.2). The equation of motion is

$$\frac{\mathrm{d}^2\theta}{\mathrm{d}t^2} + \frac{g}{l}\sin\theta = 0 \tag{5.4}$$

which is a nonlinear equation.

Although for this very simple non-linear example we could obtain the trajectory in the phase plane, we are in this book primarily concerned with linear systems. On the other hand, with the assumption that the displacements are small, we have

$$\sin\theta \simeq \theta$$

and equation (5.4) then becomes

$$\frac{\mathrm{d}^2\theta}{\mathrm{d}t^2} + \frac{g}{l}\theta = 0. \tag{5.5}$$

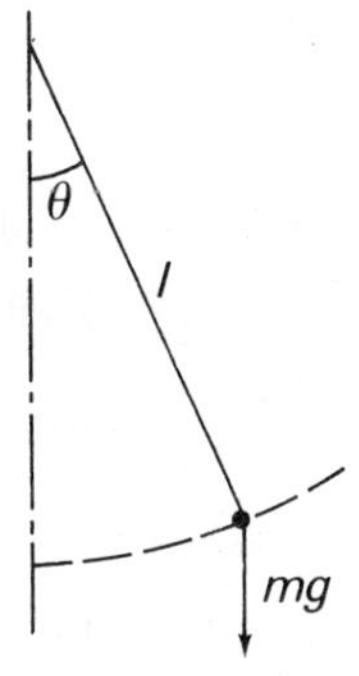

Fig. 5.2

We are again using a 'linearisation technique', first mentioned in Chapter 1. Notice that an analogous linear equation is obtained for the motion of a mass (m) on a spring of stiffness k; the equation is

$$x'' + \frac{k}{m}x = 0.$$

This is a well known linear equation of simple harmonic motion, and has the solution

$$\theta = A \sin(\sqrt{}(g/lt) + \epsilon) \tag{5.6}$$

where A and ϵ are constants determined by the initial conditions.

To write equation (5.5) in state-space form we define the variables

$$x_1 = \theta$$

and $$x_2 = \dot{\theta} \quad (= \mathrm{d}\theta/\mathrm{d}t)$$

so that $$\dot{x}_1 = x_2$$

and $$\dot{x}_2 = -g/lx_1. \tag{5.7}$$

We next find the solution to equation (5.7), in this case (using equation (5.6))

$$\left.\begin{aligned} x_1 &= A \sin(\sqrt{}(g/l)t + \epsilon) \\ x_2 &= A\sqrt{}(g/l) \cos(\sqrt{}(g/l)t + \epsilon) \end{aligned}\right\} \tag{5.8}$$

and obtain the trajectory by eliminating t in equation (5.8), it is

$$x_1^2 + \frac{x_2^2}{\sqrt{g/l}} = A^2. \tag{5.9}$$

The trajectories (for various values of A and ϵ) are ellipses as shown in Fig. 5.3.

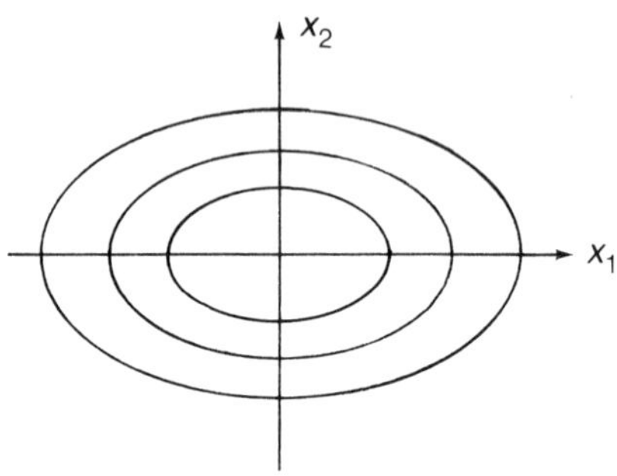

Fig. 5.3

Every trajectory shown in Fig. 5.3 is closed, showing that the pendulum, assumed to have no resistance to its motion, will swing indefinitely. A closed trajectory is typical of a periodic motion, that is, one for which the solution $\mathbf{x}(t)$ has the property

$$\mathbf{x}(t+T) = \mathbf{x}(t)$$

where T is the period of the motion.

Had we assumed a more realistic situation, when the oscillations are damped, the trajectory would terminate at the point $\mathbf{x} = \mathbf{0}$ so that it may have the form shown in Fig. 5.4. (The arrow indicates the direction of motion as t increases.)

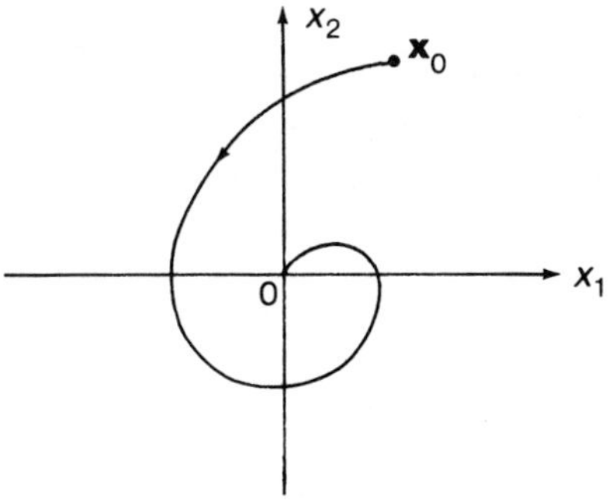

Fig. 5.4

In general autonomous system, the state equation corresponding to equation (5.7) takes the form

$$\dot{x}_1 = P(x_1, x_2) \quad \text{and} \quad \dot{x}_2 = Q(x_1, x_2) \tag{5.9}$$

The solutions (corresponding to equation (5.8)) are

$$x_1 = \phi(t) \quad \text{and} \quad x_2 = \psi(t) \tag{5.10}$$

and can be plotted to give the trajectory.

For an n-dimensional state equation, the algebra is similar, but of course the trajectory cannot be plotted.

Definition

A point (x_1^0, x_2^0), for which $P(x_1^0, x_2^0) = 0 = Q(x_1^0, x_2^0)$ is called a **critical point**.

Notice that the solution of equation (5.9) at a critical point (x_1^0, x_2^0) is

$$x_1 = (\text{constant}) = x_1^0 \quad \text{and} \quad x_2 = (\text{constant}) = x_2^0,$$

hence the point (x_1^0, x_2^0) is the trajectory; that is, the solution starting at this point, remains there. This is the reason why the solution of the state equation is said to be in **equilibrium** at a critical point.

For a linear system there is only one equilibrium point (we assume that the matrix A is non-singular) and it is the point $\mathbf{x} = \mathbf{0}$. For the linearised pendulum equation (5.7) this is obviously the case; the equilibrium point is $(\theta = 0, \dot{\theta} = 0)$.

It is the trajectory of the state vector $\mathbf{x}$, when the system is disturbed from its position of equilibrium $\mathbf{x} = \mathbf{0}$, which determines the stability or the instability of the equilibrium point and hence (in the case of a linear system) of the system. Given the initial conditions, this trajectory is completely determined by the system state equations.

In fact, we can see from the discussion of the pendulum problem, that the equilibrium point $\mathbf{x} = \mathbf{0}$ appears stable since a trajectory starting 'near' this point always remains within 'reasonable' distance of it. On the other hand, a system is asymptotically stable (that is, damped oscillations) when it is stable and the trajectory eventually approaches $\mathbf{x} = \mathbf{0}$ as $t \to \infty$. In mathematical terminology, these definitions take the form shown in Fig. 5.5.

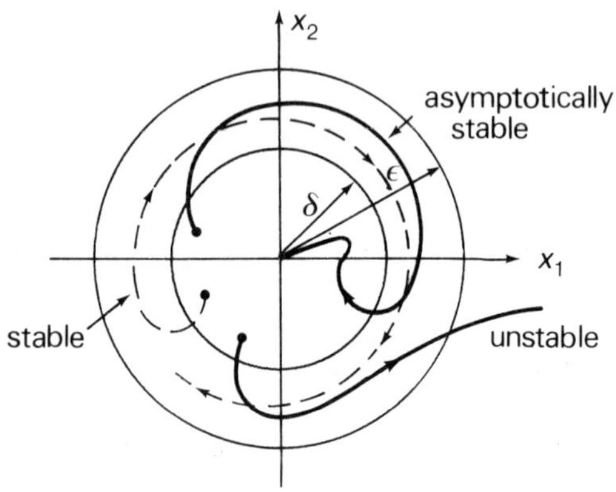

Fig. 5.5

The equilibrium point $\mathbf{x} = \mathbf{0}$ (Fig. 5.5) is said to be

(1) **stable** if, given any $t > 0$, there exists $\delta > 0$ so that every solution (equation (5.10)) which at $t = 0$ is such that

$$x_1^2 + x_2^2 = \psi(0)^2 + \phi(0)^2 < \delta^2$$

implies

$$\psi^2(t) + \phi^2(t) < \epsilon^2 \quad \text{for} \quad t \geqslant 0.$$

(2) **asymptotically stable** if

(a) it is stable, and

(b) $\lim_{t\to\infty} \psi(t) = 0$ and $\lim_{t\to\infty} \phi(t) = 0.$

The two conditions in (b) can be combined and written as

$$\lim_{t\to\infty} [\psi^2(t) + \phi^2(t)] = 0.$$

(3) **unstable** if it is not stable.

In the definitions, starting 'near' the equilibrium point is defined by the 'δ-neighbourhood', whereas the 'reasonable' distance is the 'ϵ-neighbourhood'.

If we consider an unforced system, having the state equation

$$\dot{\mathbf{x}} = A\mathbf{x} \tag{5.11}$$

we have found the solution (see equation (3.19))

$$\mathbf{x} = e^{At}\mathbf{x}(0)$$

$$= (A_1 e^{\lambda_1 t} + A_2 e^{\lambda_2 t} + \ldots + A_n e^{\lambda_n t})\mathbf{x}(0). \tag{5.12}$$

For asymptotic stability,

$$\lim_{t\to\infty} ||\mathbf{x}(t)|| = 0$$

where $||\mathbf{x}||$ is the norm of the (solution) vector $\mathbf{x}$ defined by

$$||\mathbf{x}||^2 = (\mathbf{x},\mathbf{x}) = x_1^2 + x_2^2 + \ldots + x_n^2.$$

Hence for asymptotic stability *all* the state variables $x_1, x_2, \ldots, x_n$ must decrease to zero as $t \to \infty$. From equation (5.12) we see that this is the case when $\text{Re}\,\lambda_i < 0$ $(i = 1, 2, \ldots, n)$, where λ_i are the eigenvalues of A, that is, the solutions of the characteristic equation

$$|sI - A| = 0.$$

We have, therefore, the same problem as in the previous section except that now we are examining the system state equations, whereas in the previous section we considered the system transfer function.

Returning now to Liapunov, his method depends on the state trajectory – but not directly. He suggested that there may exist functions, associated with the trajectory, whose properties may determine whether or not the trajectory terminates at the origin.

It is a well known principle in classical mechanics that the potential energy U of a conservative system has stationary values at equilibrium points and if U is a minimum at such a point, the point is stable.

We illustrate this with the pendulum example (see Fig. 5.2).

$$U = mgl(1 - \cos\theta) \simeq \tfrac{1}{2}mgl\theta^2 \quad \text{(for small } \theta\text{)}.$$

Notice that $U = 0$ (a minimum) at $\theta = \mathbf{0}$.

The kinetic energy is

$$\tfrac{1}{2}ml^2\dot{\theta}^2.$$

Hence the system energy is

$$\begin{aligned} V &\simeq \tfrac{1}{2}mgl\theta^2 + \tfrac{1}{2}ml^2\dot{\theta}^2 \\ &= \tfrac{1}{2}mglx_1^2 + \tfrac{1}{2}ml^2x_2^2 \quad \text{(see equation (5.7))} \\ &= V(x_1, x_2). \end{aligned} \tag{5.13}$$

Notice that V is zero at the point (0,0).

It follows that

$$\frac{x_1^2}{a^2V} + \frac{x_2^2}{b^2V} = 1$$

where $a^2 = \dfrac{2}{mgl}$ and $b^2 = \dfrac{2}{ml^2}$.

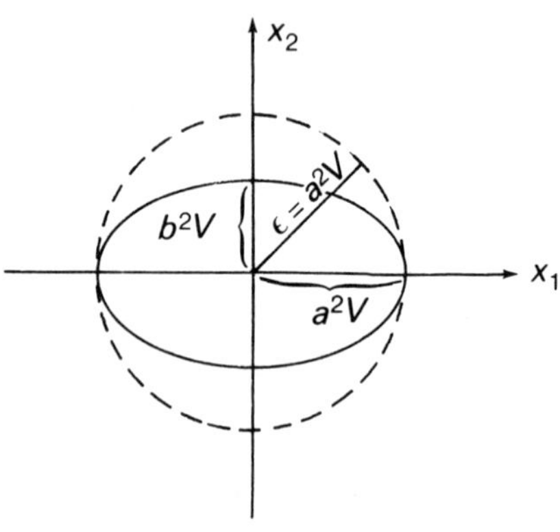

Fig. 5.6

The trajectory is (again) seen to be an ellipse (see Fig. 5.6) with major and minor axes having lengths $2a^2V$ and $2b^2V$ respectively. The nearer the initial state $\mathbf{x}(0) = (x_1, x_2)$ is to the point of equilibrium (0,0), the smaller is V and so the corresponding ellipse has smaller major and minor axes.

This fits in well with our definition of stability – once the system has been displaced from its position of equilibrium, the trajectory will remain within the $\epsilon = a^2V$ neighbourhood. The point of interest here is that we no longer consider the actual trajectory but an energy function associated with the trajectory.

Notice that the rate of change of the energy along a trajectory is (see equation (5.13)).

$$\dot{V}(\mathbf{x}) = \frac{dV(\mathbf{x})}{dt} = \frac{\partial V}{\partial x_1}\frac{dx_1}{dt} + \frac{\partial V}{\partial x_2}\frac{dx_2}{dt} \tag{5.14}$$

$$= mglx_1\frac{dx_1}{dt} + ml^2x_2\frac{dx_2}{dt}$$

$$= mglx_1x_2 + ml^2x_2(-g/l)x_1 \quad \text{(using equation (5.7))}$$

$$= 0.$$

Hence the total energy is a constant along any trajectory – as we expect for a conservative system.

Again the point of interest here is that we are able to obtain dV/dt along a trajectory without the necessity for solving the system equations – we have only used the expression for V and the state equation (5.7).

We shall not go through the mathematical manipulations involved when the oscillations are damped, that is, when the system is asymptotically stable and the trajectory terminates at the equilibrium point. V is still given by equation (5.13), but $dV/dt < 0$ showing that the energy is decreasing monotonically along every trajectory. This means that every trajectory moves towards the point of minimum energy, that is, the equilibrium point (0,0). This is in essence the Liapunov's criterion for asymptotic stability.

Liapunov's function $V(\mathbf{x})$ is similar to the energy function; it has the following properties.

(1) $V(\mathbf{x})$ and its partial derivatives $\partial V/\partial x_i$ $(i = 1, 2, \ldots, n)$ are continuous.

(2) $V(\mathbf{x}) > 0$ for $\mathbf{x} \neq \mathbf{0}$ in the neighbourhood of $\mathbf{x} = \mathbf{0}$, and $V(\mathbf{0}) = 0$.

(3) $\dot{V}(\mathbf{x}) < 0$ for $\mathbf{x} \neq \mathbf{0}$ (in the neighbourhood of $\mathbf{x} = \mathbf{0}$) and $\dot{V}(\mathbf{0}) = 0$.

$V(\mathbf{x})$ satisfying the above properties is called a **Liapunov function**.

Before discussing Liapunov's theorem we first define some of the technical terminology used.

Definition

$V(\mathbf{x})$ is **positive-definite** if

(1) $V(\mathbf{x}) > 0$ for $\mathbf{x} \neq \mathbf{0}$, and

(2) $V(\mathbf{0}) = 0$.

$V(\mathbf{x})$ is **positive-semi-definite** if

$V(\mathbf{x}) \geqslant 0$ for all $\mathbf{x}$.

$V(\mathbf{x})$ is **negative-definite** if

$(-V(\mathbf{x}))$ is positive-definite.

$V(\mathbf{x})$ is **negative-semi-definite** if

$(-V(\mathbf{x}))$ is positive-semi-definite.

Example 5.5

Classify the following functions according to the above definitions:

(a) $V(\mathbf{x}) = x_1^2 + 3x_2^2$.

(b) $V(\mathbf{x}) = -(x_1 + x_2)^2 - 2x_1^2$.

(c) $V(\mathbf{x}) = (2x_1 + x_2)^2$.

(d) $V(\mathbf{x}) = x_1^2 + 2x_2^2 - 4x_1x_2$.

Solution

(a) $V(\mathbf{x}) > 0$, for $\mathbf{x} \neq \mathbf{0}$ and $V(\mathbf{0}) = 0$, hence $V(\mathbf{x})$ is positive-definite.

(b) $V(\mathbf{x}) < 0$ for $\mathbf{x} \neq \mathbf{0}$, $V(\mathbf{0}) = 0$, hence $V(x)$ is negative-definite.

(c) Whenever $x_2 = -2x_1$, $V(\mathbf{x}) = 0$, otherwise $V(\mathbf{x}) > 0$.
Hence $V(\mathbf{x}) \geqslant 0$. $V(\mathbf{x})$ is positive-semi-definite.

(d) $V(1,1) < 0$, $V(1,3) > 0$. $V(\mathbf{x})$ assumes both positive and negative values, hence is **indefinite**.

Liapunov's Theorem

The origin (that is, the equilibrium point) is asymptotically stable if there exists a Liapunov function in the neighbourhood of the origin.

If $\dot{V}(\mathbf{x})$ is only negative-semi-definite, that is, $\dot{V}(\mathbf{x}) \leqslant 0$, then the origin is a stable point.

Proof

First consider the case when $\dot{V}(\mathbf{x}) \leqslant 0$ (the stability of the origin). Then $V(\mathbf{x}) = K$ (K is a constant) and this represents a closed curve enclosing the origin.

Also $\dot{V}(\mathbf{x}) = \dot{V}(x_1, x_2)$ in the form of equation (5.14) can be written as the scalar (the 'dot') product of two vectors, that is

$$\dot{V}(x_1, x_2) = \frac{\partial V}{\partial x_1}\frac{dx_1}{dt} + \frac{\partial V}{\partial x_2}\frac{dx_2}{dt} = \left[\frac{\partial V}{\partial x_1}, \frac{\partial V}{\partial x_2}\right] \cdot \left[\frac{dx_1}{dt}, \frac{dx_2}{dt}\right].$$

Writing x_1, x_2 in terms of their solutions equation (5.10) the vector

$$\left[\frac{dx_1}{dt}, \frac{dx_2}{dt}\right]' = \frac{d\mathbf{x}}{dt}$$

is the tangent vector to the trajectory (defined by $x_1 = \phi(t)$, $x_2 = \psi(t)$). On the other hand,

$$\left|\frac{\partial V}{\partial x_1}, \frac{\partial V}{\partial x_2}\right|' = \text{grad } V$$

is a vector normal to the surface $V(x_1, x_2) = K$.

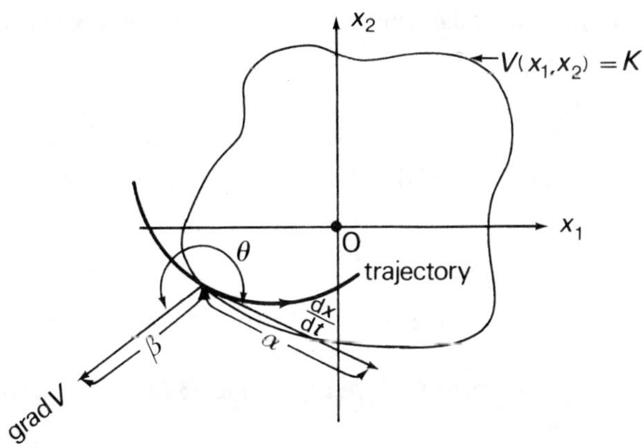

Fig. 5.7

Hence

$$\dot{V}(x_1, x_2) = \alpha\beta \cos\theta \leqslant 0$$

where α and β are the magnitudes of the two vectors involved, $d\mathbf{x}/dt$ and grad V, and θ is the angle between them.

It follows that $\frac{\pi}{2} \leqslant \theta \leqslant \frac{3\pi}{2}$.

If $\dot{V}(\mathbf{x}) = 0$ then $\theta = \pi/2$ and the direction of the motion on the trajectory is tangent (that is, along) the curve.

If $\dot{V}(\mathbf{x}) < 0$, then $\pi/2 < \theta \leqslant 3\pi/2$, that is, the direction of the motion on the trajectory is into the curve. Notice that no trajectory exists on which the motion would be outwards from the curve. It follows that once a trajectory starts inside the curve $V(\mathbf{x}) = K$, it will never cross it to move outwards. This proves that the origin is stable by the criterion discussed earlier in this section. This is sometimes referred to as **stability in the sense of Liapunov**.

If $\dot{V}(\mathbf{x})$ is negative-definite, the asymptotic stability of the origin follows by similar reasoning.

Before working through a number of examples, we consider a family of functions which are of particular usefulness to our consideration of stability of linear systems.

5.5 QUADRATIC FORMS

Definition

A quadratic form is a scalar function $V(\mathbf{x})$ of variables $\mathbf{x} = [x_1\, x_2 \ldots x_n]'$ defined by

$$V(\mathbf{x}) = \mathbf{x}'P\mathbf{x} = [x_1\, x_2 \ldots x_n] \begin{bmatrix} p_{11} & p_{12} & \cdots & p_{1n} \\ p_{21} & p_{22} & \cdots & p_{2n} \\ \vdots & & & \\ p_{n1} & p_{n2} & \cdots & p_{nn} \end{bmatrix} \begin{bmatrix} x_1 \\ x_2 \\ \vdots \\ x_n \end{bmatrix} \tag{5.15}$$

$$\begin{aligned} &= p_{11}x_1^2 + p_{12}x_1x_2 + p_{13}x_1x_3 + \ldots + p_{1n}x_1x_n \\ &\quad + p_{21}x_2x_1 + p_{22}x_2^2 + p_{23}x_2x_3 + \ldots + p_{2n}x_2x_n \\ &\quad \vdots \\ &\quad + p_{n1}x_nx_1 + p_{n2}x_nx_2 + p_{n3}x_nx_3 + \ldots + p_{nn}x_n^2 \\ &= \sum_{i=1}^{n} \sum_{j=1}^{n} p_{ij}x_ix_j. \end{aligned}$$

The matrix P is of order $n \times n$ and is symmetric, that is

$$p_{ij} = p_{ji} \quad (\text{all } i, j\,;\, i \neq j)$$

so that $\quad P' = P.$

For example, we can write

$$V(\mathbf{x}) = x_1^2 + 2x_2^2 - 4x_1x_2$$

as

$$V(\mathbf{x}) = [x_1 \; x_2] \begin{bmatrix} 1 & -2 \\ -2 & 2 \end{bmatrix} \begin{bmatrix} x_1 \\ x_2 \end{bmatrix}.$$

In problems involving more than two or three variables, the determination of whether or not a quadratic form is positive-definite can be formidable. Fortunately, there exists a procedure, known as **Sylvester's criterion**, which is very helpful. It states that the necessary and sufficient condition for a quadratic form $V(\mathbf{x}) = \mathbf{x}'P\mathbf{x}$ to be positive-definite is that all the successive principal minors of the symmetric matrix P are positive, that is,

$$p_{11} > 0, \quad \begin{vmatrix} p_{11} & p_{12} \\ p_{12} & p_{22} \end{vmatrix} > 0, \quad \begin{vmatrix} p_{11} & p_{12} & p_{13} \\ p_{12} & p_{22} & p_{23} \\ p_{13} & p_{23} & p_{33} \end{vmatrix} > 0, \quad \text{etc.}$$

Example 5.6

Determine whether

$$V(\mathbf{x}) = 4x_1^2 + 3x_2^2 + x_3^2 + 4x_1x_2 - 2x_1x_3 - 2x_2x_3$$

is positive-definite.

Solution

$V(\mathbf{x})$ can be written as

$$V(\mathbf{x}) = [x_1 \;\; x_2 \;\; x_3] \begin{bmatrix} 4 & 2 & -1 \\ 2 & 3 & -1 \\ -1 & -1 & 1 \end{bmatrix} \begin{bmatrix} x_1 \\ x_2 \\ x_3 \end{bmatrix}.$$

Since $4 > 0, \begin{vmatrix} 4 & 2 \\ 3 & 3 \end{vmatrix} = 8 > 0$ and $\begin{vmatrix} 4 & 2 & -1 \\ 2 & 3 & -1 \\ -1 & -1 & 1 \end{vmatrix} = 5 > 0.$

Sylvester's criterion assures us that $V(\mathbf{x})$ is positive-definite.

We are now in a position to show the procedure for finding a Liapunov's function, at least in the context of linear systems with which we are concerned.

5.6 DETERMINATION OF LIAPUNOV'S FUNCTIONS

As mentioned in section 5.3, the stability of a system (or more correctly, the stability of the equilibrium point) depends on the dynamics of the system, that is, on the unforced state equation (equation (5.11)) $\dot{\mathbf{x}} = A\mathbf{x}$.

Definition

The matrix A is said to be stable, if the corresponding system defined by equation (5.11) is stable.

As explained above, to determine whether A is stable we seek a symmetric matrix P so that

$$V(\mathbf{x}) = \mathbf{x}'P\mathbf{x}$$

is a positive-definite function, and

$$\begin{aligned}\dot{V}(\mathbf{x}) &= \dot{\mathbf{x}}'P\mathbf{x} + \mathbf{x}'P\dot{\mathbf{x}} \\ &= (A\mathbf{x})'P\mathbf{x} + \mathbf{x}'PA\mathbf{x} \quad \text{(using equation (5.11)} \\ &= \mathbf{x}'[A'P + PA]\mathbf{x} \end{aligned} \tag{5.16}$$

is negative-definite.

Notice that $[A'P + PA]' = [P'A + A'P']$
$= [PA + A'P]$ (since P is symmetric).

Hence $\dot{V}(\mathbf{x})$ is a quadratic form.

$$\text{Let} \quad -Q = A'P + PA; \tag{5.17}$$

for $\dot{V}(\mathbf{x})$ to be negative-definite,

$$\mathbf{x}'Q\,\mathbf{x}$$

must be positive-definite.

Liapunov's theorem can now be expressed in the following form:

> Given a positive-definite matrix Q, the matrix A is stable if the solution to the (matrix) equation (5.17) results in a positive-definite matrix P.

The solution to equation (5.17) can be very tedious. It is, of course, in our interest to make the calculations as simple as possible; for this purpose we choose Q to be simple, $Q = I$ for example.

Example 5.7

Use Liapunov's direct method to determine whether the following systems are asymptotically stable:

(a) $\dot{x}_1 = x_2$

$\dot{x}_2 = 2x_1 - x_2.$

(b) $\dot{x}_1 = x_2$

$\dot{x}_2 = -5x_1 - 2x_2.$

Solution

(a) $A = \begin{bmatrix} 0 & 1 \\ 2 & -1 \end{bmatrix}$ and equation (5.17) becomes

$$\begin{bmatrix} 0 & 2 \\ 1 & -1 \end{bmatrix}\begin{bmatrix} p_{11} & p_{12} \\ p_{12} & p_{22} \end{bmatrix} + \begin{bmatrix} p_{11} & p_{12} \\ p_{12} & p_{22} \end{bmatrix}\begin{bmatrix} 0 & 1 \\ 2 & -1 \end{bmatrix} = \begin{bmatrix} -1 & 0 \\ 0 & -1 \end{bmatrix}.$$

On solving, we obtain

$$P = \begin{bmatrix} -\frac{3}{4} & -\frac{1}{4} \\ -\frac{1}{4} & \frac{1}{4} \end{bmatrix}$$

which, by Sylvester's criterion, is *not* positive definite. Hence the system is not stable.

Note that the poles (the eigenvalues of A) are $\lambda_1 = 1$ and $\lambda_2 = -2$.

(b) In this case

$$A = \begin{bmatrix} 0 & 1 \\ -5 & -2 \end{bmatrix}$$

and on solving the equation

$$A'P + PA = -I$$

we obtain

$$P = \frac{1}{10}\begin{bmatrix} 17 & 1 \\ 1 & 3 \end{bmatrix}$$

which is positive-definite - hence the system is asymptotically stable. Note that the poles are $\lambda_1 = -1-2i$ and $\lambda_2 = -1+2i$.

Example 5.8

Determine, using Liapunov's method, the range of K for the system in Fig. 5.8 to be asymptotically stable.

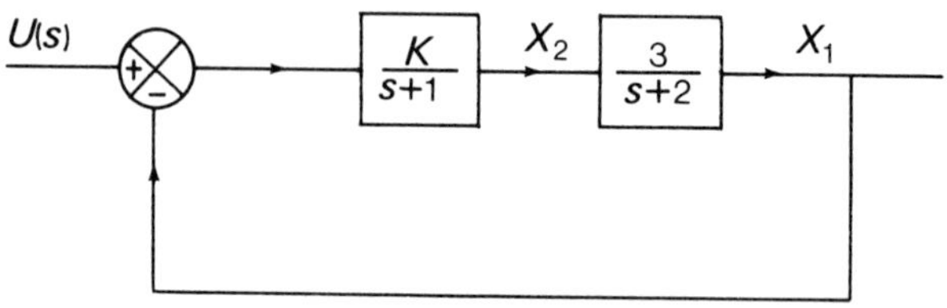

Fig. 5.8

Solution

From Fig. 5.8

$$X_1 = \frac{3}{s+2} X_2 \quad \text{hence} \quad \dot{x}_1 = -2x_1 + 3x_2$$

$$X_2 = \frac{K}{s+1}(-X_1 + U) \quad \text{hence} \quad \dot{x}_2 = -Kx_1 - x_2 + Ku.$$

The state-space equation is

$$\begin{bmatrix} \dot{x}_1 \\ \dot{x}_2 \end{bmatrix} = \begin{bmatrix} -2 & 3 \\ -K & -1 \end{bmatrix} \begin{bmatrix} x_1 \\ x_2 \end{bmatrix} + \begin{bmatrix} 0 \\ K \end{bmatrix} u.$$

Equation (5.17) becomes

$$\begin{bmatrix} -2 & -K \\ 3 & -1 \end{bmatrix} \begin{bmatrix} p_{11} & p_{12} \\ p_{12} & p_{22} \end{bmatrix} + \begin{bmatrix} p_{11} & p_{12} \\ p_{12} & p_{22} \end{bmatrix} \begin{bmatrix} -2 & 3 \\ -K & -1 \end{bmatrix} = \begin{bmatrix} -1 & 0 \\ 0 & -1 \end{bmatrix}.$$

from which, after some algebra, we deduce that

$$P = \frac{1}{18K+12} \begin{bmatrix} K^2 + 3K + 3 & 3 - 2K \\ 3 - 2K & 3K + 15 \end{bmatrix}.$$

Sylvester's criterion leads us to conclude that

$$K > -\frac{2}{3}.$$

Even in the case of 2 X 2 matrices it is seen that solving equation (5.17) requires an appreciable amount of computation. Clearly, efficient methods for this would be most desirable. A lot of work has been done on this and a good survey is found in [5].

5.7 THE NYQUIST STABILITY CRITERION

The two previous methods of stability analysis were dependent on the detailed knowledge of the system dynamics. Any change in the system required the recalculation of the stability criteria.

If the transfer function of a system is not known, it is often possible to obtain it experimentally by exciting the system with a range of sinusoidal inputs of (generally) fixed amplitude but varying frequencies, and observing the resulting changes in amplitude and phase shift at the output.

The information obtained in this way, known as the **system frequency response**, can be used in various ways for system analysis and design. In particular, a polar plot of the frequency response of the open-loop system is called a **Nyquist diagram**. The stability of the corresponding closed-loop unity feedback system is determined by a simple inspection of the Nyquist diagram and depends on the position of the frequency locus relative to the point (−1,0). An important feature of this method is that it makes use of the open-loop transfer function which is much simpler to obtain and handle than the closed-loop transfer function.

So although the mathematical justification of the method is rather cumbersome, its use is particularly simple.

5.8 THE FREQUENCY RESPONSE

Since we are dealing with linear systems, we would expect that for a sinusoidal input, the output will also be sinusoidal. We now investigate this statement.

If the system has a transfer function $G(s)$ then the output is given by (see equation (2.3))

$$Y(s) = G(s)U(s).$$

For a sinusoidal input of unit amplitude

$$u(t) = \sin \omega t,$$

so that

$$U(s) = \frac{\omega}{s^2 + \omega^2}$$

and

$$Y(s) = G(s)\frac{\omega}{s^2 + \omega^2}.$$

Assuming that $G(s)$ has poles at $s = -p_1, -p_2, \ldots, -p_m$, $Y(s)$ can be expanded in partial fractions as

$$Y(s) = \sum_{j=1}^{n} \frac{K_j}{s + p_j} + \frac{Q_1}{s + i\omega} + \frac{Q_2}{s - i\omega} \tag{5.18}$$

where $i^2 = -1$ and (see Appendix 1)

$$K_j = \left[\frac{(s + p_j)\omega G(s)}{s^2 + \omega^2}\right]_{s=-p_j}$$

$$Q_1 = \left[\frac{(s + i\omega)\omega G(s)}{s^2 + \omega^2}\right]_{s=-i\omega} = -\frac{1}{2i}G(-i\omega)$$

and

$$Q_2 = \left[\frac{(s - i\omega)\omega G(s)}{s^2 + \omega^2}\right]_{s=i\omega} = \frac{1}{2i}G(i\omega).$$

On taking the inverse Laplace transform of equation (5.18), we obtain

$$y(t) = \underbrace{\sum_{j=1}^{n} K_j e^{-p_j t}}_{\text{transient terms}} + \underbrace{Q_1 e^{-i\omega t} + Q_2 e^{i\omega t}}_{\text{steady-state terms}}. \tag{5.19}$$

If the system is stable, all the poles assumed at $s = -p_1, -p_2, \ldots, -p_n$, are in the left-half of the s-plane (that is, they have negative real parts), so that as t increases, the transient terms decrease and become negligible relative to the steady-state terms in equation (5.19).

The steady-state terms can be written as

$$\frac{1}{2i}[G(i\omega)e^{i\omega t} - G(-i\omega)e^{-i\omega t}]$$

$$= \text{Im}\{G(i\omega)e^{i\omega t}\} \tag{5.20}$$

(We use the notation Im and Rl to indicate the 'imaginary part' and the 'real part' respectively.) Expressing $G(i\omega)$ in polar form as

$$G(i\omega) = |G(i\omega)|e^{i\phi}$$

where $\phi = \tan^{-1} \dfrac{\text{Im}\{G(i\omega)\}}{\text{Rl}\{G(i\omega)\}}$

we can write equation (5.20) as

$$\text{Im}\{|G(i\omega)|e^{i(\omega t+\phi)}\} = |G(i\omega)|\sin(\omega t + \phi) \tag{5.21}$$

where we have used the identity

$$e^{ix} = \cos x + i \sin x.$$

Equation (5.21) shows that the steady-state output is also sinusoidal, and relative to the input the amplitude is multiplied by $|G(i\omega)|$. Although the frequency does not change, there is a phase shift ϕ relative to the input.

The frequency diagram is the plot of $G(i\omega)$ in the complex plane as the frequency varies from $-\infty$ to $+\infty$.

Since for the systems considered the transfer functions $G(s)$ are the ratios of two polynomials with real coefficients, it follows that

$$G(i\omega) = \text{the complex conjugate of } G(-i\omega)$$

and it is sufficient to plot $G(i\omega)$ for $0 \leqslant \omega < \infty$; the remainder of the locus is symmetrical about the real axis to this plot.

Example 5.9

Determine the frequency diagrams for the following transfer functions:

(i) $G(s) = Ks$, (ii) $G(s) = K/s$, (iii) $G(s) = K/(a + s)$

where K and a are constants.

Solution

(i) $G(i\omega) = iK\omega$, hence

$|G(i\omega)| = K\omega$ and $\phi(\omega) = 90°$.

The plot is the positive imaginary axis (for $\omega \geqslant 0$).

(ii) $G(i\omega) = K/(i\omega)$

$|G(i\omega)| = K/\omega$ and $\phi(\omega) = -90°$.

The plot is the negative imaginary axis (for $\omega \geqslant 0$).

(iii) To plot the locus it is helpful to construct a table giving $|G(i\omega)|$ and ϕ for various values of ω.

$$G(i\omega) = \frac{K}{a+i\omega} = \frac{K(a-i\omega)}{a^2+\omega^2}.$$

Hence $|G(i\omega)| = \dfrac{K}{(a^2+\omega^2)^{\frac{1}{2}}}$ and $\phi = -\tan^{-1}\omega/a$

ω	$\lvert G(i\omega)\rvert$	ϕ
0	K/a	0
a	$K/\sqrt{2}a$	$-45°$
$2a$	$K/\sqrt{5}a$	$-63°$
⋮	⋮	⋮
∞	0	$-90°$

The locus for $\omega \geqslant 0$ is the lower semicircle shown in Fig. 5.9.

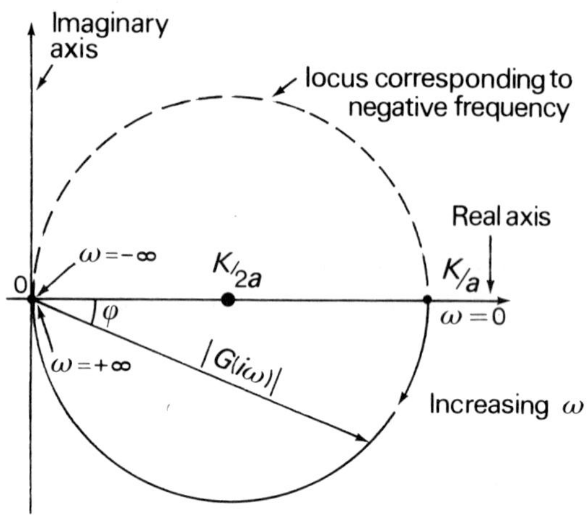

Fig. 5.9

The dashed part of the locus corresponding to negative ω is the mirror image of the lower half which is the locus for positive ω. The fact that the locus is a circle can be shown analytically.

Let $G(i\omega) = u + iv$, then

$$u = \frac{Ka}{a^2 + \omega^2} \quad \text{and} \quad v = -\frac{K\omega}{a^2 + \omega^2}.$$

It can now be verified that

$$(u - K/2a)^2 + v^2 = K^2/4a^2.$$

This is a circle with centre $(K/2a,0)$ and radius $K/2a$.

In the example (iii) above, the phase shift $\phi(\omega)$ is negative at all frequencies $\omega \geqslant 0$. A factor $1/(a + s)$ in a system transfer function is said to introduce a **lag** $\phi(\omega)$ into the system.

Similarly a differentiating term (s) introduces a 90° phase **lead**, and an integrating term $(1/s)$ introduces a 90° phase lag into the system at all frequencies.

If the transfer function involves several factors, the above procedure can be used for each factor individually, and the overall result is obtained by the following rules;

If $\quad G(s) = G_1(s)G_2(s)\ldots G_k(s),\quad$ then

$$|G(i\omega)| = |G_1(i\omega)|\,|G_2(i\omega)|\ldots|G_k(i\omega)|, \quad \text{and}$$

$$\phi(\omega) = \phi_1(\omega) + \phi_2(\omega) + \ldots + \phi_k(\omega)$$

where

$$\phi_j(\omega) = \tan^{-1}\frac{\mathrm{Im}\{G_j(i\omega)\}}{\mathrm{Rl}\{G_j(i\omega)\}} \qquad (j = 1, 2, \ldots, k)$$

Example 5.10

Construct the frequency diagram for the system having the transfer function

$$G(s) = \frac{K}{s(s+1)} \qquad (K \text{ is constant})$$

Solution

Since $G(i\omega) = (\frac{1}{i\omega})(\frac{K}{i\omega+1}) = G_1(i\omega).G_2(i\omega)$ (say), we obtain

$$|G(i\omega)| = \frac{1}{\omega}.\frac{1}{\sqrt{(1+\omega^2)}}$$

and $\quad \phi = -\pi/2 - \tan^{-1}\omega\ .$

We can now construct a frequency table:

ω	$\lvert G_1\rvert$	$\lvert G_2\rvert$	$\lvert G\rvert$	ϕ
0	∞	K	∞	-90
1	1	$K/\sqrt{2}$	$0.7K$	-135
2	$\frac{1}{2}$	$K/\sqrt{5}$	$0.22K$	-153
10	$\frac{1}{10}$	$K/101$	$0.01K$	-174
$\vdots$	$\vdots$	$\vdots$	$\vdots$	$\vdots$
∞	0	0	0	-180

The corresponding frequency diagram is shown in Fig. 5.10.

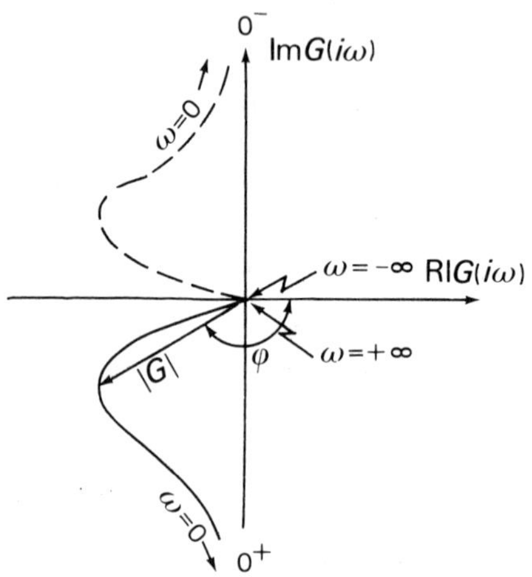

Fig. 5.10

To use the Nyquist criterion we must plot the frequency response of the system's open-loop transfer function, which will now be defined.

The system in Fig. 5.11 has the closed-loop transfer function (see equation (2.16)).

$$\frac{C(s)}{R(s)} = \frac{G(s)}{1 + G(s)H(s)} \tag{5.22}$$

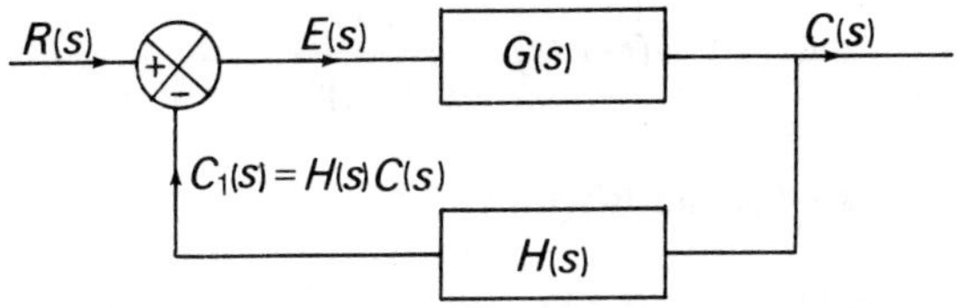

Fig. 5.11

The ratio of the feedback signal $C_1(s)$ to the actuating error signal $E(s)$ is

$$\frac{C_1(s)}{E(s)} = G(s)\,H(s)$$

and is called the system **open-loop transfer function.**

If we now consider the unity negative feedback system shown in Fig. 5.12 we again find that

$$\frac{C_1(s)}{E(s)} = G(s)\,H(s).$$

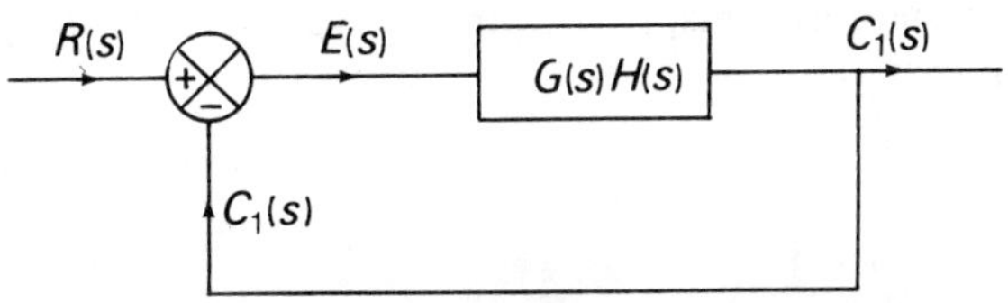

Fig. 5.12

So both the systems in Figs. 5.11 and 5.12 have the same open-loop transfer functions.

It will be assumed that if there are any factors which cancel in the product $G(s)\,H(s)$ they involve only poles and zeros in the left half s-plane.

We will consider the system to be stable so long as the closed-loop transfer function equation (5.22) has no poles in the rignt half s-plane.

We note that $s = s_1$ is a pole of equation (5.22) if

(1) $1 + G(s_1)\,H(s_1) = 0$ and $G(s_1) \neq 0$.

Since there is no cancellation of factors in the product $G(s)\,H(s)$ the condition $1 + G(s_1)\,H(s_1) = 0$ implies $G(s_1) \neq 0$ when $s = s_1$ is in the right half s-plane.

(2) $G(s) = \dfrac{A(s)}{(s - s_1)^n}$ and $1 + G(s)\,H(s) = \dfrac{B(s)}{(s - s_1)^m}$

where $n > m$, $A(s_1) \neq 0$ and $B(s_1) \neq 0$.

This condition holds only if there is a cancellation of factors of form $(s - s_1)$ in the product $G(s)\,H(s)$. By the above assumption this condition cannot hold if $s = s_1$ is in the right half s-plane.

From (1) and (2) above we conclude that the system having the transfer function equation (5.22) is stable if and only if $[1 + G(s)\,H(s)]$ has no zeros in the right half s-plane.

5.9 AN INTRODUCTION TO CONFORMAL MAPPINGS

To establish the Nyquist criterion we regard the open-loop transfer function

$$G(s)\,H(s) = F(s) \quad \text{(say)}$$

of a feedback system as a mapping from the s-plane to the $F(s)$-plane. As an example, consider

$$F(s) = \frac{1}{s + 4}.$$

To each value of s (except $s = -4$), there corresponds a unique value $F(s)$. As both s and $F(s)$ are complex, we can consider two planes:

(1) the s-plane, having a real axis, Rl$s = \sigma$ and an imaginary axis Im$s = \omega$, where

$$s = \sigma + i\omega,$$

(2) The $F(s)$-plane which has similarly a real axis Rl $F(s)$ and an imaginary axis Im $F(s)$.

We can go a step further and consider corresponding paths in the two planes. In Fig. 5.13 we have chosen a square in the s-plane, and as s takes on the values along this path we have calculated the corresponding path in the $F(s)$-plane.

It is convenient to construct a table giving some s and the corresponding $F(s)$ values.

Point	s	$F(s) = \dfrac{1}{s+4}$
A	$1+i$	$0.19-0.04i$
B	$-1+i$	$0.3\ -0.1i$
C	$-1-i$	$0.3\ +0.1i$
D	$1-i$	$0.19+0.04i$

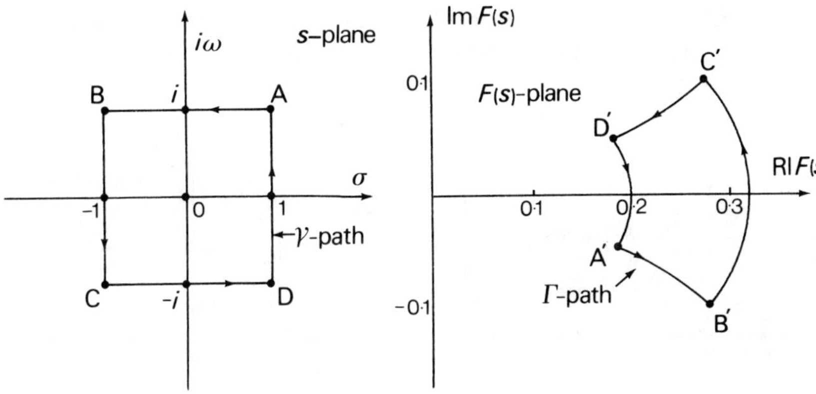

Fig. 5.13

In Fig. 5.13 it is shown that $F(s)$ traces out circular arcs as s moves along the sides of the square.

This can of course be proved analytically. For example, along DA

$$s = 1 + i\omega \qquad -1 \leqslant \omega \leqslant 1$$

and correspondingly along D′A′

$$F(s) = \frac{1}{5 + i\omega} = \frac{5 - i\omega}{25 + \omega^2} = u + iv \quad \text{(say)}$$

where $$u = \frac{5}{25+\omega^2} \quad \text{and} \quad v = -\frac{\omega}{25+\omega^2}.$$

It is now not difficult to verify the relations between u and v:

$$(u - \tfrac{1}{10})^2 + v^2 = (\tfrac{1}{10})^2.$$

This is the equation of a circle, centre $(\frac{1}{10},0)$ and radius $\frac{1}{10}$.

In Fig. 5.13 we note that

(1) the closed contour γ in the s-plane is mapped into a closed contour Γ in the $F(s)$-plane.

(2) Any point inside γ (try the origin) maps into a point inside Γ.

(3) We can define a *positive* direction (it does not matter which direction is chosen to be positive) in which s traces out its path. In Fig. 5.13 we note that as s sweeps out the contour γ in the s-plane in a positive direction, $F(s)$ sweeps out the contour Γ in the $F(s)$-plane also in the positive direction.

It is interesting to consider the direction in which $F(s)$ sweeps out its path, as s moves along a contour which encloses the pole ($s = -4$) of $F(s)$. This is shown in Fig. 5.14.

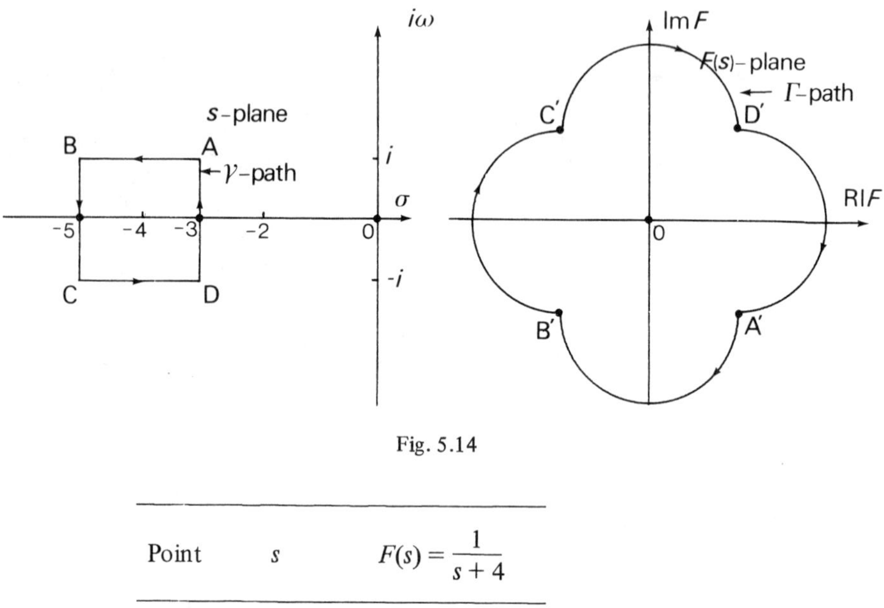

Fig. 5.14

Point	s	$F(s) = \dfrac{1}{s+4}$
A	$-3+i$	$0.5-0.5i$
B	$-5+i$	$-0.5-0.5i$
C	$-5-i$	$-0.5+0.5i$
D	$-3-i$	$0.5+0.5i$

In this case we note that as s traces the γ-contour in the positive direction, $F(s)$ traces the corresponding Γ-contour in the opposite, that is, *negative* direction. Also this time the Γ-contour encircles the origin of the $F(s)$-plane.

To explain how the direction and the encirclement of the origin of the $F(s)$-plane by the Γ-contour is affected by the γ-path, we will first consider a more formal (but not a rigorous) definition of conformal mappings.

We say that a function $F(s)$ is analytic at a point $s = s_1$ if it is differentiable at every point in the neighbourhood of s_1.† For each point in the s-plane, for which $F(s)$ is analytic (and not zero) the mapping in the $F(s)$-plane is said to be **conformal**, and it can be proved that the corresponding angles in the s and $F(s)$-planes have the same magnitude and sense. For example in Fig. 5.13, the lines AB and AD intersect at a right angle and (since $F(s)$ is analytic at A) the curves A′B′ and A′D′ also intersect at a right angle.

As for the sense of the angles, this can be determined by considering a man swimming along the path DAB. As he passes A, he will sweep out with his left hand the angle DAB in a clockwise sense. Similarly swimming along the path D′A′B′ he will sweep out with his left hand the angle D′A′B′ in the same direction.

Since in this book we consider transfer functions which are the ratios of polynomials, a function such as $F(s)$ defined above is analytic except at a finitie number of singular points, corresponding to the poles of $F(s)$.

We now consider an important result in the theory of complex variables known as **Cauchy's theorem**.

Let γ be a closed contour in the s-plane enclosing P poles and Z zeros of $F(s)$, but not passing through any poles or zeros of $F(s)$. As s traces out γ in the positive direction, $F(s)$ traces out a closed contour Γ in the $F(s)$-plane, which encircles the origin $(Z-P)$ times in the positive direction.

For proof of the theorem see MacFarlane [38].

We shall first apply this theorem to Fig. 5.13. Here γ does not enclose any poles or zeros of $F(s)$, so that

$$P = 0 = Z.$$

The origin of the $F(s)$-plane is *not* encircled by the Γ-contour.

In Fig. 5.14, γ encircles a pole of $F(s)$ so that

$$P = 1 \quad \text{and} \quad Z = 0$$

hence

$$Z - P = -1$$

and Γ encircles the origin (-1) times in the positive direction, that is, Γ encircles the origin once in the *negative* direction.

†This definition may not be strong enough for every function $F(s)$, but it is sufficient for the functions that we consider in this book.

5.10 APPLICATION OF CONFORMAL MAPPINGS TO THE FREQUENCY RESPONSE

In section 5.8 we discussed the frequency response, which required s to trace out the imaginary axis in the s-plane. In section 5.9, Cauchy's theorem calls for a closed contour in the s-plane avoiding the poles and zeros of $F(s)$. We now combine these two contours and define the **Nyquist path**. This contour consists of the entire imaginary axis, from $\omega = -\infty$ to $\omega = +\infty$, and a semicircular path of large radius R (where $R \to \infty$) in the right half of the s-plane, as shown in Fig. 5.15.

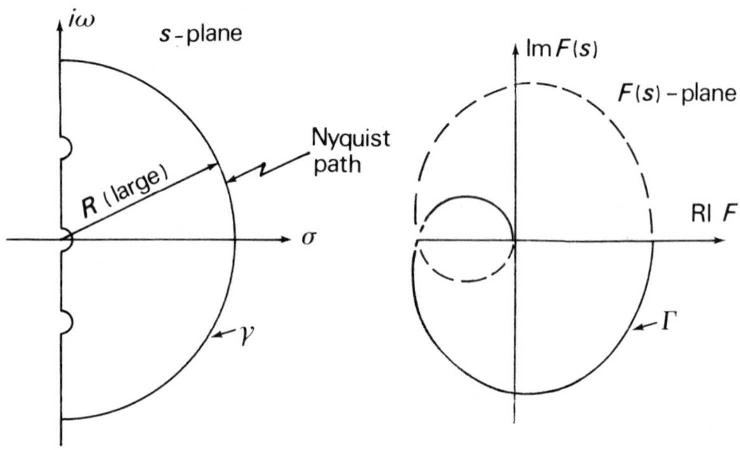

Fig. 5.15

The small semicircular indentations are used to avoid poles and zeros of $F(s)$ which lie along the imaginary axis. Normally these indentations map into arcs in the $F(s)$-plane; the nearer the indentation to the origin of the s-plane, the larger is the corresponding arc in the $F(s)$-plane.

The semi-circle in the s-plane is assumed large enough to enclose all the poles and zeros of $F(s)$ which may lie in the right half plane.

A typical map Γ of the Nyquist contour is also shown in Fig. 5.15. The dotted contour corresponds to the map for negative values of ω and is the mirror image of the contour for $\omega > 0$ (as already explained). It is, of course, not necessary to make this distinction and, having made the point about symmetry, we will no longer do so. From the discussion above, Γ encircles the origin of the $F(s)$-plane $(Z - P)$ times.

If, as we assume for the time being, the open-loop transfer function $F(s)$ is stable, it has no poles, although it may have zeros, in the right half of the s-plane. In this case (since $P = 0$), the Γ contour will encircle the $F(s)$-plane origin Z times (in the positive direction). In fact, since no poles of $F(s)$ are involved, we are not concerned with the direction of the encirclement - it can only be in the positive direction (as pointed out above, this is the direction in which s traces out the Nyquist contour in the s-plane - whether clockwise or counterclockwise is immaterial).

It follows that Z encirclements of the origin by the Γ contour imply that Z solutions of $F(s) = 0$, that is, of

$$G(s)\,H(s) = 0,$$

are in the right half of the s-plane.

We are now near to a final phase of our discussion. The plane $F(s) + 1$ is obtained from the $F(s)$-plane by moving the origin of the $F(s)$-plane to the point $(-1 + i0)$, called the **critical point**, as shown in Fig. 5.16.

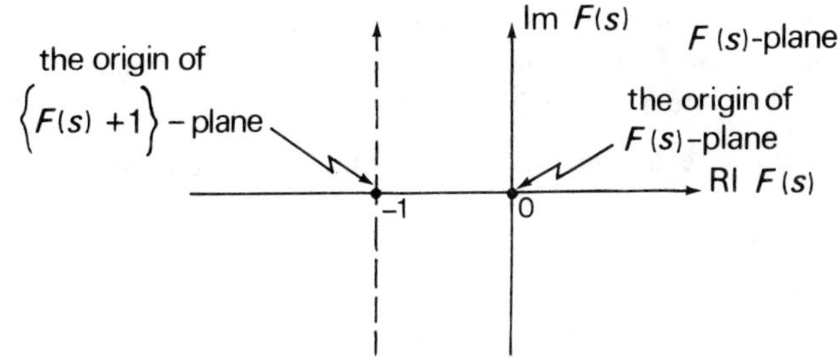

Fig. 5.16

Now, an encirclement of the origin of the $\{F(s) + 1\}$-plane indicates the existence of a right half plane solution of $F(s) + 1 = 0$, that is, of

$$G(s)\,H(s) + 1 = 0.$$

But this is the system characteristic equation, the denominator of equation (5.22).

Hence the closed-loop transfer function (equation (5.22)) has a pole in the right half of the s-plane and so the system is unstable.

Finally if the open-loop system is unstable, and $G(s)\,H(s)$ has P poles in the right half plane (see example 2.10) the situation is more complicated.

Suppose that N is the number of encirclements of the critical point in the positive direction, then by Cauchy's theorem, the number Z of right half plane zeros is

$$Z = P - N.$$

It is, of course, obvious that in this case, when the open-loop system has right-hand poles, it is impossible to obtain the frequency response experimentally.

On the other hand, if the transfer function is known analytically, the method is very useful.

We now sum up, under the heading of Nyquist criterion, the results obtained above. It is stressed that these results have not been proved here. For formal proof see various references in the *Bibliography*, for example MacFarlane [38].

The Nyquist Criterion

To determine the stability of a unity negative feedback system (Fig. 5.12) we consider the map Γ in the $F(s)$-plane of the Nyquist contour traversed in a positive direction.

The system is stable if

(1) Γ does not encircle the point $(-1,0)$ when the open-loop transfer function $G(s)\,H(s)$ has no poles in the right half plane.

(2) Γ encircles the point $(-1,0)$ P times in the positive direction when the open-loop transfer function has P poles in the right half plane.
Otherwise the system is unstable.

Fig. 5.17 shows examples of a stable and an unstable system for the case where $P = 0$.

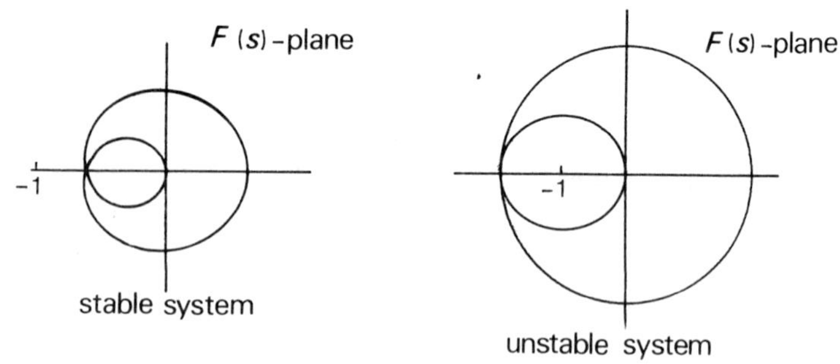

Fig. 5.17

Example 5.11

Use the Nyquist criterion to determine the stability of the closed-loop system having an open-loop transfer function

$$F(s) = \frac{K}{s(s+1)(s+2)}$$

when (1) $K = 1$ and (2) $K = 10$.

Solution

Since $F(s)$ has a pole at the origin, the Nyquist contour has a semicircular indentation as shown (greatly exaggerated in size) in Fig. 5.18(a).

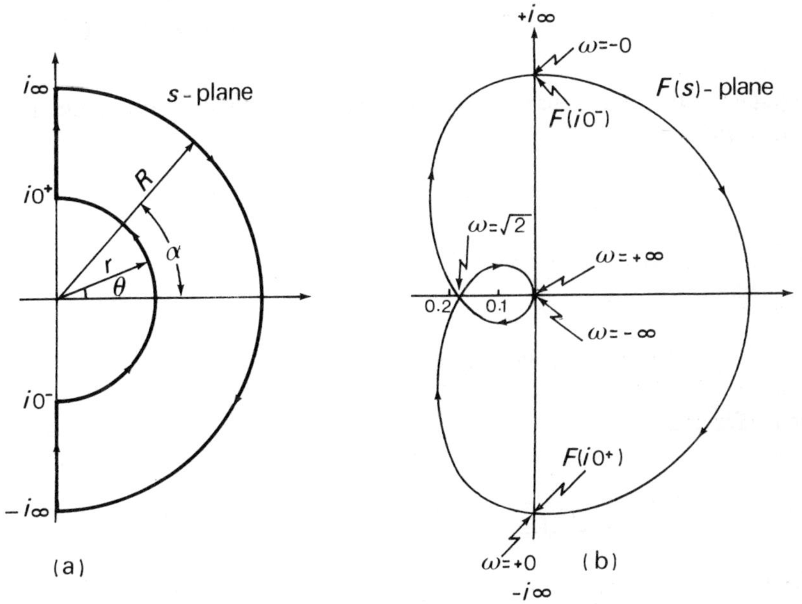

Fig. 5.18

Because this is an illustrative example we shall carefully consider the mapping by $F(s)$ as s sweeps out each section of the Nyquist contour. Such detailed analysis is not necessary after one has become used to the various approximations involved - a simple table, as used in the previous section, will be quite sufficient!

We have used the notation $(i0^-)$ and $(i0^+)$ to refer to values of the frequency just less than zero and just greater than zero respectively. Also $+i\infty$ and $-i\infty$ refer to very large positive and negative frequencies respectively.

Now, $$F(i\omega) = \frac{K}{i\omega(1+i\omega)(2+i\omega)}$$

which we can write in polar form as

$$F(i\omega) = \frac{K}{\omega(1+\omega^2)^{\frac{1}{2}}(4+\omega^2)^{\frac{1}{2}}}\,e^{i\phi} \tag{5.23}$$

where $\phi = -90° - \tan^{-1}\omega - \tan^{-1}\omega/2$,

or as

$$F(i\omega) = \frac{-3K}{(1+\omega^2)(4+\omega^2)} - i\,\frac{(\omega^2-2)K}{\omega(1+\omega^2)(4+\omega^2)} = \mathrm{Rl}\{F(i\omega)\} + i\mathrm{Im}\{F(i\omega)\}. \quad (5.24)$$

We start in the neighbourhood of zero frequency. Along the semicircular path, s can be written as

$$s = re^{i\theta}$$

where θ varies from $-90°$ to $+90°$.

As r is very small

$$1 + s \simeq 1 \quad \text{and} \quad 2 + s \simeq 2,$$

hence for small r

$$F(s) \simeq \frac{K}{re^{i\theta}.1.2} = \frac{K}{2r}\,\mathrm{e}^{-i\theta}.$$

As $r \to 0$, $K/2r \to \infty$, so that as θ varies from $-90°$ to $+90°$, $F(s)$ describes a semicircle of infinite radius, joining the points $F(i0^-)$ and $F(i0^+)$.

From equation (5.24), we find

$$F(i0^-) = +i\infty \quad \text{and} \quad F(i0^+) = -i\infty.$$

We next consider s tracing out the section of the contour from $i0^+$ to $i\infty$. We use equation (5.23) and construct a table as in the previous section.

ω	$\|F\|$	ϕ
0^+	∞	-90
$\frac{1}{2}$	0.87	-130
$\sqrt{2}$	0.17	-180
2	0.08	-200
$\vdots$	$\vdots$	$\vdots$
∞	0	-270

On the large semicircle, s can be written as

$$s = Re^{i\alpha}$$

as α varies from 90° to −90°. Hence

$$F(s) \simeq \frac{K}{R^3} e^{-3\alpha}.$$

But $K/R^3 \to 0$ as $R \to \infty$, so that this semicircle is mapped into the origin of the $F(s)$-plane.

Finally as s traces out the path from $-i\infty$ to $i0^-$, the $F(s)$ path is the mirror image of the $i0^+$ to $i\infty$ path.

The $F(s)$ contour for $K = 1$ corresponding to the Nyquist path is shown in Fig. 5.18(b).

(1) For $K = 1$, the point $(-1 + i0)$ is not encircled by the $F(s)$ contour. Since $F(s)$ has no poles in the right half s-plane ($P = 0$), we conclude that the system is stable.

(2) For $K = 10$, we note from equation (5.23) that the magnitude $|F(i\omega)|$ is 10 times larger than when $K = 1$ for each (finite) value of ω. But the phase ϕ is independent of K. This means that Fig. 5.18 is valid for $K = 10$ if we change the coordinate scales by a factor of 10.

This time the critical point $(-1 + i0)$ is encircled twice; we conclude that there are two closed-loop poles in the right half s-plane, and the system is unstable.

This example shows that as K varies, the plot contracts or dilates, but only one plot is necessary to determine the status of system stability in each case.

Of course there exists a value of K, say K_0, for which the $F(s)$ plot passes through the critical point. For values of K in the neighbourhood of K_0 the system may be just stable or just unstable, but in any case, a slight chånge in a system parameter may drive the system into an unstable state. So the distance of the $F(s)$ contour from the critical point does give a measure of what is known as **relative stability**. This is a topic which we do not explore in this book, but the interested reader will find many references to it in various publications mentioned in the *Bibliography*.

PROBLEMS

(1) the characteristic equations of two systems are as follows

(a) $s^4 + 2s^3 + 3s^2 + 5s + 1 = 0$

(b) $s^3 + s^2 + s + 2 = 0$.

In each case construct the Routh array of coefficients and hence determine whether the system is stable; if it is unstable, determine the number of roots with positive real parts.

(2) With reference to Fig. 4.14 on p. 100, the transfer function of the plant is

$$A(s) = \frac{1}{s^2 + 3s + 2}.$$

and that of the controller is

$$B(s) = K/s.$$

Determine the range of values of K for which the system is stable. What is the status of stability when $K = 6$?

(3) Determine the stability margin σ (consider σ to be an integer) of the system whose characteristic equation is

$$P(s) = s^3 + 9s^2 + 26.75s + 26.25 = 0.$$

(4) Determine which of the following quadratic forms are positive definite.

$$V_1 = 2x_1^2 + 3x_2^2 + x_3^2 - 2x_1x_2 + 2x_1x_3$$

$$V_2 = 8x_1^2 + 2x_2^2 + x_3^2 - 8x_1x_2 + 2x_1x_3 - 2x_2x_3$$

$$V_3 = x_1^2 + 2x_3^2 - 2x_1x_2 + 6x_2x_3.$$

(5) Use Liapunov's direct method to determine whether the following systems are asymptotically stable.

(a) $\dot{x}_1 = -2x_1 + x_2$

$\dot{x}_2 = -4x_2$

(b) $\dot{x}_1 = -2x_1$

$\dot{x}_2 = -x_2$

$\dot{x}_3 = x_2 - x_3.$

(6) Consider the following two systems:

(i) a particle of mass m falling under gravity

(ii) a mass M attached to a spring of stiffness k (the spring constant) moving on a frictionless plane.

For both systems, choose an appropriate state vector $\mathbf{x}(t)$ and determine the state equations; determine the trajectories of $\mathbf{x}(t)$.

CHAPTER 6

Controllability and observability

6.1 INTRODUCTION

The concepts of controllability and observability, introduced by Kalman† in 1960 are fundamental to modern control theory. The fundamental questions to be answered for a system, in particular for a multivariable system are:

(1) Can a control function $\mathbf{u}(t)$ be found which will transform the initial state $\mathbf{x}_0$ of a system to some desired final state $\mathbf{x}_f$ in finite time?

(2) Can the state of the system be determined by measuring the system output over a finite time interval?

The two concepts involved are called controllability and observability respectively. So if the answer is yes to the first question, the system is **controllable**. Similarly if the answer is yes to the second question the system is **observable**.

It should be appreciated that these questions are fundamental. For example, it is pointless to try to control a system by state-variable feedback which allows for the arbitrary positioning of the system poles, unless the system is controllable.

Similarly it is pointless to try to reconstruct the unmeasurable state variables of the system, by a device known as an **observer**, unless the system is observable.

Controllability and observability are in fact two dual concepts and are closely related to the cancellation of a pole and a zero in the system transfer function.

6.2 CONTROLLABILITY

We consider a system described by the state equations

$$\dot{\mathbf{x}} = A\mathbf{x} + B\mathbf{u} \tag{6.1}$$

$$\mathbf{y} = C\mathbf{x}$$

where A is $n \times n$, B is $n \times m$ and C is $r \times n$.

†Kalman, R. E. On the general theory of Control Systems. *Proc. IFAC.*, Moscow 1960 p. 481, Butterworth, 1961.

With the transformation (see equation (3.25))

$$\mathbf{x} = P\mathbf{z} \tag{6.2}$$

we can transform equation (6.1) into the form

$$\begin{aligned} \dot{\mathbf{z}} &= A_1\mathbf{z} + B_1\mathbf{u} \\ \mathbf{y} &= C_1\mathbf{z} \end{aligned} \tag{6.3}$$

where $A_1 = P^{-1}AP$, $B_1 = P^{-1}B$ and $C_1 = CP$. Assuming that A has distinct eigenvalues $\lambda_1, \lambda_2, \ldots, \lambda_n$ we can choose P so that A_1 is a diagonal matrix, that is,

$$A_1 = \text{diag}\{\lambda_1, \lambda_2, \ldots, \lambda_n\}.$$

If $n = m = r = 2$, the first of the equations (6.3) has the form

$$\begin{bmatrix} \dot{z}_1 \\ \dot{z}_2 \end{bmatrix} = \begin{bmatrix} \lambda_1 & 0 \\ 0 & \lambda_2 \end{bmatrix} \begin{bmatrix} z_1 \\ z_2 \end{bmatrix} + \begin{bmatrix} b_{11} & b_{12} \\ b_{21} & b_{22} \end{bmatrix} \begin{bmatrix} u_1 \\ u_2 \end{bmatrix}$$

which is written as

$$\begin{aligned} \dot{z}_1 &= \lambda_1 z_1 + \mathbf{b}_1'\mathbf{u} \\ \dot{z}_2 &= \lambda_2 z_2 + \mathbf{b}_2'\mathbf{u} \end{aligned} \tag{6.4}$$

where $\mathbf{b}_1'$ and $\mathbf{b}_2'$ are the row vectors of the matrix B_1.

The output equation is

$$\begin{bmatrix} y_1 \\ y_2 \end{bmatrix} = \begin{bmatrix} c_{11} & c_{12} \\ c_{21} & c_{22} \end{bmatrix} \begin{bmatrix} z_1 \\ z_2 \end{bmatrix}$$

which can be written as

$$\begin{aligned} y_1 &= c_{11} z_1 + c_{12} z_2 \\ y_2 &= c_{21} z_1 + c_{22} z_2 \end{aligned}$$

or

$$\begin{bmatrix} y_1 \\ y_2 \end{bmatrix} = \begin{bmatrix} c_{11} \\ c_{21} \end{bmatrix} z_1 + \begin{bmatrix} c_{12} \\ c_{22} \end{bmatrix} z_2$$

so that

$$\mathbf{y} = \mathbf{c}_1 z_1 + \mathbf{c}_2 z_2 \tag{6.5}$$

where $\mathbf{c}_1$ and $\mathbf{c}_2$ are the column vectors of C_1.

So, in general, equation (6.3) can be written in the form

$$\begin{aligned} \dot{z}_i &= \lambda_i z_i + \mathbf{b}_i'\mathbf{u} \qquad (i = 1, 2, \ldots, n) \\ \mathbf{y} &= \sum_{i=1}^{n} \mathbf{c}_i z_i \end{aligned} \tag{6.6}$$

It is seen from equation (6.6) that if $\mathbf{b}_i'$, the ith row of B_1, has all zero components, then

$$\dot{z}_i = \lambda_i \mathbf{z}_i + 0$$

and the input $\mathbf{u}(t)$ has no influence on the ith mode of the system (see section 3.3 for definition). The mode is said to be **uncontrollable**, and a system having one or more such modes is uncontrollable.

Otherwise, where all the modes are **controllable** the system is said to be **completely state controllable**, or just *controllable.*

Although it can be proved that the above statement is a necessary condition for controllability, we instead make use of the following intuitive argument.

By definition, the system equation (6.6) is controllable if and only if a control function $\mathbf{u}(t)$ can be found to transform the initial state $\mathbf{z}_0 = [\mathbf{z}_{01}, \mathbf{z}_{02}, \ldots, \mathbf{z}_{0n}]'$ to a specified state $\mathbf{z}_1 = [\mathbf{z}_{11}, \mathbf{z}_{12}, \ldots, \mathbf{z}_{1n}]'$.

It follows that for complete state controllability each of the components of $\mathbf{z}$ must be transformed by an appropriate choice of $\mathbf{u}(t)$ from $\mathbf{z}_{0y}$ to $\mathbf{z}_{1y}$ $(y = 1, 2, \ldots, n)$. If the ith mode is uncontrollable, this cannot be achieved for the ith component of $\mathbf{z}$. It follows that the system is uncontrollable.

Example 6.1

Check whether the system having the state-space representation

$$\dot{\mathbf{x}} = \begin{bmatrix} -1 & 2 \\ -3 & 4 \end{bmatrix} \mathbf{x} + \begin{bmatrix} 4 \\ 6 \end{bmatrix} u$$

$$\mathbf{y} = [1 \quad -2]\mathbf{x}$$

is controllable.

Solution

The characteristic equation is

$$|\lambda I - A| = \lambda^2 - 3\lambda + 2 = 0 \quad \text{for} \quad \lambda = 1 \quad \text{and} \quad \lambda = 2.$$

The corresponding eigenvectors are

$$\mathbf{x}_1 = [1 \quad 1]' \quad \text{and} \quad \mathbf{x}_2 = [2 \quad 3]',$$

so that the modal matrix is

$$P = \begin{bmatrix} 1 & 2 \\ 1 & 3 \end{bmatrix} \quad \text{and} \quad P^{-1} = \begin{bmatrix} 3 & -2 \\ -1 & 1 \end{bmatrix}.$$

Using the transformation $\mathbf{x} = P\mathbf{z}$, the state-equation becomes

$$\dot{\mathbf{z}} = \begin{bmatrix} 1 & 0 \\ 0 & 2 \end{bmatrix} \mathbf{z} + \begin{bmatrix} 0 \\ 2 \end{bmatrix} u$$

$$y = [-1 \quad -4]\mathbf{z}.$$

This equation shows that the first mode is uncontrollable and so the system is uncontrollable.

On the basis of the above result, we now derive an extremely useful criterion for determining whether a system is controllable. Although at this stage we consider only the necessity of this criterion, it is also a sufficient condition. The sufficiency is explored in problem 9 at the end of this chapter.

To simplify the notation and the mathematical manipulations, we consider a SISO (single-input and single-output) system, so that in equation (6.1) B is a one column matrix, that is a column vector $\mathbf{b}$ (say), and C is a row vector $\mathbf{c}_1'$ (say). The result holds for the more general case when the system is multivariable.

Equations (6.1) and (6.3) are then written as

$$\dot{\mathbf{x}} = A\mathbf{x} + \mathbf{b}u \tag{6.1a}$$

$$y = \mathbf{c}'\mathbf{x}$$

and

$$\dot{\mathbf{z}} = A_1\mathbf{z} + \mathbf{b}_1 u \tag{6.3a}$$

$$y = \mathbf{c}_1'\mathbf{z}.$$

We have chosen an indirect way of deriving the controllability criterion. It has the advantage of simplicity, but a penalty we pay for this is some loss in the logic behind the setting up of the criterion.

We have established that the necessary condition for the system defined by equation (6.1a) to be controllable is that the components of the vector

$$\mathbf{b}_1 = [\beta_1 \beta_2 \dots \beta_n]' \qquad \text{(say) in equation (6.3a) are all non-zero.}$$

In equation (6.3a) the matrix $A_1 = \text{diag}\{\lambda_1, \lambda_2, \dots, \lambda_n\}$ where the eigenvalues $\lambda_1, \lambda_2, \dots, \lambda_n$ are assumed distinct.

Hence the (Vandermonde) matrix

$$\begin{bmatrix} 1 & \lambda_1 & \dots & \lambda_1^{n-1} \\ 1 & \lambda_2 & & \lambda_2^{n-1} \\ \vdots & \vdots & & \vdots \\ 1 & \lambda_n & & \lambda_n^{n-1} \end{bmatrix}$$

has linearly independent columns, so that it is non-singular.

It follows that the necessary condition for the system to be controllable is that the (partitioned) matrix

$$Q_1 = [\mathbf{b}_1 \,\vdots\, A_1 \mathbf{b}_1 \,\vdots\, \cdots \,\vdots\, A_1^{n-1} \mathbf{b}_1] = \begin{bmatrix} \beta_1 & \lambda_1 \beta_1 & \dots & \lambda_1^{n-1} \beta_1 \\ \beta_2 & \lambda_2 \beta_2 & & \lambda_2^{n-1} \beta_2 \\ \vdots & \vdots & & \vdots \\ \beta_n & \lambda_n \beta_n & & \lambda_n^{n-1} \beta_n \end{bmatrix} \tag{6.7}$$

is non-singular. (As already mentioned, the above is also a sufficient condition for controllability.)

Since $A_1 = P^{-1}AP$ and $\mathbf{b}_1 = P^{-1}\mathbf{b}$, we have

$$A_1 \mathbf{b}_1 = P^{-1}A\mathbf{b},\; A_1^2 \mathbf{b}_1 = P^{-1}A^2\mathbf{b}, \dots,\; A_1^{n-1}\mathbf{b}_1 = P^{-1}A^{n-1}\mathbf{b},$$

so that

$$Q_1 = P^{-1}[\mathbf{b} \,\vdots\, A\mathbf{b} \,\vdots\, \dots \,\vdots\, A^{n-1}\mathbf{b}] = P^{-1}Q$$

where

$$Q = [\mathbf{b} \,\vdots\, A\mathbf{b} \,\vdots\, \dots \,\vdots\, A^{n-1}\mathbf{b}]. \tag{6.8}$$

Since Q_1 (for a controllable system) and P^{-1} are both non-singular, Q is also non-singular.

As Q is non-singular, its n columns are linearly independent. So that the rank of the matrix Q written as $r(Q)$ is n.

The Controllability Criterion

It follows that if the rank of the matrix Q in equation (6.8) is n (Q is non-singular), the system is controllable. If the rank is less than n (Q is singular), the system is uncontrollable. (For a multivariable system, $Q = [B \,\vdots\, AB \,\vdots\, \ldots \, A^{m-1}B]$, B being the matrix defined in equation (6.1)).

Definition

The matrix Q of equation (6.8) is called the system **controllability matrix**.

Example 6.2

Using the controllability criterion verify that the system examined in example 6.1 is uncontrollable.

Solution

For this system

$$\mathbf{b} = \begin{bmatrix} 4 \\ 6 \end{bmatrix} \quad \text{and} \quad A\mathbf{b} = \begin{bmatrix} -1 & 2 \\ -3 & 4 \end{bmatrix}\begin{bmatrix} 4 \\ 6 \end{bmatrix} = \begin{bmatrix} 8 \\ 12 \end{bmatrix}, \quad \text{so that}$$

$$r(Q) = r[\mathbf{b} \quad A\mathbf{b}] = r\begin{bmatrix} 4 & 8 \\ 6 & 12 \end{bmatrix} = 1.$$

Since the rank of Q is less than 2, the system is uncontrollable.

Example 6.3

Determine whether the system, governed by the equation

$$\begin{bmatrix} \dot{x}_1 \\ \dot{x}_2 \end{bmatrix} = \begin{bmatrix} 0 & 6 \\ -1 & -5 \end{bmatrix}\begin{bmatrix} x_1 \\ x_2 \end{bmatrix} + \begin{bmatrix} 3 & 6 \\ -1 & -2 \end{bmatrix}\begin{bmatrix} u_1 \\ u_2 \end{bmatrix},$$

is controllable.

Solution

$$r(Q) = r[B \; AB] = r\left[\begin{array}{cc:cc} 3 & 6 & -6 & -12 \\ -1 & -2 & 2 & 4 \end{array}\right].$$

It is obvious that the rank of this matrix is 1. The system is therefore uncontrollable.

6.3 OBSERVABILITY

Using the transform $\mathbf{x} = P\mathbf{z}$, as in the previous section, we end up with the system state equations in the form of equation (6.3), that is

$$\dot{\mathbf{z}} = A_1\mathbf{z} + B_1\mathbf{u}$$

$$\mathbf{y} = C_1\mathbf{z}.$$

If a row of the matrix C_1 is zero, the corresponding mode of the system will not appear in the output $\mathbf{y}$. In this case the system is **unobservable**, since we cannot determine the state variable corresponding to the row of zeros in C_1 from $\mathbf{y}$.

Example 6.4

Determine whether the system having the state equations

$$\begin{bmatrix} \dot{x}_1 \\ \dot{x}_2 \end{bmatrix} = \begin{bmatrix} -5 & 4 \\ -6 & 5 \end{bmatrix} \begin{bmatrix} x_1 \\ x_2 \end{bmatrix} + \begin{bmatrix} 1 \\ 2 \end{bmatrix} u$$

$$y = [3 \quad -2] \begin{bmatrix} x_1 \\ x_2 \end{bmatrix}$$

is observable.

Solution

The characteristic equation for this system is $|\lambda I - A| = \lambda^2 - 1 = 0$ for $\lambda = -1$ and $\lambda = 1$. The corresponding eigenvectors are $[1 \quad 1]'$ and $[2 \quad 3]'$. The modal matrix is

$$P = \begin{bmatrix} 1 & 2 \\ 1 & 3 \end{bmatrix} \quad \text{so that} \quad P^{-1} = \begin{bmatrix} 3 & -2 \\ -1 & 1 \end{bmatrix}.$$

The transformation $\mathbf{x} = P\mathbf{z}$, transforms the state-equations into

$$\begin{bmatrix} \dot{z}_1 \\ \dot{z}_2 \end{bmatrix} = \begin{bmatrix} -1 & 0 \\ 0 & 1 \end{bmatrix} \begin{bmatrix} z_1 \\ z_2 \end{bmatrix} + \begin{bmatrix} -1 \\ 1 \end{bmatrix} u$$

$$y = [1 \quad 0] \begin{bmatrix} z_1 \\ z_2 \end{bmatrix}.$$

This result shows that the system is unobservable since the output y is not influenced by the state variable z_2. This is very unfortunate in this particular case since the corresponding mode is seen to be unstable, whereas the output y would indicate that the system is (apparently) stable!

The above example illustrates the importance of the observability concept. In this case we have a non-stable system, whose instability is not observed in the output measurement.

The dual controllability concept is of equal theoretical importance. An uncontrollable system has one or more modes which are not influenced by the input.

We now derive a criterion for observability in a similar manner to that used to derive the controllability criterion.

Again for simplicity we consider a SISO system, but the result holds for the more general multivariable system.

We have seen that the necessary conditions for systems defined by equation (6.1a) to be observable is that the components of the vector

$$\mathbf{c}_1' = [\gamma_1, \gamma_2, \ldots, \gamma_n] \quad \text{(say) in equation (6.3a)}$$

are all non-zero. It follows that for a controllable system the matrix

$$R_1 = \begin{bmatrix} \mathbf{c}_1' \\ \mathbf{c}_1' A_1 \\ \vdots \\ \mathbf{c}_1' A_1^{n-1} \end{bmatrix} = \begin{bmatrix} \gamma_1 & \gamma_2 & \ldots & \gamma_n \\ \gamma_1 \lambda_1 & \gamma_2 \lambda_2 & \ldots & \gamma_n \lambda_n \\ \\ \gamma_1 \lambda_1^{n-1} & \gamma_2 \lambda_2^{n-1} & \ldots & \gamma_n \lambda_n^{n-1} \end{bmatrix} \tag{6.9}$$

must be non-singular.

Since $A_1 = P^{-1}AP$ and $\mathbf{c}_1' = \mathbf{c}'P$, we have

$$\mathbf{c}_1' A_1 = \mathbf{c}'AP, \quad \mathbf{c}_1' A_1^2 = \mathbf{c}'A^2P, \ \ldots, \ \mathbf{c}_1' A^{n-1} = \mathbf{c}'A^{n-1}P,$$

so that

$$R_1 = \begin{bmatrix} \mathbf{c}' \\ \mathbf{c}'A \\ \vdots \\ \mathbf{c}'A^{n-1} \end{bmatrix} P = RP$$

where

$$R = \begin{bmatrix} \mathbf{c}' \\ \mathbf{c}'A \\ \vdots \\ \mathbf{c}'A^{n-1} \end{bmatrix}. \tag{6.10}$$

Since R_1 (for an observable system) and P are non-singular, R must also be non-singlular.

The Observability Criterion

If the rank of the matrix R in equation (6.10) is n (R is non-singular), the system is observable. If the rank is less than n (R is singular), the system is unobservable. For a multivariable system, the matrix R becomes the (partitioned) matrix in which the vector $\mathbf{c}'$ is replaced by the matrix C defined in equation (6.1).

Definition

The matrix R of equation (6.10) is called the system **observability matrix**.

Example 6.5

Using the observability criterion, verify that the system examined in example 6.4 is unobservable.

Solution

For this system

$$\mathbf{c}' = [3 \quad -2], \quad A = \begin{bmatrix} -5 & 4 \\ -6 & 5 \end{bmatrix} \quad \text{hence}$$

$$R = \begin{bmatrix} 3 & -2 \\ -3 & 2 \end{bmatrix}, \quad \text{so that} \quad r(R) = 1.$$

It follows that the system is unobservable.

6.4 DECOMPOSITION OF THE SYSTEM STATE

From the discussion in the two previous sections it is clear that in general the state variables (or equivalently, the corresponding modes) of a linear system can be classified into four exclusive groups as follows:

Group 1: Controllable and observable
Group 2: Controllable but unobservable
Group 3: Uncontrollable but observable
Group 4: Uncontrollable and unobservable.

Assuming that the system matrix A has distinct eigenvalues, by appropriate transformations the state equations can be reduced to the form below,

$$\begin{bmatrix} \dot{\mathbf{x}}_1 \\ \dot{\mathbf{x}}_2 \\ \dot{\mathbf{x}}_3 \\ \dot{\mathbf{x}}_4 \end{bmatrix} = \begin{bmatrix} A_1 & 0 & 0 & 0 \\ 0 & A_2 & 0 & 0 \\ 0 & 0 & A_3 & 0 \\ 0 & 0 & 0 & A_4 \end{bmatrix} \begin{bmatrix} \mathbf{x}_1 \\ \mathbf{x}_2 \\ \mathbf{x}_3 \\ \mathbf{x}_4 \end{bmatrix} + \begin{bmatrix} B_1 \\ B_2 \\ 0 \\ 0 \end{bmatrix} u \tag{6.11}$$

$$\mathbf{y} = [C_1 \;\; 0 \;\; C_3 \;\; 0] \begin{bmatrix} \mathbf{x}_1 \\ \mathbf{x}_2 \\ \mathbf{x}_3 \\ \mathbf{x}_4 \end{bmatrix}.$$

The (transformed) system matrix A has been put into a 'block-diagonal' form where each A_i ($i = 1,2,3,4$) is in a diagonal form. The suffix i of the state-variable vector $\mathbf{x}_i$ implies that the elements of this vector are the state variables corresponding to the ith group defined above.

Example 6.6

Classify the state variables in the system defined by the following state equations:

$$\begin{bmatrix} \dot{x}_1 \\ \dot{x}_2 \\ \dot{x}_3 \\ \dot{x}_4 \\ \dot{x}_5 \\ \dot{x}_6 \end{bmatrix} = \begin{bmatrix} -2 & 0 & 0 & 0 & 0 & 0 \\ 0 & -1 & 0 & 0 & 0 & 0 \\ 0 & 0 & 1 & 0 & 0 & 0 \\ 0 & 0 & 0 & -3 & 0 & 0 \\ 0 & 0 & 0 & 0 & 0 & 0 \\ 0 & 0 & 0 & 0 & 0 & -4 \end{bmatrix} \begin{bmatrix} x_1 \\ x_2 \\ x_3 \\ x_4 \\ x_5 \\ x_6 \end{bmatrix} + \begin{bmatrix} 1 & -1 \\ 0 & 0 \\ 1 & 1 \\ 0 & 0 \\ 1 & 0 \\ 0 & 2 \end{bmatrix} \begin{bmatrix} u_1 \\ u_2 \end{bmatrix}$$

$$\begin{bmatrix} y_1 \\ y_2 \end{bmatrix} = \begin{bmatrix} 0 & 0 & -1 & 2 & 0 & 1 \\ 1 & 0 & 1 & -1 & 0 & 1 \end{bmatrix} \begin{bmatrix} x_1 \\ x_2 \\ x_3 \\ x_4 \\ x_5 \\ x_6 \end{bmatrix}.$$

Solution

By inspection we can classify the state variables into the four groups as follows:
Controllable and observable, x_1, x_3 and x_6
Controllable and unobservable, x_5
Uncontrollable and observable, x_4
Uncontrollable and unobservable, x_2.
Notice that x_5 is marginally stable, but is unobservable. On the other hand x_3 is unstable, but fortunately it is both observable and controllable!

We can represent the decompositions of the state variables into four groups by a diagram (Fig. 6.1) showing the system divided into four subsystems each having state variables belong to one group only as indicated by the suffix i of S_i.

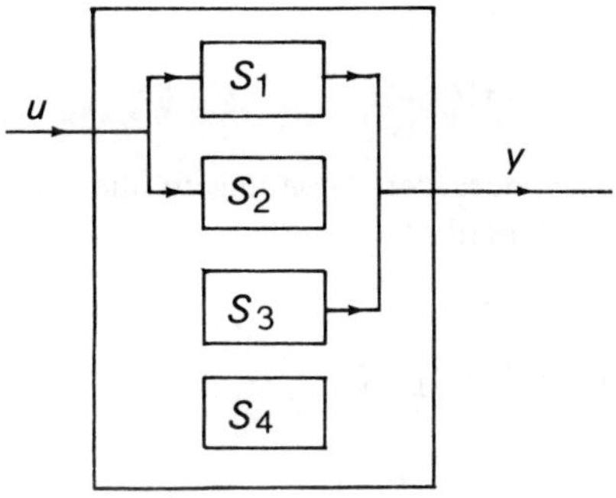

Fig. 6.1

This insight into the system structure explains the difference which may exist between the form of the system transfer function calculated from the system differential equations and that obtained by experimentation (that is, by obtaining the system frequency response.

We have defined (see equation (3.31)) the transfer function $G(s)$ as the ratio of the Laplace transforms of the output $Y(s)$ to the input $U(s)$. The transfer function obtained from the differential equations (or equivalently from the system state equations) includes all the state variables (or modes) of the system. But the transfer function found by experimentation involves that part of the system which is influenced by the input *and* which influences the output. From Fig. 6.1 it is seen that it is the transfer function of the subsystem S_1 that is found in this manner and it includes only the state variables (or modes) which are both controllable and observable.

So, in general, a transfer function $G(s)$ represents only the subsystem S_1 of the system considered, and indeed on adding to S_1 the subsystems S_2, S_3 and S_4 will not affect $G(s)$.

Example 6.7

Illustrate the above discussion with the systems considered in examples 6.1 and 6.4.

Solution

In example 6.1, the state equations were transformed into the diagonal form

$$\dot{\mathbf{z}} = \begin{bmatrix} 1 & 0 \\ 0 & 2 \end{bmatrix} \mathbf{z} + \begin{bmatrix} 0 \\ 2 \end{bmatrix} u$$

$$y = [-1 \quad -4]\mathbf{z}.$$

The system has two modes corresponding to the poles $\lambda = 1$ and $\lambda = 2$. The equations can be represented by Fig. 6.2.

From equation (3.31)

$$G(s) = C[sI - A]^{-1}B$$

$$= [-1 \quad -4] \begin{bmatrix} \dfrac{1}{s-2} & 0 \\ 0 & \dfrac{1}{s-1} \end{bmatrix} \begin{bmatrix} 0 \\ 2 \end{bmatrix} = \frac{-8}{s-2}.$$

It is seen that the transfer function involves only the mode corresponding to $\lambda = 2$ – the controllable and observable one (see Fig. 6.2). The uncontrollable mode ($\lambda = 1$) does not appear in the transfer function.

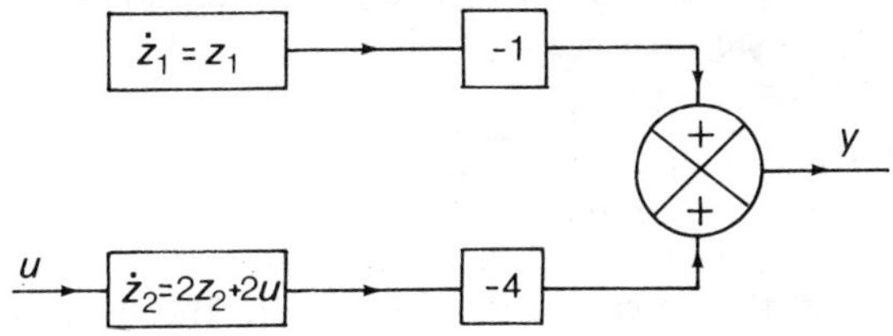

Fig. 6.2

In example 6.4 the transformed state equations are:

$$\dot{\mathbf{z}} = \begin{bmatrix} -1 & 0 \\ 0 & 1 \end{bmatrix} \mathbf{z} + \begin{bmatrix} -1 \\ 1 \end{bmatrix} u$$

$$y = [1 \quad 0]\mathbf{z}.$$

These equations can be represented by Fig. 6.3.

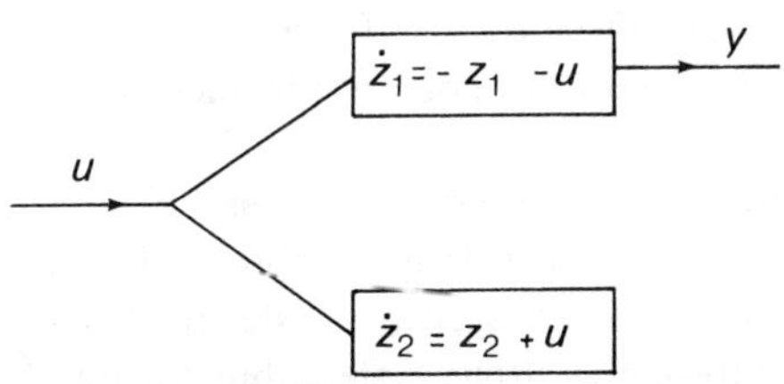

Fig. 6.3

In this case

$$G(s) = [1 \quad 0] \begin{bmatrix} \frac{1}{s+1} & 0 \\ 0 & \frac{1}{s-1} \end{bmatrix} \begin{bmatrix} -1 \\ 1 \end{bmatrix} = -\frac{1}{s+1}.$$

Again it is seen that the unobservable mode (corresponding to the pole $\lambda = 1$) is not involved in the transfer function.

In the above example it was noted that the uncontrollable or the unobservable mode of a system is absent in its transfer function. This fact is worth exploring further, and it gives rise to another criterion for a system to be controllable and/or observable. The matrix A is considered to be of order $n \times n$ having distinct eigenvalues.

The matrix $[sI-A]^{-1}$ in the equation of the system transfer function (equation (3.31) can be written as

$$[sI-A]^{-1} = \frac{\mathrm{adj}[sI-A]}{|sI-A|}$$

where $|sI-A|$ is the determinant of $[sI-A]$. The elements of the matrix adjoint $[sI-A]$ are polynomials of degree $(n-1)$ or less in s.$|sI-A|$ is the characteristic polynomial of the matrix A; it has the form

$$|sI-A| = \prod_{r=1}^{n} (s-\lambda_r)$$

where $\lambda_1, \lambda_2, \ldots, \lambda_n$ are the distinct eigenvalues of A (that is, the system poles).

It can be shown that in this case (distinct eigenvalues) $|sI-A|$ and $\mathrm{adj}[sI-A]$ can have no factors in common.

The controllability of a system is independent of the output (that is independent of the matrix C). When the system is uncontrollable, at least one of the system modes is missing from the transfer function $G(s)$. We conclude that there must be a cancellation between $\mathrm{adj}[sI-A]B$ and $|sI-A|$, that is between a zero and a pole of the transfer function.

Similarly when a system is unobservable there is a cancellation between $C\,\mathrm{adj}[sI-A]$ and $|sI-A|$.

So we have here another criterion for system controllability and observability. If a system is controllable (and/or observable) there is no cancellation between a zero and a pole of the system transfer functions. The converse is also true for a single input-single output system, but for multivariable systems the situation is more complicated - cancellations can occur yet the system may be both controllable and observable.

The conclusion of the above discussion is that when a system is both controllable and observable, we can obtain unique state equations from the transfer function, but this is not the case otherwise. If a system is uncontrollable and/or unobservable, the transfer function represents only the subsystems S_1 (see Fig. 6.1) so that in this case there are infinite possible state equations. This leads to the following definition.

Definition

The state equations equation (6.1)

$$\dot{\mathbf{x}} = A\mathbf{x} + B\mathbf{u}$$

$$\mathbf{y} = C\mathbf{x}$$

are said to be a **realisation** of the transfer function $G(s)$ if

$$G(s) = C[sI - A]^{-1}B.$$

The number of uncontrollable and unobservable modes which could be involved in a realisation of a transfer function, will influence the order of the system, that is, the order of the matrix A. But since all the controllable and observable modes of the system are involved in $G(s)$, the order of the system cannot be less than the number of such modes, hence:

Definition

A realisation equation (6.1) of a transfer function $G(s)$ is said to be **minimal** if it is both controllable and observable.

The above definition effectively states that a realisation is minimal if it is the realisation of the transfer function of the subsystem S_1 - the controllable and observable part of the system. A minimal realisation of a system is of course not unique, since any (non-singular) linear transformation leads to the same transfer function.

Example 6.8

Obtain a minimal realisation of a system having the transfer function

$$G(s) = \frac{1}{(s+1)(s+2)}\begin{bmatrix} -s & -(s+1) \\ 3s+2 & 2(s+1) \end{bmatrix}.$$

Solution

The minimal realisations must have two modes corresponding to the zeros of the characteristic equation (the denominator of $G(s)$). Hence A is a 2×2 matrix.

Also B and C are 2×2 matrices. In its diagonal form, A is the matrix $A = \Lambda = \text{diag}\{-1, -2\}$, so that

$$G(s) = C[sI - \Lambda]^{-1}B$$

$$= \begin{bmatrix} c_{11} & c_{12} \\ c_{21} & c_{22} \end{bmatrix}\begin{bmatrix} \frac{1}{s+1} & 0 \\ 0 & \frac{1}{s+2} \end{bmatrix}\begin{bmatrix} b_{11} & b_{12} \\ b_{21} & b_{22} \end{bmatrix}$$

$$= \begin{bmatrix} \frac{c_{11}b_{11}}{s+1} + \frac{c_{12}b_{21}}{s+2} & \frac{c_{11}b_{12}}{s+1} + \frac{c_{12}b_{22}}{s+2} \\ \frac{c_{21}b_{11}}{s+1} + \frac{c_{22}b_{21}}{s+2} & \frac{c_{21}b_{12}}{s+1} + \frac{c_{22}b_{22}}{s+2} \end{bmatrix}.$$

The given transfer function can be put in the form

$$\begin{bmatrix} \dfrac{1}{s+1} - \dfrac{2}{s+2} & -\dfrac{1}{s+2} \\ -\dfrac{1}{s+1} + \dfrac{4}{s+2} & \dfrac{2}{s+2} \end{bmatrix}$$

so that on comparing the elements in the two matrices representing the transfer function, we obtain

$$\begin{aligned} c_{11}b_{11} &= 1 \qquad & c_{12}b_{21} &= -2 \\ c_{11}b_{12} &= 0 & c_{12}b_{22} &= -1 \\ c_{21}b_{11} &= -1 & c_{22}b_{21} &= 4 \\ c_{21}b_{12} &= 0 & c_{22}b_{22} &= 2\ . \end{aligned}$$

Hence $b_{12} = 0$; taking arbitrarily $b_{11} = 1$, $b_{21} = 2$ and $b_{22} = 1$ we obtain $c_{11} = 1$, $c_{21} = -1$, $c_{12} = -1$ and $c_{22} = 2$. A minimal realisation is therefore

$$\dot{\mathbf{x}} = \begin{bmatrix} -1 & 0 \\ 0 & -2 \end{bmatrix} \mathbf{x} + \begin{bmatrix} 1 & 0 \\ 2 & 1 \end{bmatrix} u$$

$$\mathbf{y} = \begin{bmatrix} 1 & -1 \\ -1 & 2 \end{bmatrix} \mathbf{x}\ .$$

Notice that if T is any non-singular matrix of order 2×2, then

$$\dot{\mathbf{x}} = T^{-1} \begin{bmatrix} -1 & 0 \\ 0 & -2 \end{bmatrix} T\mathbf{x} + T^{-1} \begin{bmatrix} 1 & 0 \\ 2 & 1 \end{bmatrix} \mathbf{u}$$

$$\mathbf{y} = \begin{bmatrix} 1 & -1 \\ -1 & 2 \end{bmatrix} T\mathbf{x}$$

is also a minimal realisation of $G(s)$ (see the solution to example 3.10).

Finally, one non-minimal realisation is

$$\dot{\mathbf{x}} = \begin{bmatrix} -1 & 0 & 0 \\ 0 & -2 & 0 \\ 0 & 0 & -3 \end{bmatrix} \mathbf{z} + \begin{bmatrix} 1 & 0 \\ 2 & 1 \\ 0 & 0 \end{bmatrix} \mathbf{u}$$

$$\mathbf{y} = \begin{bmatrix} 1 & -1 & 0 \\ -1 & 2 & -1 \end{bmatrix} \mathbf{z}$$

(this can be verified by calculating the transfer function).

6.5 A TRANSFORMATION INTO THE COMPANION FORM

It was mentioned in section 3.5 that there exists a transformation, say

$$\mathbf{x} = T\mathbf{z} \tag{6.12}$$

transforming the system

$$\dot{\mathbf{x}} = A\mathbf{x} + \mathbf{b}u \tag{6.13}$$

where A is $n \times n$ matrix assumed to have n distinct eigenvalues, into

$$\dot{\mathbf{z}} = C\mathbf{z} + \mathbf{d}u \tag{6.14}$$

where

$$C = T^{-1}AT$$

is in a companion form, that is

$$C = \begin{bmatrix} 0 & 1 & 0 \ldots & 0 \\ 0 & 0 & 1 \ldots & 0 \\ \vdots & & & \\ 0 & 0 & 0 \ldots & 1 \\ -a_0 & -a_1 & -a_2 \ldots & -a_{n-1} \end{bmatrix}, \text{ and } \mathbf{d} = T^{-1}\mathbf{b} = \begin{bmatrix} 0 \\ 0 \\ 0 \\ \vdots \\ 1 \end{bmatrix} \tag{6.15}$$

Also (see Chapter 3)

$$|\lambda I - A| = \lambda^n + a_{n-1}\lambda^{n-1} + \ldots + a_0 = |\lambda I - C| = 0$$

is the characteristic equation of A (and C).

We construct T in the following way:

Let $\mathbf{p}$ be a vector of order $n \times 1$. Then $\mathbf{p}'A, \mathbf{p}'A^2, \ldots, \mathbf{p}'A^{n-1}$ are n row vectors, which we collect to make up the n rows of the matrix T^{-1}, that is

$$T^{-1} = \begin{bmatrix} \mathbf{p}' \\ \mathbf{p}'A \\ \vdots \\ \mathbf{p}'A^{n-1} \end{bmatrix}. \tag{6.16}$$

Assume for the moment that T^{-1} is a non-singular matrix, so that T exists and has the partitioned form (say)

$$T = [\mathbf{q}_1\,\mathbf{q}_2 \ldots \mathbf{q}_n] \tag{6.17}$$

so that $T^{-1}T = I$, hence

$$\begin{bmatrix} \mathbf{p}'\mathbf{q}_1 & \mathbf{p}'\mathbf{q}_2 & \ldots & \mathbf{p}'\mathbf{q}_n \\ \mathbf{p}'A\mathbf{q}_1 & \mathbf{p}'A\mathbf{q}_2 & \ldots & \mathbf{p}'A\mathbf{q}_n \\ \vdots & & & \\ \mathbf{p}'A^{n-1}\mathbf{q}_1 & \mathbf{p}'A^{n-1}\mathbf{q}_2 & \ldots & \mathbf{p}'A^{n-1}\mathbf{q}_n \end{bmatrix} = \begin{bmatrix} 1 & 0 & 0 \ldots 0 \\ 0 & 1 & 0 \ldots 0 \\ \vdots & & \\ 0 & 0 & 0 \ldots 1 \end{bmatrix}. \tag{6.18}$$

On taking the product

$$T^{-1}AT = \begin{bmatrix} \mathbf{p}' \\ \mathbf{p}'A \\ \vdots \\ \mathbf{p}'A^{n-1} \end{bmatrix} A[\mathbf{q}_1\,\mathbf{q}_2 \ldots \mathbf{q}_n],$$

we obtain the matrix

$$\begin{bmatrix} \mathbf{p}'A\mathbf{q}_1 & \mathbf{p}'A\mathbf{q}_2 & \ldots & \mathbf{p}'A\mathbf{q}_n \\ \mathbf{p}'A^2\mathbf{q}_1 & \mathbf{p}'A^2\mathbf{q}_2 & \ldots & \mathbf{p}'A^2\mathbf{q}_n \\ \vdots & & & \\ \mathbf{p}'A^n\mathbf{q}_1 & \mathbf{p}'A^n\mathbf{q}_2 & \ldots & \mathbf{p}'A^n\mathbf{q}_n \end{bmatrix}. \tag{6.19}$$

Note that the first $(n-1)$ rows of equation (6.19) are the same (row by row) as the last $(n-1)$ rows of the matrix equation (6.18).

So the matrix equation (6.19) is in the companion form, and (see equation (6.15)) the vectors $\mathbf{q}_1, \mathbf{q}_2, \ldots, \mathbf{q}_n$ are such that

$$-a_0 = \mathbf{p}'A^n\mathbf{q}_1,\ -a_1 = \mathbf{p}'A^n\mathbf{q}_2,\ \ldots\ -a_{n-1} = \mathbf{p}'A^n\mathbf{q}_n. \tag{6.20}$$

We must also have (equation (6.15))

$$\begin{bmatrix} \mathbf{p}' \\ \mathbf{p}'A \\ \vdots \\ \mathbf{p}'A^{n-1} \end{bmatrix} \mathbf{b} = \begin{bmatrix} 0 \\ 0 \\ \vdots \\ 1 \end{bmatrix}$$

or $\mathbf{p}'\mathbf{b} = 0$, $\mathbf{p}'A\mathbf{b} = 0, \ldots, \mathbf{p}'A^{n-1}\mathbf{b} = 1$, that is

$$\mathbf{p}'[\mathbf{b}\ \ A\mathbf{b}\ \ \ldots\ \ A^{n-1}\ \mathbf{b}] = \mathbf{d}' \tag{6.21}$$

Equation (6.21) is a system of n simultaneous equations in n unknowns (the n components of $\mathbf{p}$) and will have a unique solution if and only if the rank of the system controllability matrix is n, that is if

$$r[\mathbf{b}, A\mathbf{b}, \ldots, A^{n-1}\mathbf{b}] = n \tag{6.22}$$

(then the system is controllable).

Further, if the condition equation (6.22) holds, the matrix T^{-1} is non-singular. Indeed, assume there exist scalars $c_1, c_2, \ldots, c_n$ not all zero such that

$$c_1\mathbf{p}' + c_2\mathbf{p}'A + \ldots + c_n\mathbf{p}'A^{n-1} = \mathbf{0}. \tag{6.23}$$

Multiplying equation (6.23) by $\mathbf{b}$ and using equation (6.21) we obtain

$$c_1\mathbf{p}'\mathbf{b} + c_2\mathbf{p}'A\mathbf{b} + \ldots + c_n\mathbf{p}'A^{n-1}\mathbf{b} = c_n\mathbf{p}'A^{n-1}\mathbf{b} = \mathbf{0}$$

and so

$$c_n = 0.$$

Similarly on multiplying equation (6.23) by $A\mathbf{b}, A^2\mathbf{b}, \ldots, A^{n-1}\mathbf{b}$ we obtain respectively

$$c_{n-1} = \ldots = c_2 = c_1 = 0.$$

Hence the vectors $\mathbf{p}', \mathbf{p}'A, \ldots, \mathbf{p}'A^{n-1}$ are linearly independent, so that T^{-1} is non-singular. Finally, this proves that a matrix T exists such that the transformation equation (6.12) transforms the system state equations into one in which the state matrix is in companion form.

Example 6.9

The state equation of a system is

$$\dot{\mathbf{x}} = \begin{bmatrix} 13 & 35 \\ -6 & -16 \end{bmatrix} \mathbf{x} + \begin{bmatrix} -2 \\ 1 \end{bmatrix} u.$$

Find the transformation so that the state matrix of the transformed system is in a companion form.

Solution

Using equations (6.21), we obtain

$$[p_1 \; p_2] \begin{bmatrix} -2 \\ 1 \end{bmatrix} = 0 \quad \text{and} \quad [p_1 \; p_2] \begin{bmatrix} 13 & 35 \\ -6 & -16 \end{bmatrix} \begin{bmatrix} -2 \\ 1 \end{bmatrix} = 1$$

that is,

$$-2p_1 + p_2 = 0$$

$$9p_1 - 4p_2 = 1.$$

Hence $\mathbf{p}' = [p_1 \; p_2] = [1 \; 2]$, so that (equation (6.16))

$$T^{-1} = \begin{bmatrix} 1 & 2 \\ 1 & 3 \end{bmatrix} \quad \text{and} \quad T = \begin{bmatrix} 3 & -2 \\ -1 & 1 \end{bmatrix}.$$

We can then verify that

$$T^{-1}AT = \begin{bmatrix} 0 & 1 \\ -2 & -3 \end{bmatrix} \quad \text{and} \quad T^{-1}\mathbf{b} = \begin{bmatrix} 0 \\ 1 \end{bmatrix}.$$

The required transformation is

$$\mathbf{x} = \begin{bmatrix} 3 & -2 \\ -1 & 1 \end{bmatrix} \mathbf{z}.$$

PROBLEMS

(1) Determine whether the systems having the following representations are
(a) controllable and (b) observable.
Indicate which mode is uncontrollable and/or unobservable.

(i) $$\dot{\mathbf{x}} = \begin{bmatrix} 3 & 2 \\ -1 & 0 \end{bmatrix} \mathbf{x} + \begin{bmatrix} 4 \\ -2 \end{bmatrix} u$$
$$y = [-2 \quad -3]\mathbf{x}$$

(ii) $$\dot{\mathbf{x}} = \begin{bmatrix} 5 & 6 \\ -1 & 0 \end{bmatrix} \mathbf{x} + \begin{bmatrix} -2 \\ 1 \end{bmatrix} u$$
$$y = [-1 \quad -4]\mathbf{x}$$

(iii) $$\dot{\mathbf{x}} = \begin{bmatrix} -4 & -3 \\ 6 & 5 \end{bmatrix} \mathbf{x} + \begin{bmatrix} 2 \\ -4 \end{bmatrix} u$$
$$y = [-1 \quad 1]\mathbf{x}$$

(2) For each system in problem 1 find both the controllability and the observability matrices and determine their ranks.

(3) Calculate the transfer functions for each system in problem 1.

(4) Determine the values of p and q for which the following system is uncontrollable and/or unobservable.

$$\dot{\mathbf{x}} = \begin{bmatrix} 1 & 12 \\ 1 & 0 \end{bmatrix} \mathbf{x} + \begin{bmatrix} p \\ -1 \end{bmatrix} u$$
$$y = [q \quad 1]\mathbf{x}.$$

(5) Obtain a minimal realization of a system having the transfer matrix

$$G(s) = \frac{1}{(s+2)(s+3)} \begin{bmatrix} 1 & -s-4 \\ -3s-7 & 4s+10 \end{bmatrix}.$$

(6) Given the system defined by

$$\dot{\mathbf{x}} = \begin{bmatrix} 1 & 2 & 0 \\ 0 & -1 & 1 \\ 1 & 0 & 1 \end{bmatrix} \mathbf{x} + \begin{bmatrix} 1 \\ 0 \\ 1 \end{bmatrix} u.$$

Find the matrix T which will transform the state equations so that the resulting system matrix is in a companion form.

(7) Show that for a controllable system whose states at time $t = 0$ and $t = t_1$ are $\mathbf{x}_0$ and $\mathbf{x}_1$ respectively, where (see equation (3.20))

$$\mathbf{x}_1 = e^{At_1}\left[\mathbf{x}_0 + \int_0^{t_1} e^{-At}\mathbf{b}u(t)\mathrm{d}t\right]$$

no vector $\mathbf{w} \neq \mathbf{0}$ exists such that

$$[e^{-At}b]^{-}\mathbf{w} = \mathbf{0} \quad \text{for} \quad 0 \leqslant t \leqslant t_1.$$

Hint: assume the above and then show that the controllability criterion is violated.

(8) Determine the (constant) vector $\mathbf{w}$ so that

$$u = [e^{-At}\mathbf{b}]^{-}\mathbf{w}$$

is the input function which will transform the state of the system in problem 7 from $\mathbf{x}_0$ to $\mathbf{x}_1$ in time $t_1 > 0$.

(9) Based on the solutions to problems 7 and 8 above conclude why the controllability criterion $r(Q) = n$ (see equation (6.8)) is a sufficient condition for system controllability.

(10) The state equation of a system is

$$\dot{\mathbf{x}} = \begin{bmatrix} 0 & 2 \\ 0 & 0 \end{bmatrix}\mathbf{x} + \begin{bmatrix} 0 \\ 1 \end{bmatrix}u$$

and $\mathbf{x}(0) = \mathbf{x}_0 = \mathbf{0}$.

(i) Show that the system is controllable.

(ii) Find the input $u(t)$ necessary to transfer the system state from $\mathbf{x} = \mathbf{0}$

to $\mathbf{x} = \begin{bmatrix} 1 \\ 2 \end{bmatrix}$ in time $t = 3$ seconds.

CHAPTER 7

Multivariable feedback and pole location

7.1 INTRODUCTION

We have considered various aspects of feedback control, especially in section 4.6. We have in particular discussed how to achieve an adequate steady-state system response, but have said nothing about the transient response. Under certain circumstances, the feedback techniques we have considered for single-input-single-output systems can be applied to multivariable systems. This is possible when the system is decoupled, that is when each input affects only one output. If the system is not decoupled, state variable feedback can be used to achieve this objective.

But it is the attainment of a desirable system transient response which will be our main concern in this chapter. In fact we shall be even less ambitious. In this book we have been more concerned with the underlying principles and less with actual design procedures; we continue with this basic philosophy. We have previously noted that the transient response depends on the location of the system poles, that is on the eigenvalues of the closed-loop system matrix (see Chapter 4). The problem we now investigate, possibly the most important one in linear control theory, is whether and under what circumstances can we use feedback to assign the eigenvalues in any desired configuration in the s-plane for a linear system.

7.2 STATE FEEDBACK OF A SISO SYSTEM

Consider a SISO system defined by

$$\dot{\mathbf{x}} = A\mathbf{x} + \mathbf{b}u. \tag{7.1}$$

The dynamics of this system are determined by the poles, that is, the eigenvalues $\lambda_1, \lambda_2, \ldots, \lambda_n$ - (assumed distinct) of the matrix A.

We know (see section 6.5) that if the system is controllable, there exists a matrix T such that

$$\mathbf{x} = T\mathbf{z}$$

transforms the system defined by equation (7.1) to

$$\dot{\mathbf{z}} = C\mathbf{z} + \mathbf{d}u \tag{7.2}$$

where

$$C = T^{-1}AT = \begin{bmatrix} 0 & 1 & 0 & \dots & 0 \\ 0 & 0 & 1 & \dots & 0 \\ \vdots & & & & \\ 0 & 0 & 0 & \dots & 1 \\ -a_1 & -a_2 & -a_3 & \dots & -a_n \end{bmatrix} \quad \text{and} \quad \mathbf{d} = T^{-1}\mathbf{b} = \begin{bmatrix} 0 \\ 0 \\ \vdots \\ 0 \\ 1 \end{bmatrix}.$$

Now apply a feedback control of the form

$$u(t) = \mathbf{k}'\mathbf{z} \tag{7.3}$$

where $\mathbf{k}' = [k_1, k_2, \dots, k_n]$.

Substituting (7.3) into equation (7.2) results in

$$\dot{\mathbf{z}} = [C + \mathbf{d}\mathbf{k}']\mathbf{z} \tag{7.4}$$

which is the system whose dynamics are determined by the eigenvalues of $[C + \mathbf{d}\mathbf{k}']$, that is by the eigenvalues of

$$\begin{bmatrix} 0 & 1 & 0 & \dots & 0 \\ 0 & 0 & 1 & \dots & 0 \\ \vdots & & & & \\ 0 & 0 & 0 & \dots & 1 \\ -a_1 & -a_2 & -a_3 & \dots & -a_n \end{bmatrix} + \begin{bmatrix} 0 & 0 & 0 & \dots & 0 \\ 0 & 0 & 0 & \dots & 0 \\ \vdots & & & & \\ 0 & 0 & 0 & \dots & 0 \\ k_1 & k_2 & k_3 & \dots & k_n \end{bmatrix}$$

$$= \begin{bmatrix} 0 & 1 & 0 & \dots & 0 \\ 0 & 0 & 1 & \dots & 0 \\ \vdots & & & & \\ 0 & 0 & 0 & \dots & 1 \\ -r_1 & -r_2 & -r_3 & \dots & -r_n \end{bmatrix}$$

where $\quad -r_i = k_i - a_i \quad (i = 1, 2, \dots, n).$ (7.5)

The characteristic equation of $[C + \mathbf{d}'\mathbf{k}]$ is

$$s^n + r_n s^{n-1} + \ldots + r_2 s + r_1 = 0.$$

By an appropriate choice of the components $k_1, k_2, \ldots, k_n$ of $\mathbf{k}$ in equation (7.5) we can assign arbitrary values to the coefficients $r_1, r_2, \ldots, r_n$ and so achieve any desired pole configurations.

Having obtained the desired feedback law when the system matrix is in a companion form, we must now determine the law in terms of the original parameters.

Since $\mathbf{z} = T^{-1}\mathbf{x}$, it follows that

$$u(t) = \mathbf{k}'\mathbf{z} = \mathbf{k}'T^{-1}\mathbf{x}. \tag{7.6}$$

Example 7.1

Given the system

$$\dot{\mathbf{x}} = \begin{bmatrix} -13 & 8 \\ -25 & 15 \end{bmatrix} \mathbf{x} + \begin{bmatrix} 1 \\ 2 \end{bmatrix} u$$

find the feedback law so that the resulting closed-loop system has poles at $\lambda = -2$ and $\lambda = -3$.

Solution

We proceed as in example 6.9 to find

$$T^{-1} = \begin{bmatrix} 2 & -1 \\ -1 & 1 \end{bmatrix}, \quad \text{so that} \quad T = \begin{bmatrix} 1 & 1 \\ 1 & 2 \end{bmatrix}.$$

With the transformation $\mathbf{x} = T\mathbf{z}$, the state equation becomes

$$\dot{\mathbf{z}} = T^{-1}AT\mathbf{z} + T^{-1}\mathbf{b}u, \quad \text{that is}$$

$$\dot{\mathbf{z}} = \begin{bmatrix} 0 & 1 \\ -5 & 2 \end{bmatrix} \mathbf{z} + \begin{bmatrix} 0 \\ 1 \end{bmatrix} u.$$

Notice that the characteristic equation in this case is

$$\lambda^2 - 2\lambda + 5 = 0.$$

The system poles are at $\lambda = 1 \pm 2i$ so that the system is unstable!

We apply feedback in the form

$$u = \mathbf{k}'\mathbf{z} \quad \text{where} \quad \mathbf{k}' = [k_1, k_2].$$

Since the system poles are to be at $\lambda_1 = -2$ and $\lambda_2 = -3$ the corresponding characteristic equation is

$$\lambda^2 + 5\lambda + 6 = 0.$$

Making use of equation (7.5) we obtain

$$k_1 = a_1 - r_1 = 5 - 6 = -1 \text{ and } k_2 = a_2 - r_2 = -2 - 5 = -7,$$

hence $\mathbf{k}' = [-1, -7]$.

In terms of the original coordinates

$$u(t) = \mathbf{k}'T^{-1}\mathbf{x} = [-1 \quad -7]\begin{bmatrix} 2 & -1 \\ -1 & 1 \end{bmatrix}\mathbf{x} = [5 \quad -6]\mathbf{x}.$$

We now check whether the system with the above feedback control has the assigned poles,

$$\dot{\mathbf{x}} = A\mathbf{x} + \mathbf{b}u = [A + \mathbf{b}\mathbf{k}'T^{-1}]\mathbf{x}$$

and

$$A + \mathbf{b}\mathbf{k}'T^{-1} = \begin{bmatrix} -13 & 8 \\ -25 & 15 \end{bmatrix} + \begin{bmatrix} 1 \\ 2 \end{bmatrix}[5 \quad -6]$$

$$= \begin{bmatrix} -8 & 2 \\ -15 & 3 \end{bmatrix}.$$

The corresponding characteristic equation is

$$\lambda^2 + 5\lambda + 6 = 0,$$

as desired!

7.3 MULTIVARIABLE SYSTEMS

As one would expect, the pole assignment for a multivariable system is more complicated than for a scalar system since a multitude of inputs are to be controlled, and not just one! The irony is that apparently there are too many degrees of freedom which we are not sure how to exploit.

We shall discuss an interesting method (see MacFarlane [39]) which severely restricts the number of degrees of freedom so that it is by no means a general method, but it has the great advantage of being relatively simple.

To demonstrate the method we make use of a relation involving the determinants of matrices.

If P and Q are matrices of orders $m \times n$ and $n \times m$ respectively, then

$$|I_m - PQ| = |I_n - QP| \tag{7.7}$$

where the notation $|A|$ stands for the determinant of A.

I_m and I_n are the unit matrices of order $m \times m$ and $n \times n$ respectively.

This relation is by no means obvious, and we first discuss its derivation. If $n = m$ and P is non-singular, the derivation makes use of the result $|AB| = |BA|$ and is quite simple.

$$|I - PQ| = |P[P^{-1} - Q]| = |[P^{-1} - Q]P| = |I - QP|$$

In the more general case when $n \neq m$, we can make use of the determinant of a partitioned matrix as explained below. If A is a matrix of order $n \times n$, then the determinant of the (partitioned square) matrix

$$\left|\begin{array}{c|c} I & 0 \\ \hline 0 & A \end{array}\right| = |A|.$$

This is obvious when we consider the expansion of the determinant by its first column. For example

$$\left|\begin{array}{cc|cc} 1 & 0 & 0 & 0 \\ 0 & 1 & 0 & 0 \\ \hline 0 & 0 & a_{11} & a_{12} \\ 0 & 0 & a_{21} & a_{22} \end{array}\right| = \left|\begin{array}{ccc} 1 & 0 & 0 \\ 0 & a_{11} & a_{12} \\ 0 & a_{21} & a_{22} \end{array}\right| = \left|\begin{array}{cc} a_{11} & a_{12} \\ a_{21} & a_{22} \end{array}\right|.$$

Similarly

$$\left|\begin{array}{c|c} A & 0 \\ \hline 0 & I \end{array}\right| = |A|.$$

On the other hand the determinant of the partitioned matrix

$$\left|\begin{array}{c|c} I & A \\ \hline 0 & I \end{array}\right| = 1.$$

Again this is obvious when we consider the expansion of the determinant by its first column. Similar considerations lead to the conclusion that

$$\left|\begin{array}{c|c} I & 0 \\ \hline A & I \end{array}\right| = 1.$$

We now make use of the above and of the standard result that the determinant of a product of matrices equals the product of the determinants of the matrices, that is

$$|ABC| = |A||B||C|$$

to consider the determinants of partitioned matrices of more complicated forms. For example, if A is non-singular, then since

$$\left[\begin{array}{c|c} A & B \\ \hline 0 & C \end{array}\right] = \left[\begin{array}{c|c} A & 0 \\ \hline 0 & I \end{array}\right]\left[\begin{array}{c|c} I & 0 \\ \hline 0 & C \end{array}\right]\left[\begin{array}{c|c} I & A^{-1}B \\ \hline 0 & I \end{array}\right]$$

it follows that

$$\left|\begin{array}{c|c} A & B \\ \hline 0 & C \end{array}\right| = |A|.|C|.1.$$

Similarly if C is non-singular, then since

$$\left[\begin{array}{c|c} A & 0 \\ \hline B & C \end{array}\right] = \left[\begin{array}{c|c} A & 0 \\ \hline 0 & I \end{array}\right]\left[\begin{array}{c|c} I & 0 \\ \hline 0 & C \end{array}\right]\left[\begin{array}{c|c} I & 0 \\ \hline C^{-1}B & I \end{array}\right]$$

it follows that

$$\left|\begin{array}{c|c} A & 0 \\ \hline B & C \end{array}\right| = |A|.|C|.1.$$

We now use those last two results to prove equation (7.7).

Since

$$\left[\begin{array}{c|c} I_m & 0 \\ \hline -Q & I_n \end{array}\right]\left[\begin{array}{c|c} I_m & P \\ \hline Q & I_n \end{array}\right] = \left[\begin{array}{c|c} I_m & P \\ \hline 0 & I_n - QP \end{array}\right]$$

it follows that

$$\left|\begin{array}{c|c} I_m & P \\ \hline Q & I_n \end{array}\right| = |I_n - QP| \qquad (7.8)$$

(the value of determinant of the first matrix on the left-hand side is unity). Similarly, since

$$\left[\begin{array}{c|c} I_m & P \\ \hline Q & I_n \end{array}\right]\left[\begin{array}{c|c} I_m & 0 \\ \hline -Q & I_n \end{array}\right] = \left[\begin{array}{c|c} I_m - PQ & P \\ \hline 0 & I_n \end{array}\right]$$

it follows that

$$\left|\begin{array}{c|c} I_m & P \\ \hline Q & I_n \end{array}\right| = |I_m - PQ|.$$

Comparing equations (7.8) and (7.9), we obtain equation (7.7).

The following example illustrates equation (7.7).

Example 7.2

Given $P = [p_1 \;\; p_2]$ and $Q = \begin{bmatrix} q_1 \\ q_2 \end{bmatrix}$, show that

$$|I_m - PQ| = |I_n - QP|.$$

Solution

Notice that since P is a one-row matrix, PQ is also a one-row matrix, so that in the above equation

$$I_m = [1].$$

It follows that

$$|I_m - PQ| = 1-(p_1q_1+p_2q_2).$$

Also

$$[I_n - QP] = \begin{bmatrix} 1 & 0 \\ 0 & 1 \end{bmatrix} - \begin{bmatrix} q_1p_1 & q_1p_2 \\ q_2p_1 & q_2p_2 \end{bmatrix} = \begin{bmatrix} 1-q_1p_1 & -q_1p_2 \\ -q_2p_1 & 1-q_2p_2 \end{bmatrix}$$

hence

$$\begin{aligned} |I_n - QP| &= (1-q_1p_1)(1-q_2p_2)-q_1p_2q_2p_1 \\ &= 1-(p_1q_1+p_2q_2). \end{aligned}$$

We can now proceed with the main theme of this section.

We consider a multivariable system having n state variables and m inputs. To simplify the mathematical manipulations we shall initially assume that the state matrix is in a diagonal form, that is the state equation has the form

$$\dot{\mathbf{z}} = \Lambda\mathbf{z} + B_1\mathbf{u} \tag{7.10}$$

where $\Lambda = \text{diag}\{\lambda_1, \lambda_2, \ldots, \lambda_n\}$

B_1 is of order $n \times m$

$\mathbf{z}$ is $n \times 1$ and $\mathbf{u}$ is $m \times 1$.

The system open loop characteristic polynomial is

$$|sI - \Lambda| = (s-\lambda_1)(s-\lambda_2)\ldots(s-\lambda_n) \tag{7.11}$$

where the system poles (also known as the **open loop characteristic frequencies**) $\lambda_1, \lambda_2, \ldots, \lambda_n$ are assumed to be distinct. To correspond to equation (7.2) we assume a feedback law of the form

$$\mathbf{u} = K_1\mathbf{z} \tag{7.12}$$

where K_1, known as a **proportional-gain** matrix, is of order $m \times n$.

Combining equations (7.10) and (7.12), we obtain

$$\dot{\mathbf{z}} = [\Lambda + B_1K_1]\mathbf{z}. \tag{7.13}$$

The objective in this procedure is to choose K_1 so that the closed loop system poles, say $\rho_1, \rho_2, \ldots, \rho_n$, have prescribed values.

These poles are the solutions of the closed loop characteristic equation, that is the roots of

$$|sI - \Lambda - B_1K_1| = (s - \rho_1)(s - \rho_2) \ldots (s - \rho_n) = 0. \tag{7.14}$$

(The poles, $\rho_1, \rho_2, \ldots, \rho_n$, are also called the **closed loop characteristic frequencies.**)

Even without a detailed mathematical analysis, it is clear that if a solution to this problem exists - and this is something that we consider later on in this section - there must be many proportional gain matrices such that the left-hand side of equation (7.14) is satisfied by prescribed values of $\rho_1, \rho_2, \ldots, \rho_n$.

A procedure of great merit [4] because of its relative simplicity and elegance is to select K_1 to have the so-called **dyadic** form, a product of a column and a row vector

$$K_1 = \mathbf{f}\mathbf{d}' \tag{7.15}$$

where $\mathbf{f} = [f_1, f_2, \ldots f_m]'$ and $\mathbf{d}' = [d_1, d_2, \ldots, d_n]$.

With this choice for K_1, equation (7.12) becomes

$$\mathbf{u} = \mathbf{f}\mathbf{d}'\mathbf{z} \tag{7.16}$$

which shows that this control law requires the feedback to be the proportions $f_1, f_2, \ldots, f_m$ of the fixed combination

$$\mathbf{d}'\mathbf{z} = d_1z_1 + d_2z_2 + \ldots + d_nz_n$$

so that the components of the input $\mathbf{u} = [u_1, u_2, \ldots, u_m]'$ have the form $u_1 = f_1\mathbf{d}'\mathbf{z}$, $u_2 = f_2\mathbf{d}'\mathbf{z}, \ldots, u_m = f_m\mathbf{d}'\mathbf{z}$.

After substituting for K_1 of equation (7.15) into equation (7.14) we can then write

$$[sI - \Lambda - B_1\mathbf{f}\mathbf{d}'] = [sI - \Lambda][I - (sI - \Lambda)^{-1}B_1\mathbf{f}\mathbf{d}']$$

On taking the determinants of both sides and using equations (7.11) and (7.14), we obtain

$$\prod_{i=1}^{n} (s - \rho_i) = \prod_{i=1}^{n} (s - \lambda_i)|I - (sI - \Lambda)^{-1}B_1\mathbf{f}\mathbf{d}'|. \tag{7.17}$$

Let $P = [sI - \Lambda]^{-1}B_1\mathbf{f}$, a matrix of order $n \times 1$ and $Q = \mathbf{d}'$, a matrix of order $1 \times n$. We now use equation (7.7), noting that the order of the matrix $m \times n$ in that equation, corresponds to $n \times 1$ above, so that

$$I_n = [1].$$

It follows that

$$|I - (sI - \Lambda)^{-1}B_1\mathbf{f}\mathbf{d}'| = |1 - \mathbf{d}'(sI - \Lambda)^{-1}B_1\mathbf{f}|. \tag{7.18}$$

Let

$$B_1\mathbf{f} = \mathbf{r} = \begin{bmatrix} r_1 \\ r_2 \\ \vdots \\ r_n \end{bmatrix}, \text{ say} \tag{7.19}$$

we then have

$$\mathbf{d}'(sI - \Lambda)^{-1}B_1\mathbf{f} = [d_1, d_2, \ldots, d_n] \begin{bmatrix} \frac{1}{s-\lambda_1} & 0 & \ldots 0 \\ 0 & \frac{1}{s-\lambda_2} & \ldots 0 \\ \vdots & & \\ 0 & 0 & \cdots \frac{1}{s-\lambda_n} \end{bmatrix} \begin{bmatrix} r_1 \\ r_2 \\ \vdots \\ r_n \end{bmatrix}$$

$$= \frac{d_1 r_1}{s-\lambda_1} + \frac{d_2 r_2}{s-\lambda_2} + \ldots + \frac{d_n r_n}{s-\lambda_n}. \tag{7.20}$$

Substituting equations (7.18) and (7.20) into equation (7.17), we obtain

$$\prod_{i=1}^{n} (s - \rho_i) = \prod_{i=1}^{n} (s - \lambda_i) \left[1 - \sum_{i=1}^{n} \frac{d_i r_i}{s - \lambda_i} \right].$$

Hence,

$$\frac{\prod_{i=1}^{n} (s - \lambda_i) - \prod_{i=1}^{n} (s - \rho_i)}{\prod_{i=1}^{n} (s - \lambda_i)} = \sum_{i=1}^{n} \frac{d_i r_i}{s - \lambda_i}. \tag{7.21}$$

The numerator of the left-hand side of equation (7.21) is the difference of the two characteristic polynomials both of degree n and both **monic** (which implies that the coefficient of s^n is unity). It follows that the numerator is a polynomial of degree $(n-1)$ or less, so that the expression can be expanded into partial fractions as

$$\frac{\prod_{i=1}^{n}(s-\lambda_i) - \prod_{i=1}^{n}(s-\rho_i)}{\prod_{i=1}^{n}(s-\lambda_i)} = \sum_{i=1}^{n}\frac{R_i}{s-\lambda_i} \tag{7.22}$$

where the residues R_j at the eigenvalue λ_j $(j = 1,2,\ldots,n)$ can be evaluated as (see Appendix 1)

$$R_j = \lim_{s\to\lambda_j}\frac{(s-\lambda_j)\left[\prod_{i=1}^{n}(s-\lambda_i) - \prod_{i=1}^{n}(s-\rho_i)\right]}{\prod_{i=1}^{n}(s-\lambda_i)}. \tag{7.23}$$

We now discuss the procedure for the pole assignment.

Using equation (7.22) we calculate the residues R_j $(j = 1, 2, \ldots, n)$. Comparing equations (7.21) and (7.22) we have

$$R_j = d_j r_j. \tag{7.24}$$

With certain restrictions which we consider later on (see *Notes* on p. 186) we can specify the matrix $\mathbf{f}$ of equation (7.15), and this will enable us to compute (see equation (7.19))

$$\mathbf{r} = B_1\mathbf{f}$$

and then use equation (7.24) to determine $\mathbf{d}'$.

To illustrate the procedure we begin with an example of a single input system.

Example 7.3

Given a system defined by the state equation

$$\dot{\mathbf{x}} = \begin{bmatrix} -1 & 0 \\ 0 & -2 \end{bmatrix}\mathbf{x} + \begin{bmatrix} 2 \\ 1 \end{bmatrix}u$$

find a proportional gain matrix K_1 such that the closed loop system poles are both at $\lambda = -3$.

Solution

In this case $n = 2$ and $m = 1$ so that

$$K_1 = \mathbf{fd}'$$

is a one-row matrix, and $\mathbf{f}$ is a one-element matrix.

The open loop characteristic polynomial is

$$(s-\lambda_1)(s-\lambda_2) = (s+1)(s+2) = s^2+3s+2.$$

The closed loop characteristic polynomial is

$$(s-\rho_1)(s-\rho_1) = (s+3)^2 = s^2+6s+9.$$

The residuals are

$$R_1 = \lim_{s \to -1} \frac{(s+1)(-3s-7)}{(s+1)(s+2)} = -4$$

and

$$R_2 = \lim_{s \to -2} \frac{(s+2)(-3s-7)}{(s+1)(s+2)} = 1.$$

Hence (equation 7.24)

$$r_1 d_1 = -4 \qquad r_2 d_2 = 1.$$

From equation (7.19)

$$\begin{bmatrix} 2 \\ 1 \end{bmatrix} f = \begin{bmatrix} r_1 \\ r_2 \end{bmatrix} \quad \text{so that} \quad \begin{bmatrix} 2f \\ f \end{bmatrix} = \begin{bmatrix} r_1 \\ r_2 \end{bmatrix}.$$

Choose $f = 1$, then $r_1 = 2$ and $r_2 = 1$

so that $\quad d_1 = -2 \quad$ and $\quad d_2 = 1 \quad$ and

$$K_1 = [-2, \ 1].$$

Check

The closed loop characteristic polynomial is

$$|sI - \Lambda - B_1K_1|$$

$$= \left| \begin{bmatrix} s & 0 \\ 0 & s \end{bmatrix} - \begin{bmatrix} -1 & 0 \\ 0 & -2 \end{bmatrix} - \begin{bmatrix} 2 \\ 1 \end{bmatrix} [-2, \ 1] \right|$$

$$= \left| \begin{bmatrix} s & 0 \\ 0 & s \end{bmatrix} - \begin{bmatrix} -1 & 0 \\ 0 & -2 \end{bmatrix} - \begin{bmatrix} -4 & 2 \\ -2 & 1 \end{bmatrix} = \begin{bmatrix} s & 0 \\ 0 & s \end{bmatrix} - \begin{bmatrix} -5 & 2 \\ -2 & -1 \end{bmatrix} \right|$$

$$= \begin{vmatrix} s+5 & -2 \\ 2 & s+1 \end{vmatrix} = (s+3)^2$$

We now consider the general case, when the state matrix is not necessarily in a diagonal form. Assume that

$$\dot{\mathbf{x}} = A\mathbf{x} + B\mathbf{u}. \tag{7.25}$$

Using the transformation

$$\mathbf{x} = T\mathbf{z}$$

equation (7.25) becomes

$$\begin{aligned} \dot{\mathbf{z}} &= T^{-1}AT\mathbf{z} + T^{-1}B\mathbf{u} \\ &= \Lambda z + B_1 u \end{aligned} \tag{7.26}$$

where $\Lambda = T^{-1}AT$ and $B_1 = T^{-1}B$.

Notice that equation (7.26) is identical to equation (7.10).

Having transformed the state equation into a convenient diagonal form, we can compute the proportional gain matrix K_1, in the manner indicated above.

Now applying the inverse transformation,

$$\mathbf{z} = T^{-1}\mathbf{x},$$

to the resulting state equation (equation 7.13), we obtain

$$\dot{\mathbf{x}} = T\Lambda T^{-1}\mathbf{x} + TB_1K_1T^{-1}\mathbf{x}$$

or

$$\dot{\mathbf{x}} = A\mathbf{x} + BK\mathbf{x} \tag{7.27}$$

where

$$K = K_1T^{-1} \tag{7.28}$$

is the proportional gain matrix associated with the state of the system specified by equation (7.25).

Example 7.4

Given

$$\dot{\mathbf{x}} = \begin{bmatrix} 4 & 10 \\ -3 & -7 \end{bmatrix}\mathbf{x} + \begin{bmatrix} 9 \\ -5 \end{bmatrix}u$$

Find the proportional gain matrix K such that the closed loop system poles are both at $\lambda = -3$.

Solution

On applying the transformation

$$\mathbf{x} = T\mathbf{z} = \begin{bmatrix} 2 & 5 \\ -1 & -3 \end{bmatrix}\mathbf{z}$$

for which $T^{-1} = \begin{bmatrix} 3 & 5 \\ -1 & -2 \end{bmatrix}$

it is found that the system is the same as the one considered in example 7.3. Hence

$$K_1 = [-2,\ 1], \quad \text{so that (equation (7.28))}$$

$$K = K_1T^{-1} = [-2,\ 1]\begin{bmatrix} 3 & 5 \\ -1 & -2 \end{bmatrix} = [-7,\ -12],$$

so the feedback control

$$u = [-7,\ 12]\mathbf{x}$$

will achieve the desired pole assignment.

Check

The new system matrix is

$$A + BK = \begin{bmatrix} 4 & 10 \\ -3 & -7 \end{bmatrix} + \begin{bmatrix} -63 & -108 \\ 35 & 60 \end{bmatrix} = \begin{bmatrix} -59 & -98 \\ 32 & 53 \end{bmatrix}.$$

Hence the closed loop system poles are the roots of the characteristic equation

$$\begin{vmatrix} s+59 & 98 \\ -32 & s-53 \end{vmatrix} = s^2 + 6s + 9 = 0,$$

as required.

We have, in the above two examples, applied a method developed for multivariable system to scalar ones. This was done to illustrate as simply as possible the concepts involved. In the next example we consider a multi-input system.

Example 7.5

Consider the system defined by

$$\dot{\mathbf{x}} = \begin{bmatrix} 4 & 10 \\ -3 & -7 \end{bmatrix}\mathbf{x} + \begin{bmatrix} 1 & 1 \\ 2 & 0 \end{bmatrix}\mathbf{u}.$$

Find the proportional-gain matrix K for the above system so that the closed loop system poles are both at $\lambda = -3$.

Solution

Using the transformation of the previous example, that is

$$\mathbf{x} = T\mathbf{z} = \begin{bmatrix} 2 & 5 \\ -1 & -3 \end{bmatrix}\mathbf{z},$$

the system state equation becomes

$$\dot{\mathbf{z}} = \begin{bmatrix} -1 & 0 \\ 0 & -2 \end{bmatrix} \mathbf{z} + \begin{bmatrix} 13 & 3 \\ -5 & -1 \end{bmatrix} \mathbf{u}.$$

We can now use the data from example 7.3,

$$r_1 d_1 = -4 \quad \text{and} \quad r_2 d_2 = 1.$$

We choose $\mathbf{f} = [-1 \;\; 1]'$, and obtain from equation (7.24)

$$\begin{bmatrix} r_1 \\ r_2 \end{bmatrix} = \begin{bmatrix} 13 & 3 \\ -5 & -1 \end{bmatrix} \begin{bmatrix} -1 \\ 1 \end{bmatrix} = \begin{bmatrix} -10 \\ 4 \end{bmatrix}$$

so that

$$\mathbf{d}' = [\tfrac{2}{5}, \tfrac{1}{4}], \quad \text{and}$$

$$K_1 = \begin{bmatrix} -1 \\ 1 \end{bmatrix} [\tfrac{2}{5} \;\; \tfrac{1}{4}] = \begin{bmatrix} -\tfrac{2}{5} & -\tfrac{1}{4} \\ \tfrac{2}{5} & \tfrac{1}{4} \end{bmatrix}$$

$$K = K_1 T^{-1} = \tfrac{1}{20} \begin{bmatrix} -19 & -30 \\ 19 & 30 \end{bmatrix}.$$

Check

$$A + BK = \begin{bmatrix} 4 & 10 \\ -3 & -7 \end{bmatrix} + \begin{bmatrix} 0 & 0 \\ -1.9 & -3 \end{bmatrix} = \begin{bmatrix} 4 & 10 \\ -4.9 & -10 \end{bmatrix}.$$

Hence the closed loop characteristic equation is

$$\lambda^2 + 6\lambda + 9 = 0,$$

as required.

Notes

(1) If the system is uncontrollable then at least one row (say the jth) of B_1 has all the components equal to zero. The equation

$$\mathbf{r} = B_1 \mathbf{f}$$

shows that the corresponding $r_j = 0$, whatever the choice of $\mathbf{f}$, and hence (equations (7.21) and (7.22)) $R_j = 0$.

Finally from equation (7.23) we deduce that in this case, the root $s = \lambda_j$ is common to both the open loop and the closed loop characteristic equations. This implies that we cannot assign an arbitrary value to a system pole corresponding to an uncontrollable mode.

(2) There is a restriction on the choice of the matrix **f**. As (equation (7.19))

$$r_j = (j^{\text{th}} \text{ row of } B_1)\,\mathbf{f}$$

must (in general) be non-zero since $R_j \neq 0$, we must choose **f** so that it is *not orthogonal* to any row of B_1.

(3) For each suitable matrix **f**, d' is uniquely determined (from equation (7.24)), and hence so is a proportional-gain matrix K. It follows that an infinite number of proportional-gain matrices can be computed to assign the poles of the closed loop system at pre-assigned values.

Example 7.6

Use the data of the previous example to illustrate the last point (3) above.

Solution

With the data of example 7.5, we shall choose a different **f**, say

$$\mathbf{f} = [1, \ -3]'.$$

This time

$$\mathbf{r} = [4 \ \ -2]' \quad \text{so that} \quad \mathbf{d}' = [-1 \ \ -\tfrac{1}{2}]$$

and

$$K_1 = \begin{bmatrix} -1 & -\tfrac{1}{2} \\ 3 & \tfrac{3}{2} \end{bmatrix}.$$

The proportional-gain matrix is

$$K = \tfrac{1}{2}\begin{bmatrix} -5 & -8 \\ 15 & 24 \end{bmatrix}$$

and is different from the one computed in example 7.5.

Although the method discussed above enables us to assign the system's closed loop poles it is stressed that this is not sufficient to fully control the dynamic behaviour of the system. This behaviour is determined by the system's transfer function, and the assignment of the closed loop poles only specifies the poles of this function (see section 3.6); the zeros remain unspecified.

7.4 OBSERVERS

In the previous two sections it has been assumed that the entire state vector $\mathbf{x}(t)$ of the system is available for feedback to implement a specified control law. But in many systems it is not possible to measure the entire state vector, although the measurement of some of the state variables is practical. Fortunately, Luenberger† has shown that a satisfactory estimate of the state vector can be obtained by using the available system inputs and outputs to drive a second linear system known as an **observer**.

The observer is a dynamic system whose state vector, say $\mathbf{z}(t)$, is an approximation to a linear transformation of the state $\mathbf{x}(t)$, that is

$$\mathbf{z}(t) \simeq T\mathbf{x}(t). \tag{7.29}$$

The entire state vector $\mathbf{z}$ is of course, available for feedback.

To show that it is possible to design an observer having a state defined by equation (7.29) we consider the system

$$\dot{\mathbf{x}} = A\mathbf{x} + B\mathbf{u} \tag{7.30}$$

$$y = \mathbf{c}'\mathbf{x}$$

where A is $n \times n$, B is $n \times m$, $\mathbf{c}'$ is $1 \times n$.

To simplify the mathematical manipulation, the system has been assumed to have a single output. But the results are applicable to multi-output systems.

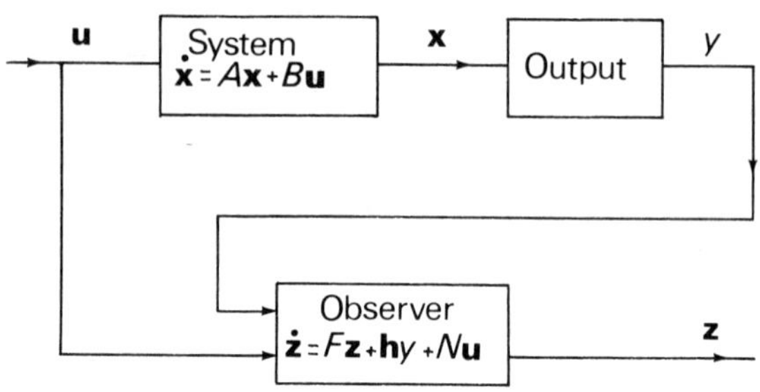

Fig. 7.1

If the dynamics of the observer when regarded as a free system (that is when the input is zero) are defined by

$$\dot{\mathbf{z}} = F\mathbf{z}$$

†An Introduction to Observers, *IEEE Trans. Automat. Contr.* **AC-16**, pp. 596–602, Dec. 1971.

then, when the inputs are applied (see Fig. 7.1), its behaviour is determined by an equation of the form

$$\dot{\mathbf{z}} = F\mathbf{z} + \mathbf{h}y + N\mathbf{u} \tag{7.31}$$

where $\mathbf{h}$ is an arbitrary vector (of order $n \times 1$). Using equations (7.30) and (7.31), we form the difference

$$\begin{aligned}\dot{\mathbf{z}} - T\dot{\mathbf{x}} &= F\mathbf{z} + \mathbf{h}y + N\mathbf{u} - TA\mathbf{x} - TB\mathbf{u} \\ &= F[\mathbf{z} - T\mathbf{x}] + [\mathbf{h}\mathbf{c}' - TA + FT]\mathbf{x} + [N - TB]\mathbf{u} \qquad (7.32)\end{aligned}$$

where the vector $FT\mathbf{x}$ has been added and subtracted on the right-hand side of equation (7.32).

If we now choose

$$\mathbf{h}\mathbf{c}' = TA - FT \tag{7.33}$$

and $\quad N = TB$

then equation (7.32) becomes

$$\frac{\mathrm{d}}{\mathrm{d}t}[\mathbf{z} - T\mathbf{x}] = F[\mathbf{z} - T\mathbf{x}]$$

which has the solution

$$\mathbf{z} - T\mathbf{x} = \mathrm{e}^{Ft}[\mathbf{z}(0) - T\mathbf{x}(0)]$$

or

$$\mathbf{z} = T\mathbf{x} + \mathrm{e}^{Ft}[\mathbf{z}(0) - T\mathbf{x}(0)]. \tag{7.34}$$

If it is possible to choose

$$\mathbf{z}(0) = T\mathbf{x}(0)$$

then $\mathbf{z} = T\mathbf{x}$ for all $t \geqslant 0$.

Generally it is more realistic to assume that $\mathbf{z}(0)$ is only approximately equal to $T\mathbf{x}(0)$, in which case it is required that e^{Ft} approaches the zero matrix as rapidly as possible. As we know from Chapter 3 (see equation (3.19)), this can be achieved by choosing the eigenvalues of F to have large negative real parts. There are good practical reasons why it is not advisable for the eigenvalues to have very large negative real parts. The practice is to place the observer eigenvalues so that their (negative) real parts are slightly larger (in modulus) than the real parts of the system eigenvalues.

Having obtained an estimate of $T\mathbf{x}$, it is now necessary to construct a system having the transfer function T^{-1}, to obtain $\mathbf{x}$. This can be avoided if we choose T to be the identity transformation I. The resulting system is called an **identity observer**.

In this case equations (7.33) become

$$\mathbf{h}\mathbf{c}' = A - F \tag{7.35}$$

$$N = B.$$

From equation (7.35) we obtain

$$F = A - \mathbf{h}\mathbf{c}' \tag{7.36}$$

so that the observer eigenvalues are seen to be determined by the choice of $\mathbf{h}$.

In fact Luenberger (see p. 188 above) shows that the identity observer can be designed to have arbitrary eigenvalues if the system defined by equation (7.30) is completely observable.

Although the identity observer is obviously convenient, there is the disadvantage that it reproduces the entire state vector $\mathbf{x}(t)$, whereas, as mentioned above, some of the state variables are already available. There is then a certain degree of redundancy in the design of this type of observer, which is both expensive and results in an unnecessarily complex system. This redundancy can be eliminated by the design of the so called **reduced order observer**. It can be shown that if the original system of order n has m independent outputs, then an adequate observer need be of order $(n-m)$ only. We do not prove this result, but the interested reader will find detailed discussion of this concept in various references of the *Bibliography*.

Example 7.7

Design an identity observer having both eigenvalues at $\lambda = -3$ for the system defined by the following state equations:

$$\dot{\mathbf{x}} = \begin{bmatrix} -1 & 1 \\ 0 & -2 \end{bmatrix} \mathbf{x} + \begin{bmatrix} 1 & -1 \\ 0 & 1 \end{bmatrix} \mathbf{u}$$

$$y = [1 \quad 0]\mathbf{x}.$$

Solution

Notice that only the state variable x_1 is available at the output. We need x_2 also. Using equation (7.36)

$$F = A - \mathbf{hc}' = \begin{bmatrix} -1 & 1 \\ 0 & -2 \end{bmatrix} - \begin{bmatrix} h_1 \\ h_2 \end{bmatrix} [1 \quad 0]$$

$$= \begin{bmatrix} -1-h_1 & 1 \\ -h_2 & -2 \end{bmatrix}.$$

The observer characteristic equation is

$$|\lambda I - F| = 0$$

that is (after simplification)

$$\lambda^2 + (h_1+3)\lambda + (2h_1+h_2+2) = 0.$$

Since the observer eigenvalues are at $\lambda = -3$, the above equation is identical to

$$\lambda^2 + 6\lambda + 9 = 0,$$

hence

$$h_1 + 3 = 6 \quad \text{and} \quad 2h_1 + h_2 + 2 = 9,$$

so that

$$h_1 = 3 \quad \text{and} \quad h_2 = 1.$$

It follows that the state matrix of the required observer is

$$F = \begin{bmatrix} -4 & 1 \\ -1 & -2 \end{bmatrix}.$$

Check

On computing the difference $\dot{\mathbf{z}} - \dot{\mathbf{x}}$, we find

$$\dot{\mathbf{z}} - \dot{\mathbf{x}} = \begin{bmatrix} -4 & 1 \\ -1 & -2 \end{bmatrix}\mathbf{z} + \begin{bmatrix} 3 & 0 \\ 1 & 0 \end{bmatrix}\mathbf{x} + \begin{bmatrix} 1 & -1 \\ 0 & 1 \end{bmatrix}\mathbf{u}$$

$$- \begin{bmatrix} -1 & 1 \\ 0 & -2 \end{bmatrix}\mathbf{x} - \begin{bmatrix} 1 & -1 \\ 0 & 1 \end{bmatrix}\mathbf{u}$$

$$= \begin{bmatrix} -4 & 1 \\ -1 & -2 \end{bmatrix} [\mathbf{z} - \mathbf{x}]$$

as required!

PROBLEMS

(1) A system is defined by

$$\dot{\mathbf{x}} = \begin{bmatrix} 15 & -20 \\ 12 & -16 \end{bmatrix} \mathbf{x} + \begin{bmatrix} 1 \\ 2 \end{bmatrix} u.$$

(i) Find the modes of the system.
(ii) Find the transformation $\mathbf{x} = T\mathbf{z}$ so that the resulting state matrix is in a companion form.
(iii) With the result from (ii) determine the feedback law which will result in the closed loop system having modes at $\lambda = -2$ and $\lambda = -3$.

(2) Compensate the system defined in problem 1 above, by determining a proportional gain matrix which will result in the closed-loop modes being at $\lambda = -2$ and $\lambda = -3$.

(3) Find a proporional gain matrix to compensate each of the following systems so that in each case the closed-loop modes are at $\lambda = -1$, $\lambda = -2$ and $\lambda = -3$.

(a) $$\dot{\mathbf{x}} = \begin{bmatrix} 1 & 0 & 0 \\ 0 & -1 & 0 \\ 0 & 0 & -4 \end{bmatrix} \mathbf{x} + \begin{bmatrix} 1 \\ 1 \\ -1 \end{bmatrix} u$$

(b) $$\dot{\mathbf{x}} = \begin{bmatrix} 1 & 0 & 0 \\ 0 & -1 & 0 \\ 0 & 0 & -4 \end{bmatrix} \mathbf{x} + \begin{bmatrix} 1 & 1 \\ 1 & -1 \\ -1 & 0 \end{bmatrix} \mathbf{u}.$$

(4) Write down the state equation of an identity observer associated with the multivariable system defined by

$$\dot{\mathbf{x}} = A\mathbf{x} + B\mathbf{u}$$
$$\mathbf{y} = C\mathbf{x}.$$

(5) A system is defined by

$$\dot{\mathbf{x}} = \begin{bmatrix} 0 & 1 \\ -3 & -2 \end{bmatrix} \mathbf{x} + \begin{bmatrix} 1 & 0 \\ 1 & 1 \end{bmatrix} \mathbf{u}$$

$$y = \begin{bmatrix} 1 & 0 \\ 0 & 0 \end{bmatrix} \mathbf{x}.$$

Design an identity observer having modes at $\lambda = -3$ and $\lambda = -4$.

Part II – Optimal Control

CHAPTER 8

Introduction to optimal control

8.1 CONTROL AND OPTIMAL CONTROL

So far in this book we have been concerned with the problem of controlling a system. In what follows we will take this one stage further and consider the problem of controlling the system in some 'best' way. The optimal control strategy will of course depend on what is defined as the best way. This is usually achieved in terms of a performance index or criterion as discussed in Chapter 4.

As a simple example, we can consider the problem of a rocket launching a satellite into an orbit about the earth. A *control* problem would be that of choosing the thrust attitude angle and rate of emission of the exhaust gases so that the rocket takes the satellite into its prescribed orbit. An associated *optimal control* problem is to choose the controls to effect the transfer with, for example, minimum expenditure of fuel, or in minimum time.

So, in what follows, we will assume that the system is controllable, and we will be looking for the controls which effect a prescribed transfer of the system in some best way. We will first look at a number of specific examples, before stating the general form for optimal control problems.

8.2 EXAMPLES

8.2.1 Economic Growth

We first consider a mathematical model of a highly simplified economy in which the rate of output, $Y(t)$, is assumed to depend on the rates of input of capital, $K(t)$ (for example in the form of machinery) and labour force, $L(t)$; that is.

$$Y = F(K, L) \tag{8.1}$$

where F is called the **production function**. This function is assumed to obey a 'return to scale' property so that

$$F(\alpha K, \alpha L) = \alpha F(K, L). \tag{8.2}$$

With $\alpha = 1/L$, and defining the output per worker, $y = Y/L$, and capital per worker, $k = K/L$, we have

$$y = Y/L = F(K,L)/L = F(k,1) = f(k), \quad \text{say.}$$

A typical example of the form of f is illustrated in Fig. 8.1; we note that $f'(k) > 0$, but $f''(k) < 0$.

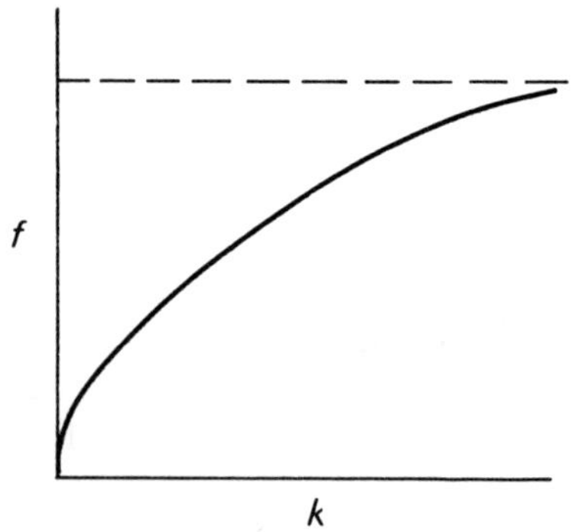

Fig. 8.1 Typical production function

Output is either consumed or invested, so that

$$Y = C(t) + I(t) \tag{8.3}$$

where C and I are the rates of consumption and investment respectively.

The investment is used to increase the capital stock and replace machinery, that is,

$$I = \frac{dK}{dt} + \mu K \tag{8.4}$$

Here μ is called the **rate of depreciation**. Defining $c = C/L$ as the consumption rate per worker, we see that from (8.3),

$$y = f(k) = c + \frac{1}{L}\frac{dK}{dt} + \mu k,$$

and since

$$\frac{d}{dt}\left(\frac{K}{L}\right) = \frac{1}{L}\frac{dK}{dt} - \frac{1}{L}\frac{dL}{dt}k,$$

we have

$$f(k) = c + \frac{dk}{dt} + \frac{\dot{L}}{L}k + \mu k.$$

Assuming that labour grows exponentially, that is $L = L_0 e^{\lambda t}$, then

$$\frac{dk}{dt} = f(k) - (\lambda + \mu)k - c(t) \qquad (8.5)$$

which is the governing equation of this economic growth model.

The consumption rate per worker, $c(t)$, is the control for this problem. The central planner's problem is to choose c on a time interval $0 \leqslant t \leqslant T$ in some best way. But what are the desired economic objectives? One method of quantifying the best way is to introduce a 'utility' function $U = U(c)$, which is a measure of the value attached to consumption. The function U will normally satisfy

$$U'(c) > 0,\ U''(c) < 0,\ \lim_{c \to 0} U'(c) = \infty,\ \lim_{c \to \infty} U'(c) = 0, \qquad (8.6)$$

which means that a fixed increment in consumption will be valued increasingly highly with decreasing consumption level. This is illustrated in Fig. 8.2. We also need to optimise consumption for all time in $[0, T]$ but with some discounting for future time. So the central planner wishes to maximise the 'welfare' integral

$$W = \int_0^T e^{-\delta t} U(c) dt,$$

where δ is known as the discount rate, which is a measure of preference for earlier rather than later consumption. If $\delta = 0$, then there is no time discounting, and consumption is valued equally at all times; as δ increases, so does the discounting of consumption and utility at future times.

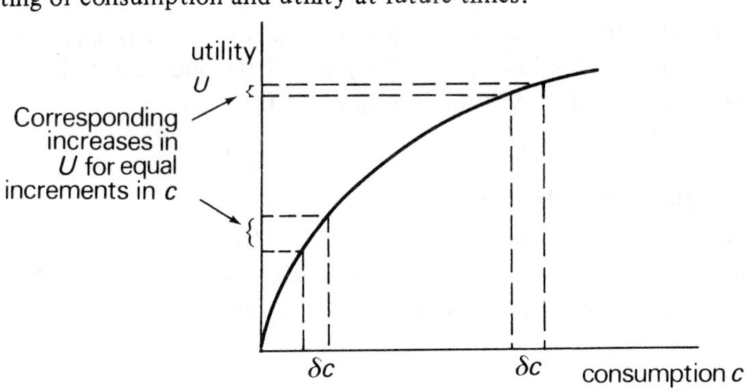

Fig. 8.2 Utility function $U(c)$

The mathematical problem has now been reduced to finding the optimal consumption path $c(t)$, $0 \leqslant t \leqslant T$, which maximises W subject to the constraint (8.5), and, for example, with $k(0) = k_0$.

8.2.2 Resource Depletion

A related problem, and one of much recent interest, is to determine the optimal depletion rate for a non-renewable resource (for example oil or coal). Suppose that the output rate now depends on two inputs, capital $K(t)$ and resource depletion rate $R(t)$, that is,

$$Y = F(K, R). \tag{8.7}$$

Now suppose output is either consumed or used to increase the capital stock giving

$$Y = C + \frac{\mathrm{d}K}{\mathrm{d}t};$$

that is, $$\mathrm{d}K/\mathrm{d}t = F(K, R) - C \tag{8.8}$$

where $C = C(t)$ is the consumption rate. If $S(t)$ denotes the resource stock, then we clearly have

$$\mathrm{d}S/\mathrm{d}t = -R. \tag{8.9}$$

As in the first example, we wish to maximise the welfare integral

$$W = \int_0^T \mathrm{e}^{-\delta t} U(C)\mathrm{d}t \tag{8.10}$$

by finding the optimal paths for resource depletion rate, R, and consumption rate, C, subject to the constraints (8.8) and (8.9), and also with some initial conditions, for example, $K = K_0$, $S = S_0$ at $t = 0$.

8.2.3 Exploited Populations

Many resources are to some extent renewable (for example, fish populations, grazing land, forests) and a vital proplem is their optimal management. With no harvesting, the resource population $x = x(t)$ is assumed to obey a growth law of the form

$$\mathrm{d}x/\mathrm{d}t = F(x). \tag{8.11}$$

A typical example for F is the Verhulst model

$$F(x) = rx(1-x/K), \tag{8.12}$$

where K is the saturation level of population, and r is a positive constant. With harvesting, (8.11) is modified to

$$dx/dt = F(x) - h(t) \tag{8.13}$$

where h is called the harvesting rate. Now h will depend on the fishing effort, E, (for example, size of nets, number of trawlers, number of fishing days) as well as the population level so that, with appropriate units, we assume

$$h = Ex. \tag{8.14}$$

Optimal management will seek to maximise the **economic rent** defined by

$$\begin{aligned} \text{economic rent} &= \text{revenue} - \text{costs} \\ &= ph - cE, \end{aligned}$$

assuming the costs to be proportional to the effort and where p is the unit price. The basic problem is to maximise the discounted economic rent over all time, called the present value, PV, that is,

$$PV = \int_0^\infty e^{-\delta t}(pEx - cE)dt, \tag{8.15}$$

subject to the constraint (8.13), and initial condition $x = x_0$.

8.2.4 Advertising Policies

Another management problem, but a vastly different one, is to determine the optimal advertising policy for a particular product. The 'classical' model for sales response to advertising is given by

$$ds/dt = rA(t)(1-s/M) - \lambda s. \tag{8.16}$$

Here $s(t)$ is the *sales rate*, $A(t)$ the *advertising rate*, M the *saturation sales level*, λ the constant rate of decay of sales in the absence of advertising, and r is a positive constant.

A typical problem is to maximise the sales over some period $[0,T]$ by choosing the advertising optimally on this time interval, given a fixed advertising budget. In mathematical terms, we wish to choose $A(t), 0 \leqslant t \leqslant T$, so that

$$S = \int_0^T s(t)\mathrm{d}t \tag{8.17}$$

is maximised, subject to (8.16) and

$$B = \int_0^T A(t)\mathrm{d}t \tag{8.18}$$

where B, the total spent on advertising, is fixed. We also take $s = s_0$ at $t = 0$ and assume that there is some maximum advertising rate, say $\bar{A}$; so that

$$0 \leqslant A \leqslant \bar{A}. \tag{8.19}$$

One possible advertising strategy satisfying (8.18) and (8.19) is to take A as a constant. Using (8.18) the level of advertising is given by

$$A = B/T, \quad 0 \leqslant t \leqslant T \tag{8.20}$$

(provided $B/T \leqslant \bar{A}$). But this is only one possible advertising strategy, and it will not necessarily maximise the sales S. A second possible strategy is given by

$$A = \begin{cases} 2B/T & 0 \leqslant t < T/4 \\ 0 & T/4 < t < 3T/4 \\ 2B/T & 3T/4 < t \leqslant T. \end{cases} \tag{8.21}$$

These are illustrated in Fig. 8.3.

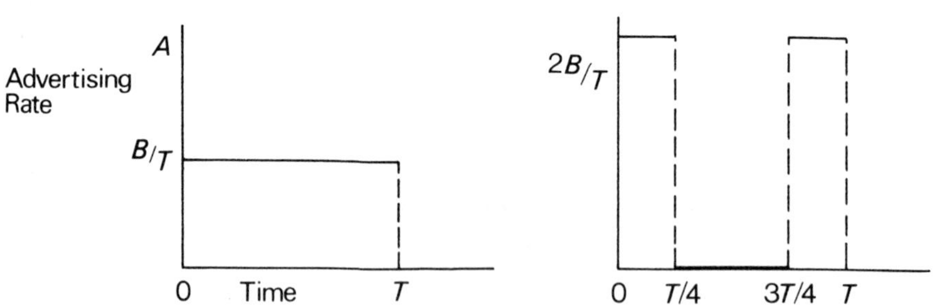

Fig. 8.3 Possible Advertising Strategies

8.2.5 Rocket Trajectories

The governing equation of rocket motion is

$$m \frac{\mathrm{d}\mathbf{v}}{\mathrm{d}t} = m\mathbf{F} + \mathbf{F}_{\text{ext}}. \tag{8.22}$$

Here m is the rocket's mass, $\mathbf{v}$ its velocity, $\mathbf{F}$ the thrust produced by the rocket motor and $\mathbf{F}_{\text{ext}}$ is the external force. It can be shown that (see Burghes and Downes [11])

$$|\mathbf{F}| = \frac{c}{m}\beta \tag{8.23}$$

where c is the relative exhaust speed, $\beta = -\mathrm{d}m/\mathrm{d}t$, the burning rate, so that if ϕ is the thrust attitude angle (the angle between the rocket axis and the horizontal, as in Fig. 8.4), then the equations of motion are

$$\frac{\mathrm{d}v_1}{\mathrm{d}t} = \frac{c\beta}{m}\cos\phi \tag{8.24}$$

$$\frac{\mathrm{d}v_2}{\mathrm{d}t} = \frac{c\beta}{m}\sin\phi - g.$$

Here $\mathbf{v} = (v_1, v_2)$ and we have taken $\mathbf{F}_{\text{ext}} = (0, -mg)$.

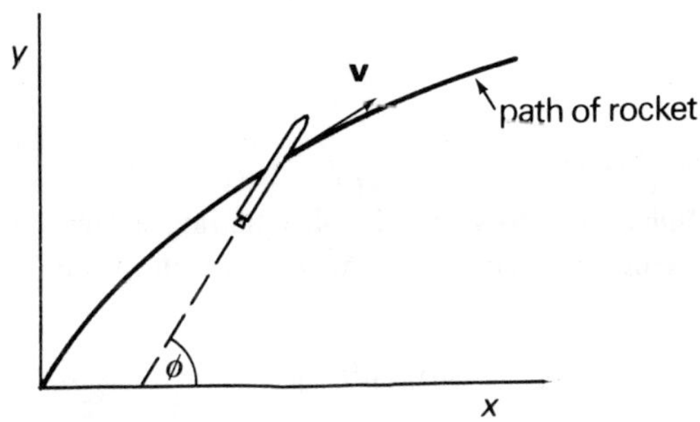

Fig. 8.4 Rocket Flight Path

The minimum fuel problem is to choose the controls, β and ϕ, so as to take the rocket from initial position to a prescribed height, say, $\bar{y}$, in such a way as to minimise the fuel used. The fuel consumed is

$$\int_0^T \beta\,\mathrm{d}t, \tag{8.25}$$

where T is the time at which $\bar{y}$ is reached.

8.2.6 Servo Problem

A control surface on an aircraft is to be kept at rest at a fixed position. Disturbances move the surface, and if not corrected it would behave as a damped harmonic oscillator (see section 4.3), for example

$$\frac{d^2\theta}{dt^2} + a\frac{d\theta}{dt} + w^2\theta = 0 \tag{8.26}$$

where θ is the angle from the desired position (that is, $\theta = 0$). The disturbance gives initial values $\theta = \theta_0$, $d\theta/dt = \theta_0'$, but a servomechanism applies a restoring torque, so that (8.26) is modified to

$$\frac{d^2\theta}{dt^2} + a\frac{d\theta}{dt} + w^2\theta = u. \tag{8.27}$$

The problem is to choose $u(t)$, which will be bounded by $|u(t)| \leqslant c$, so as to bring the system to $\theta = 0$, $d\theta/dt = 0$ in *minimum* time. We can write this time as

$$T = \int_0^T 1.dt, \tag{8.28}$$

and so this integral is required to be minimised.

8.3 FUNCTIONALS

All the examples in section 8.2 involve finding extremum values of integrals, subject to varying constraints. The integrals are all of the form

$$J = \int_{t_0}^{t_1} F(x, dx/dt, t)dt \tag{8.29}$$

where F is a given function of the function $x(t)$, its derivative dx/dt and the independent variable t. The path $x(t)$ is defined for $t_0 \leqslant t \leqslant t_1$. For a given path, say $x = x_1(t)$, (8.29) gives the corresponding value of J, say $J = J_1$. For a second path, say $x = x_2(t)$, (8.29) again gives a value of J, say $J = J_2$. In general, $J_1 \neq J_2$, and we call integrals of the form (8.29) **functionals**.

As a simple example, consider the integral

$$J = \int_0^1 \left(t^2 + t\frac{dx}{dt}\right) dt$$

where $x = x(t)$, $0 \leqslant t \leqslant 1$, is some path defined from $t = 0$ to $t = 1$. It is clear that the value of J depends on the prescribed path $x = x(t)$, $0 \leqslant t \leqslant 1$. For example,

$$\text{(i) } x = t, 0 \leqslant t \leqslant 1 \text{ gives } J = \int_0^1 (t^2 + t)\mathrm{d}t = \tfrac{5}{6};$$

$$\text{(ii) } x = \mathrm{e}^t, 0 \leqslant t \leqslant 1 \text{ gives } J = \int_0^1 (\mathrm{e}^{2t} + t\mathrm{e}^t)\mathrm{d}t = \mathrm{e}^2/2 + 1 .$$

Even if we fix the initial and final positions of x, the value of this integral will vary. For example, if $x(0) = 0$, $x(1) = 1$, then the path

$$\text{(iii) } x = t^2, 0 \leqslant t \leqslant 1 \text{ gives } J = \int_0^1 (t^4 + 2t^2)\mathrm{d}t = \tfrac{13}{15},$$

which is not equal to the value found in path (i) above. The paths are illustrated in Fig. 8.5.

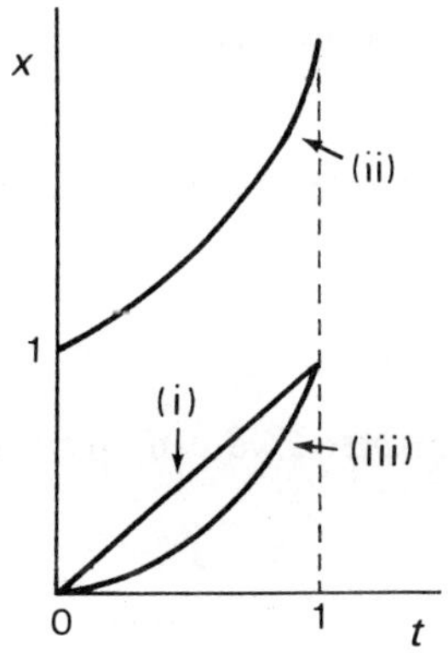

Fig. 8.5 Paths in x–t plane

On the other hand, some functionals do not vary in value for different paths between two fixed points in the x–t plane. For example,

$$J = \int_0^1 (x + t\frac{\mathrm{d}x}{\mathrm{d}t})\mathrm{d}t$$

with $x(0) = 0$, $x(1) = 1$ always has the value 1, independent of the path taken between $(0, 0)$ and $(1, 1)$ in the t–x plane. This can be verified by using the paths (i) and (iii) above, and can in fact be readily proved. The result follows since

$$J = \int_0^1 (x + t\frac{\mathrm{d}x}{\mathrm{d}t})\mathrm{d}t = \int_0^1 \frac{\mathrm{d}}{\mathrm{d}t}(xt)\mathrm{d}t = x(t)t \Big|_0^1$$

$$= x(1) - 0 = 1.$$

So the value of the integral in this case is always 1, no matter what path is taken from (0, 0) to (1, 1).

8.4 THE BASIC OPTIMAL CONTROL PROBLEM

All the examples in section 8.2 can be put in the following form. As in earlier chapters the state of the system is characterised by n variables, $x_1, x_2, \ldots, x_n$ and we write the **state vector x** as

$$\mathbf{x} = (x_1, x_2, \ldots, x_n)'. \tag{8.30}$$

These variables satisfy the coupled first order differential equations

$$\frac{\mathrm{d}x_i}{\mathrm{d}t} = f_i(x_1, x_2, \ldots, x_n; u_1, u_2, \ldots, u_m; t) \;\; (1 \leqslant i \leqslant n) \tag{8.31}$$

on $[0, T]$ and where the m variables $u_1, u_2, \ldots, u_m$ form the **control vector u**; that is

$$\mathbf{u} = (u_1, u_2, \ldots, u_m)' \tag{8.32}$$

As we have seen in earlier chapters, we can conveniently write the system in the form

$$\dot{\mathbf{x}} = \mathbf{f}(\mathbf{x}, \mathbf{u}, t), \quad \mathbf{f} = (f_1, f_2, \ldots, f_n)'$$

and if the system is linear

$$\dot{\mathbf{x}} = A\mathbf{x} + B\mathbf{u} \tag{8.33}$$

where A is an $n \times n$ matrix, B an $n \times m$ matrix.

We assume that $\mathbf{u} \in U$, a given control region in m-dimensional space and the u_j are assumed continuous, except possibly at a finite number of discontinuities. It is assumed that the f_i are continuous with continuous partial derivatives. The initial values of x_i are specified (for example $\mathbf{x} = \mathbf{x}_0$ at time $t = 0$) which means that specifying u_j, $0 \leqslant t \leqslant T$ $(j = 1, 2, \ldots, m)$ determines the x_i $(i = 1, 2, \ldots, n)$ from (8.31).

The basic control problem is to choose the control vector $\mathbf{u} \in U$ such that the state vector $\mathbf{x}$ is transferred from $\mathbf{x}_0$ to a terminal point at time T where some (or all or none) of the state variables are specified; for example, suppose x_l are specified at $t = T$ for $l = 1, 2, \ldots, q$ $(\leqslant n)$. The region U is called the **admissible control region**.

If the transfer can be accomplished, the problem in optimal control is to effect the transfer so that the functional

$$J = \int_0^T f_0(\mathbf{x}, \mathbf{u}, t) \tag{8.34}$$

is maximised (or minimised). Here f_0 is a function of $x_1, x_2, \ldots, x_n, u_1, u_2, \ldots, u_m$ and t and is continuous with continuous partial derivatives.

Using this notation, we can identify some of the optimal control problems in section 8.2.

8.2.1 $n = 1$, $m = 1$, $q = 0$, $x_1 = k$, $u_1 = c$, $f_0 = \mathrm{e}^{-\delta t}U(c)$,

$$f_1 = f(k) - (\lambda + \mu)k - c.$$

8.2.2 $n = 2$, $m = 2$, $x_1 = S$, $x_2 = K$, $u_1 = R$, $u_2 = c$ and if we take $S = 0$ at the final time, T, then $q = 1$, and

$$f_0 = \mathrm{e}^{-\delta t}U(C),\ f_1 = f(K, R) - C,\ f_2 = -R.$$

In the next three chapters we will develop the required theory which will then be applied to the problems introduced in section 8.2. We first start with variational calculus which readily leads on to continuous optimal control. We complete the theory by giving a non-rigorous proof for Pontryagin's maximum principle, which solves the problem stated above, and develop dynamic programming.

PROBLEMS

(1) (i) Solve the governing equation of economic growth, (8.5), when $f = bk$ and $c = \bar{c}$, constant.

(ii) Show that when $f = bk^{\alpha}$ $(0 < \alpha < 1)$ there are two possible equilibrium solutions. Determine their stability. Also show that the *maximum* sustainable value of the consumption occurs when the capital is given by $k = [(\lambda + \mu)/\alpha b]^{1/(\alpha-1)}$. Find the corresponding value of c. (These values of k and c are known as the **golden-rule levels** of capital and consumption.)

(2) The dynamic equation for an exploited population is (8.13). Show that the population can be in equilibrium for $0 \leqslant h \leqslant h_{\max}$. For the Verhulst population growth, show that

$$h_{\max} = \frac{rk}{4}.$$

[This is known as the **maximum sustainable yield**, M.S.Y.]

If harvesting and effort are related by $h = Ex$, determine the equilibrium population level when revenue and costs are equated. Also determine the equilibrium population level if the economic rent is maximised.

(3) Using the sales response to advertising model, described in section 8.2.4, find the total sales in the time interval $0 \leqslant t \leqslant T$ for the advertising strategy described by equation (8.20).

(4) Evaluate the integral

$$J = \int_0^1 [x^2 + (\frac{dx}{dt})^2]dt$$

along the following paths from (0,0) to (1,1)

(i) $x = t$
(ii) $x = t^2$
(iii) $x = (e^t - 1)/(e - 1)$.

(5) Repeat Exercise 4 for the integral

$$J = \int_0^1 (2xt + t^2 \frac{dx}{dt})dt.$$

Can you explain the results?

(6) Using the terminology of section 8.3, identify the parameters n, m and q and variables x_i, f_i, u_j, for the problem in section 8.2.3.

CHAPTER 9

Variational calculus

9.1 THE BRACHISTOCHRONE PROBLEM

The calculus of variations originated from a problem posed by the Swiss mathematician Johann Bernoulli (1667–1748). He required the form of the curve joining two fixed points A and B in a vertical plane such that a body sliding down the curve (under constant gravity and no friction) travels from A to B in minimum time. This problem does not have a trivial solution. The straight line from A to B is *not* the solution, as can be found experimentally by sliding beads down wires in appropriate shapes.

To pose the problem in mathematical terms, we introduce coordinates as in Fig. 9.1, so that A is the point (0,0) and B is (a, b). Assuming that the particle is released from rest at A, conservation of energy gives

$$\tfrac{1}{2}mv^2 - mgx = 0 \tag{9.1}$$

taking zero potential energy level as $x = 0$. Here v is the particle's speed, and since velocity $\mathbf{v} = (\dot{x}, \dot{y})$, we have

$$v = [\dot{x}^2 + \dot{y}^2]^{\frac{1}{2}}.$$

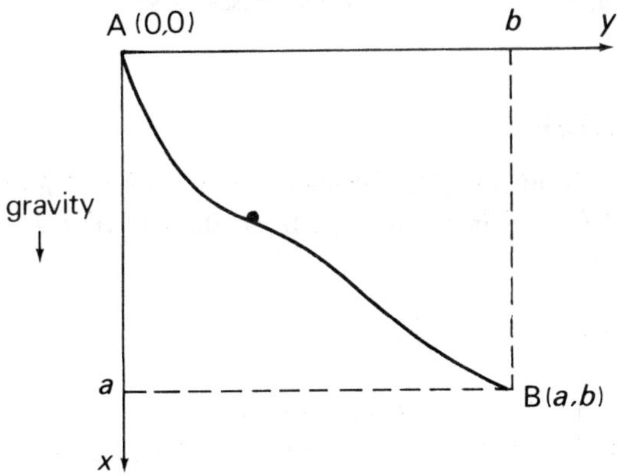

Fig. 9.1 Brachistochrane Problem

From Fig. 9.2, we can see that an element of arc length, δs, is given approximately by

$$\delta s \simeq (\delta x^2 + \delta y^2)^{\frac{1}{2}}.$$

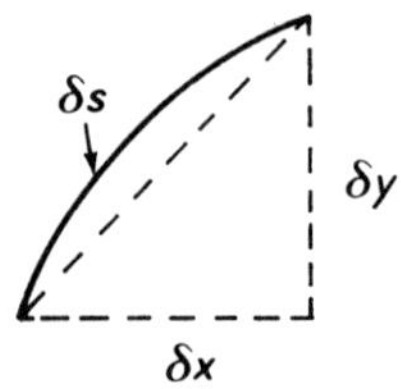

Fig. 9.2 Arc length Element

Dividing by δt, and letting $\delta t \to 0$ gives

$$\frac{ds}{dt} = \left[\left(\frac{dx}{dt}\right)^2 + \left(\frac{dy}{dt}\right)^2\right]^{\frac{1}{2}} = v \tag{9.2}$$

where s is arc length along the curve. From (9.1) and (9.2),

$$dt/ds = (2gx)^{-\frac{1}{2}}$$

so that on integrating, the time of descent is given by

$$t = \int \frac{ds}{(2gx)^{\frac{1}{2}}} = \frac{1}{(2g)^{\frac{1}{2}}} \int_0^a \left[\frac{1 + (dy/dx)^2}{x}\right]^{\frac{1}{2}} dx \tag{9.3}$$

having used $ds/dx = [1 + (dy/dx)^2]^{\frac{1}{2}}$. Our problem is to find the path $y = y(x)$ such that $y(0) = 0$, $y(a) = b$ and which minimises (9.3).

9.2 EULER EQUATION

The mathematical problem described above does not have an obvious answer. For example, the straight line from A to B has the equation $y = bx/a$, and this gives a time of descent

$$t_1 = [2(a^2 + b^2)/ag]^{\frac{1}{2}} \tag{9.4}$$

which we will later find not to be the minimum time.

The problem posed above is a special case of finding extremum values of functionals of the form

$$J = \int_a^b F(y, y', x)\mathrm{d}x \tag{9.5}$$

where F is a given function of y, y' $(= \mathrm{d}y/\mathrm{d}x)$ and x and $y = y(x)$ is some path between $A(a, \alpha)$ and $B(b, \beta)$; that is, $y(a) = \alpha$ and $y(b) = \beta$. (Not that we are using y and x in place of x and t used in section 8.3.)

Two such paths are illustrated in Fig. 9.3.

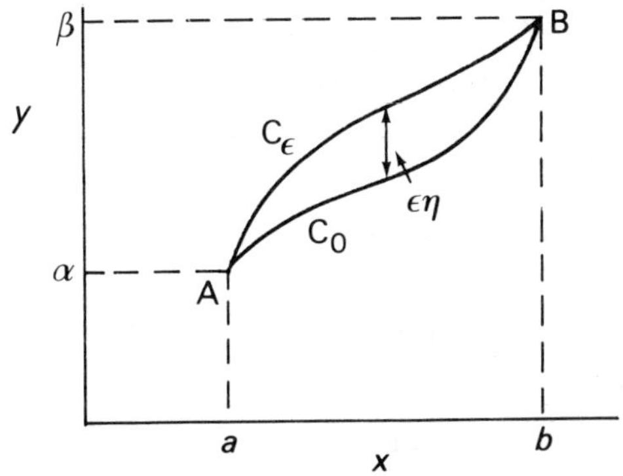

Fig. 9.3 Possible paths between A and B

Suppose we are looking for the path $y = y(x)$ which gives an extremum value for the functional (9.5). Let C_o be the required curve, say $y_o(x)$, and let C_ϵ be a neighbouring curve defined by

$$y_\epsilon(x) = y_o(x) + \epsilon\eta(x) \tag{9.6}$$

where ϵ is a small parameter and $\eta(x)$ is a differential function of x, arbitrary except for the restriction $\eta(a) = \eta(b) = 0$, which is required to ensure that both curves start at A and finish at B. Thus equation (9.6) generates a whole class of neighbouring curves C_ϵ, depending on the value of ϵ. The value $\epsilon = 0$ gives the optimal curve C_o.

As a simple example, consider the problem of finding the curve of shortest length between two points, say $(0, 0)$ and $(1, 1)$. We know the optimal curve C_o is given by

$$y_o(x) = x, \quad 0 \leqslant x \leqslant 1.$$

Taking $\eta(x) = x(1 - x)$, we can construct the family of neighbouring curves, C_ϵ, given by

$$y_\epsilon(x) = x + \epsilon x(1 - x), 0 \leqslant x \leqslant 1$$

which are illustrated in Fig. 9.4.

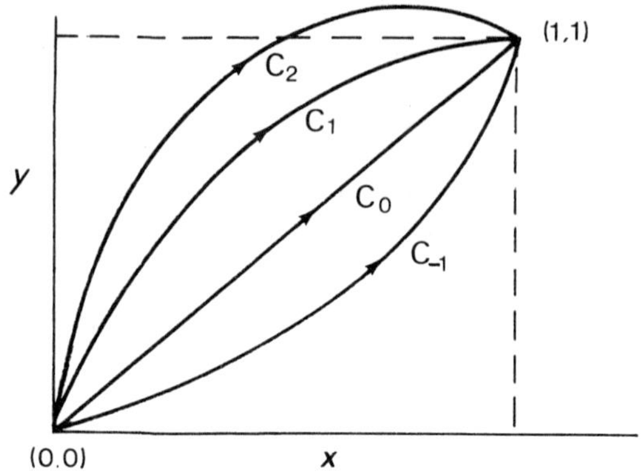

Fig. 9.4 Family of curves C_ϵ, $\epsilon = -1, 0, 1, 2$

Now the value of J for the path C_ϵ for the general problem is given by

$$J = \int_a^b F(y_0 + \epsilon\eta, y_0' + \epsilon\eta', x)\mathrm{d}x$$

where $y' = \mathrm{d}y/\mathrm{d}x$ and $\eta' = \mathrm{d}\eta/\mathrm{d}x$. For a given function $y(x)$, J will be a function of the parameter ϵ, that is, $J = J(\epsilon)$. For extremum values of J, we must have $\mathrm{d}J/\mathrm{d}\epsilon = 0$. But our choice of C_ϵ gives the extremum value at $\epsilon = 0$. Thus the condition can be written as

$$\mathrm{d}J/\mathrm{d}\epsilon = 0 \quad \text{at} \quad \epsilon = 0. \tag{9.7}$$

Taking the differentiation through the integral sign,

$$\frac{\mathrm{d}J}{\mathrm{d}\epsilon} = \int_a^b \{\eta F_y(y_0 + \epsilon\eta, y_0' + \epsilon\eta', x) + y'F_{y'}(y_0 + \epsilon\eta, y_0' + \epsilon\eta', x)\}\mathrm{d}x$$

where F_y means $\partial F/\partial y$, $F_{y'}$ means $\partial F/\partial y'$; here we have used the chain-rule for partial differentiation,

$$\frac{\mathrm{d}F}{\mathrm{d}\epsilon} = \frac{\partial F}{\partial y}\frac{\mathrm{d}y}{\mathrm{d}\epsilon} + \frac{\partial F}{\partial y'}\frac{\mathrm{d}y'}{\mathrm{d}\epsilon} = \eta F_y + \eta' F_{y'}$$

since $y = y_0 + \epsilon\eta$, $y' = y_0' + \epsilon\eta'$. Applying condition (9.7) we obtain

$$\int_a^b \{\eta F_y(y_0, y_0', x) + \eta' F_{y'}(y_0, y_0', x)\}\mathrm{d}x = 0.$$

From now on all the partial derivations of F are evaluated on the optimal path C_0, so we will leave off the (y_0, y'_0, x) dependence. With this notation, on the optimal path,

$$\int_a^b (\eta F_y + \eta' F_{y'})\mathrm{d}x = 0$$

and integrating the second term by parts gives

$$\eta F_{y'}|_a^b + \int_a^b [\eta F_y - \eta \frac{\mathrm{d}}{\mathrm{d}x}(F_{y'})]\mathrm{d}x = 0$$

$$\eta(b)F_{y'}|_{x=b} - \eta(a)F_{y'}|_{x=a} + \int_a^b \eta\{F_y - \frac{\mathrm{d}}{\mathrm{d}x}(F_{y'})\}\mathrm{d}x = 0. \tag{9.8}$$

But $\eta(a) = \eta(b) = 0$, so we are left with

$$\int_a^b \eta\{F_y - \frac{\mathrm{d}}{\mathrm{d}x}(F_{y'})\} = 0. \tag{9.9}$$

This must hold for *all* differentiable functions $\eta(x)$ such that

$$\eta(a) = \eta(b) = 0.$$

We now require the following Lemma.

Lemma

Let $\eta(x)$ be a differentiable function for $a \leqslant x \leqslant b$ such that $\eta(a) = \eta(b) = 0$. If f is continuous on $a \leqslant x \leqslant b$ and

$$\int_a^b f(x)\,\eta(x)\,\mathrm{d}x = 0$$

for all such $\eta(x)$, then $f = 0$ on $a \leqslant x \leqslant b$.

Proof

We prove this result by contradiction. Suppose $f > 0$ for some point in c in $[a, b]$. Since f is continuous, then $f > 0$ in a neighbourhood of c, say $[c_1, c_2]$, with $a < c_1 < c < c_2 < b$. Define the function

$$\eta(x) = \begin{cases} 0 & a \leqslant x \leqslant c_1 \\ (x-c_1)^2(x-c_2)^2 & c_1 \leqslant x \leqslant c_2 \\ 0 & c_2 \leqslant x \leqslant b. \end{cases}$$

Then $\eta(x)$ is a differentiable function and

$$\int_a^b f\,\eta\,\mathrm{d}x = \int_{c_1}^{c_2} f(x-c_1)^2(x-c_2)^2\mathrm{d}x > 0$$

since the integrand is strictly positive in $[c_1, c_2]$. This gives a contradiction against the hypothesis of the lemma. A similar contradiction is obtained if it is assumed that $f < 0$, and so we conclude that $f = 0$ on $[a, b]$.

Applying the result of this lemma to equation (9.9) immediately gives

$$F_y - \frac{\mathrm{d}}{\mathrm{d}x}(F_{y'}) = 0 \tag{9.10}$$

on the optimal path. This is known as **Euler's equation** (or the **Euler-Lagrange equation**) and must be satisfied by paths $y(x)$ which yield extremum values of the functional J.

Example 1

Find the curve $y(x)$ which gives an extremum value to the functional

$$J = \int_0^1 (y'^2 + 1)\,\mathrm{d}x$$

with $y(0) = 1$, $y(1) = 2$.

Solution

Here the integrand $F = y'^2 + 1$, and so $F_y = 0$. Hence the Euler equation (9.10), becomes

$$\frac{\mathrm{d}}{\mathrm{d}x}(F_{y'}) = 0$$

and, on integrating, $F_{y'} = A$, constant; that is

$$2y' = A.$$

Integrating again,

$$y = Ax/2 + B$$

where B is a constant. To satisfy the conditions $y(0) = 1$, $y(1) = 2$ requires $B = 1$ and $A = 2$, and so the optimal curve is the straight line

$$y = x + 1$$

as illustrated in Fig. 9.5.

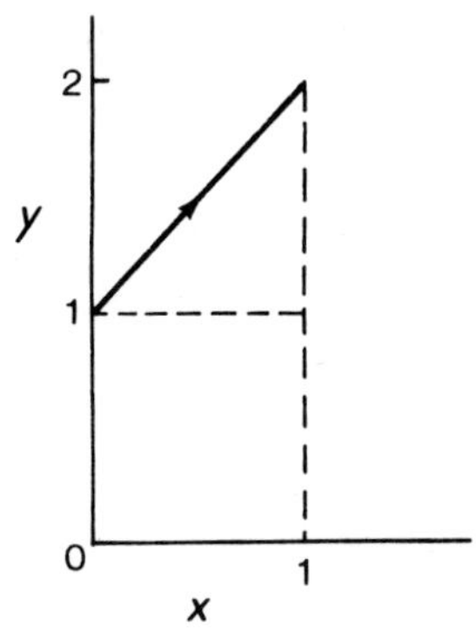

Fig. 9.5 Optimal Path

The corresponding extrememum value of J is

$$J = \int_0^1 2 \, \mathrm{d}x = 2,$$

and this is in fact a *minimum* value for J, although we have not proved this.

Example 2
Find the curve which has minimum length between $(0, 0)$ and $(1, 1)$.

Solution
Let $y = y(x)$, $0 \leqslant x \leqslant 1$, be any curve between $(0, 0)$ and $(1, 1)$. The length, L, is given by

$$L = \int_0^1 (1 + y'^2)^{\frac{1}{2}} \, \mathrm{d}x.$$

The integrand $F = (1 + y'^2)^{\frac{1}{2}}$ is independent of y, that is, $F_y = 0$. Hence (9.10) becomes

$$\frac{\mathrm{d}}{\mathrm{d}x}(F_{y'}) = 0,$$

and, on integrating $F_{y'} = A$, constant. Thus

$$y'(1 + y'^2)^{-\frac{1}{2}} = A,$$

which implies that $y' =$ constant, say B. Hence $y = Bx + C$, where C is a constant. To pass through (0, 0) and (1, 1), we obtain $y = x$ (as expected).

Example 3 (The Brachistochrone problem)

Determine the minimum value of the functional (9.3), that is

$$t = \frac{1}{(2g)^{\frac{1}{2}}} \int_0^a \left[\frac{1 + y'^2}{x}\right]^{\frac{1}{2}} \mathrm{d}x$$

such that $y(0) = 0$ and $y(a) = b$.

Solution

As in example 1, the integrand, $F = \{(1 + y'^2)/x\}^{\frac{1}{2}}$, is independent of y and so the Euler equation when integrated becomes

$$F_{y'} = c, \quad \text{constant.}$$

Hence

$$\frac{y'}{[x(1 + y'^2)]^{\frac{1}{2}}} = c$$

which gives, after squaring, $y'^2 = c^2 x(1 + y'^2)$. Hence

$$y' = \left[\frac{c^2 x}{1 - c^2 x}\right]^{\frac{1}{2}}$$

and, integrating,

$$y = c \int \frac{x \,\mathrm{d}x}{(x - c^2 x^2)^{\frac{1}{2}}}.$$

We can evaluate this integral using the substitution

$$x = \frac{1}{2c^2}(1 - \cos\theta)$$

so that $dx/d\theta = (1/2c^2)\sin\theta$ and, after substitution,

$$y = \frac{1}{2c^2}\int(1 - \cos\theta)d\theta$$

that is, $$y = \frac{1}{2c^2}(\theta - \sin\theta) + A$$

where A is a constant. Now initially $x = y = 0$, so that if we choose the parameter θ so that $\theta = 0$ corresponds to $(0, 0)$, we have $A = 0$. Thus the curve of quickest descent has parametric equations

$$\left.\begin{aligned} x &= \frac{1}{2c^2}(1 - \cos\theta) \\ y &= \frac{1}{2c^2}(\theta - \sin\theta) \end{aligned}\right\} \quad (9.11)$$

This is the equation of a **cycloid**, the constant c being chosen so that the curve passes through the point (a, b). A typical example of such a curve is shown in Fig. 9.6.

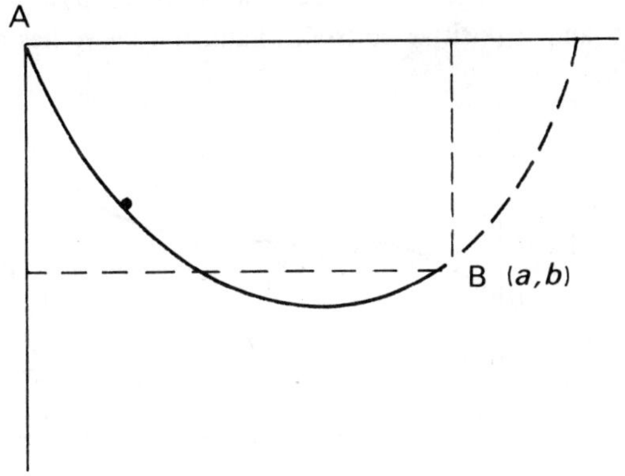

Fig. 9.6 (a, b) Cycloid from A and B

Note: Our theory has only shown us that we have a stationary value of t on this path. The path found does in fact correspond to minimum t.

The Euler equation, (9.9), takes special forms when the integrand F is independent of either y or x:

(i) F independent of y

As we have already seen, (9.10) gives $\frac{d}{dx}(F_{y'}) = 0$ which integrates to give

$$F_{y'} = \text{constant} \tag{9.12}$$

(ii) F independent of x

Since, in general,

$$\frac{dF}{dx} = \frac{\partial F}{\partial y}\frac{dy}{dx} + \frac{\partial F}{\partial y'}\frac{dy'}{dx} + \frac{\partial F}{\partial x}$$

$$= y'F_y + y''F_{y'},$$

F being independent of x. Using the Euler equation (9.10) to eliminate F_y now gives

$$\frac{dF}{dx} = y'\frac{d}{dx}(F_{y'}) + y''F_{y'} = \frac{d}{dx}(y'F_{y'}).$$

Integrating, we see that

$$F - y'F_{y'} = \text{constant}.$$

Example 4 (Minimum Surface Area)

A plane curve, $y = y(x)$ passes through the points $(0, b)$ and (a, c) $(b \neq c)$ in the xy-plane (a, b and c positive constants) and the curve lies above the x-axis as shown in Fig. 9.7.

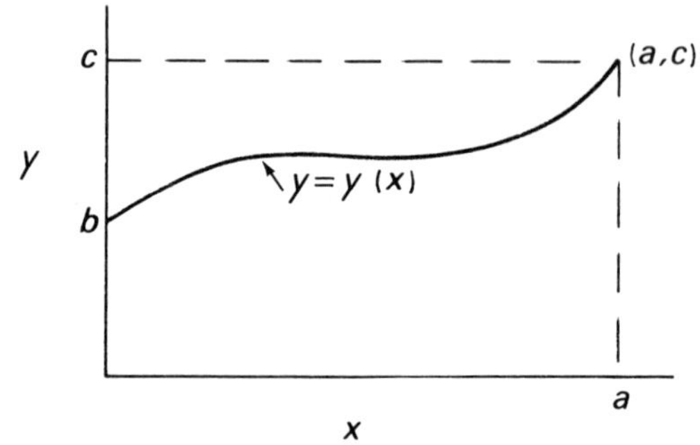

Fig. 9.7 Curve rotates about x-axis

Determine the curve $y(x)$ which, when rotated about the x-axis, yields a surface of revolution with minimum surface area.

Solution

The area of the surface of revolution is given by

$$A = \int 2\pi y \,\mathrm{d}s = 2\pi \int_0^a y[1 + (\mathrm{d}y/\mathrm{d}x)^2]^{\frac{1}{2}}\,\mathrm{d}x;$$

so we wish to find the curve $y(x)$ which minimises this integral. The integrand $F = y(1 + y'^2)^{\frac{1}{2}}$. This is independent of x and so we can apply equation (9.13) to give

$$y(1 + y'^2)^{\frac{1}{2}} - y'^2 y/(1 + y'^2)^{\frac{1}{2}} = A, \text{ constant.}$$

Simplifying

$$y^2 = A^2(1 + y'^2)$$

or

$$y' = \frac{\mathrm{d}y}{\mathrm{d}x} = \frac{1}{A}(y^2 - A^2)^{\frac{1}{2}},$$

that is,

$$\int \frac{\mathrm{d}y}{(y^2 - A^2)^{\frac{1}{2}}} = \int \frac{1}{A}\,\mathrm{d}x.$$

Integrating

$$\cosh^{-1}(y/A) = x/A + B \qquad (B \text{ constant})$$

so that

$$y = A \cosh(x/A + B). \tag{9.14}$$

The constants are chosen so that the curve passes through $(0, b)$ and (a, c). The curve is the **catenary** (which is also the shape assumed by a heavy chain of uniform density when suspended freely from its ends).

9.3 FREE END CONDITIONS

In this section we generalise the result obtained in section 9.2 for finding extremum values of functionals. We now suppose that we wish to find the path $y = y(x)$ $a \leqslant x \leqslant b$ which gives extremum values to the functional

$$J = \int_a^b F(y, y', x)\mathrm{d}x \tag{9.15}$$

but with no restriction on the value of y at either end point.

The analysis used in section 9.2 follows through directly, except that we *no* longer have the restriction $\eta(a) = \eta(b) = 0$. The optimal curve, C_o, and neighbouring curves, C_ϵ, are illustrated in Fig. 9.8.

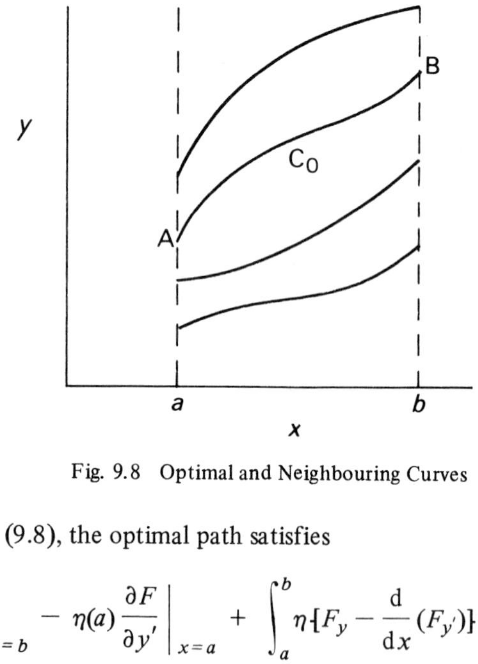

Fig. 9.8 Optimal and Neighbouring Curves

Rewriting, (9.8), the optimal path satisfies

$$\eta(b)\frac{\partial F}{\partial y'}\bigg|_{x=b} - \eta(a)\frac{\partial F}{\partial y'}\bigg|_{x=a} + \int_a^b \eta\{F_y - \frac{\mathrm{d}}{\mathrm{d}x}(F_{y'})\}\mathrm{d}x = 0. \tag{9.16}$$

As the free end point solution curve will also be the solution to the fixed end point problem which has the same end point values as the free end point solution (A, B in Fig. 9.8), it will also satisfy the Euler equation

$$F_y - \frac{\mathrm{d}}{\mathrm{d}x}(F_{y'}) = 0. \tag{9.17}$$

Equation (9.16) now becomes

$$\eta(b)\frac{\partial F}{\partial y'}\bigg|_{x=b} - \eta(a)\frac{\partial F}{\partial y'}\bigg|_{x=a} = 0.$$

This result must be true for all differentiable functions $\eta(x)$; in fact for the class of all differentiable functions $\eta(x)$ with the restriction

$$\eta(a) = 0,$$

which means that

$$\eta(b)\frac{\partial F}{\partial y'}\bigg|_{x=b} = 0.$$

Since the value of $\eta(b)$ is arbitrary, we conclude that

$$\frac{\partial F}{\partial y'} = 0 \text{ at } x = b. \tag{9.18}$$

Similarly

$$\frac{\partial F}{\partial y'} = 0 \text{ at } x = a; \tag{9.19}$$

and if y is only free at one end point, then it can be shown that

$$\partial F/\partial y' = 0$$

at that end point. We can summarise these results by

$$\frac{\partial F}{\partial y'} = 0 \quad \text{at end point at which } y \text{ is not specified.} \tag{9.20}$$

Example 5

Find the curve $y = y(x)$ which minimises the functional

$$J = \int_0^1 (y'^2 + 1)\mathrm{d}x$$

where $y(0) = 1$ and y is free at $x = 1$.

Solution

We have already seen, from example 1, that the Euler equation gives

$$y = Ax/2 + B.$$

Since x is no longer given at $x = 1$, we can apply (9.20) to give

$$\frac{\partial F}{\partial y'} = 0 \text{ at } x = 1 \qquad \text{that is,} \qquad 2y' = 0 \text{ at } x = 1.$$

Thus $y'(1) = 0$ implies $A = 0$, and so

$$y(x) = B.$$

But $y(0) = 1$ implies $B = 1$, and the optimal path is

$$y(x) = 1 \quad 0 \leqslant x \leqslant 1$$

and the corresponding value of J is $J = 1$, which is clearly a minimum.

Note that the value of J obtained above is less than the value obtained in example 1, where both end points were fixed. This is to be expected, as there is an extra restriction on possible curves in the first case.

9.4 CONSTRAINTS

The results obtained in the last two sections can be further generalised, firstly to the case where the integrand is a function of more than one independent variable, and secondly to the case of optimisation with constraints.

If we wish to find extremum values of the functional

$$J = \int_a^b F(x_1, x_2, \ldots, x_n; \dot{x}_1, \dot{x}_2, \ldots, \dot{x}_n; t)\,\mathrm{d}t \tag{9.21}$$

where $x_1, x_2, \ldots, x_n$ are independent functions of the variable t and the dot denotes differentiation with respect to t (note that, up to now in this chapter, we have been working with functions y of x; the independent variable is now t instead of x), then following a similar analysis to that in section 9.2, we obtain n Euler equations to be satisfied by the optimum path; that is,

$$\frac{\partial F}{\partial x_i} - \frac{\mathrm{d}}{\mathrm{d}t}\left(\frac{\partial F}{\partial \dot{x}_i}\right) = 0 \qquad (i = 1, 2, \ldots, n). \tag{9.22}$$

Also at any end point where x_i is free

$$\frac{\partial F}{\partial \dot{x}_i} = 0. \tag{9.23}$$

Example 6

Find the extremals of the functional

$$J = \int_0^{\pi/2} (\dot{x}_1^2 + \dot{x}_2^2 + 2x_1 x_2)\, \mathrm{d}t$$

subject to boundary conditions $x_1(0) = 0$, $x_1(\pi/2) = 1$, $x_2(0) = 0$, $x_2(\pi/2) = 1$.

Solution

Evaluating (9.22) for $n = 1$ and 2, and with $F = \dot{x}_1^2 + \dot{x}_2^2 + 2x_1x_2$ gives

$$\frac{\partial F}{\partial x_1} - \frac{\mathrm{d}}{\mathrm{d}t}\left(\frac{\partial F}{\partial \dot{x}_1}\right) = 0, \quad \text{that is,} \quad 2x_2 + \frac{\mathrm{d}}{\mathrm{d}t}(2\dot{x}_1) = 0$$

$$\frac{\partial F}{\partial x_2} - \frac{\mathrm{d}}{\mathrm{d}t}\left(\frac{\partial F}{\partial \dot{x}_2}\right) = 0, \quad \text{that is,} \quad 2x_1 + \frac{\mathrm{d}}{\mathrm{d}t}(2\dot{x}_2) = 0.$$

Hence $x_2 + \ddot{x}_1 = 0$ and $x_1 + \ddot{x}_2 = 0$, and eliminating x_2 gives

$$\frac{\mathrm{d}^4 x_1}{\mathrm{d}t^4} - x_1 = 0.$$

This has solution

$$x_1 = A\mathrm{e}^t + B\mathrm{e}^{-t} + C \sin t + D \cos t$$

and

$$x_2 = \ddot{x}_1 = A\mathrm{e}^t + B\mathrm{e}^{-t} - C \sin t - D \cos t$$

(A, B, C and D being constants). Applying the boundary conditions gives $A = \mathrm{e}^{-\pi/2}/(1 - \mathrm{e}^{-\pi}) = -B$, and $C = D = 0$; hence the optimal path is

$$x_1 = x_2 = \mathrm{e}^{-\pi/2}(\mathrm{e}^t - \mathrm{e}^{-t})/(1 - \mathrm{e}^{-\pi}), \; 0 \leqslant t \leqslant \pi/2.$$

In many optimisation problems, we have constraints on the relevant variables. We first deal with extremum values of a function, say $f(x_1, x_2)$ subject to a constraint of the form $g(x_1, x_2) = 0$ (for example, an algebraic relationship between x_1 and x_2). Assuming that it is not possible to eliminate one of x_1 or x_2 using the constraint, we introduce a Lagrange multiplier, λ, and form the augmented function

$$f^* = f(x_1, x_2) + \lambda g(x_1, x_2). \tag{9.24}$$

We now look for extremum values of f^*, that is,

$$\frac{\partial f^*}{\partial x_1} = \frac{\partial f^*}{\partial x_2} = 0. \tag{9.25}$$

Solving these equations, x_1 and x_2 will be functions of the parameter λ, that is, $x_1(\lambda), x_2(\lambda)$. But we also require the constraint to be satisfied, so the appropriate value of λ is found by satisfying $g(x_1(\lambda), x_2(\lambda)) = 0$. Since f^* now equals f, we have the condition for extremum values for f.

Example 7

Find the extremum value of

$$f(x_1, x_2) = x_1^2 + x_2^2$$

subject to $x_1 + 3x_2 = 4$.

Solution

We first note that we can easily eliminate x_1, say, from the constraint equation to give

$$f = (4 - 3x_2)^2 + x_2^2$$

and we can then solve the problem by evaluating $\mathrm{d}f/\mathrm{d}x_2 = 0$.

Tackling the problem by using a Lagrange multiplier, we consider the augmented function

$$f^* = x_1^2 + x_2^2 + \lambda(x_1 + 3x_2 - 4)$$

(note that the constraint is now in the form $g(x_1, x_2) = x_1 + 3x_2 - 4 = 0$). For extremum values of f^*,

$$\frac{\partial f^*}{\partial x_1} = 0, \quad \text{that is,} \quad 2x_1 + \lambda = 0$$

$$\frac{\partial f^*}{\partial x_2} = 0, \quad \text{that is,} \quad 2x_2 + 3\lambda = 0.$$

Hence $x_1 = -\lambda/2$, $x_2 = -3\lambda/2$, and to satisfy the constraint

$$-\lambda/2 + 3(-3\lambda/2) = 4.$$

This gives $\lambda = -4/5$ and so the extremum value of f occurs at

$$x_1 = 2/5, \quad x_2 = 6/5.$$

We can now return to the main problem. For example, suppose we wish for extremum values of (9.21), $J = \int_a^b F(\mathbf{x},\dot{\mathbf{x}},t)\mathrm{d}t$, subject to a constraint between $x_1, x_2, \ldots, x_n, \dot{x}_1, \ldots, \dot{x}_n, t$ which we write as

$$g(\mathbf{x}, \dot{\mathbf{x}}, t) = 0, \tag{9.26}$$

then we introduce a Lagrange multiplier, λ, and form the augmented functional

$$J^* = \int_a^b (F + \lambda g)\,\mathrm{d}t. \tag{9.27}$$

We now find extremum values of J^* in the usual way, regarding the variables as independent. The Lagrange multiplier can then be eliminated to yield the required optimum path. Note that, since the constraint (9.26) must hold all along the optimum path, the multiplier will in general be a function (rather than a constant as in *static* optimisation problems subject to constraints, e.g. example 7).

Example 8

Minimise the cost functional

$$J = \frac{1}{\sqrt{(2g)}}\int_0^a \frac{(1 + \dot{y}\dot{z})^{\frac{1}{2}}}{x^{\frac{1}{2}}}\,\mathrm{d}x$$

where the variables $y = y(x)$, $z = z(x)$ are subject to the constraint

$$y = z + 1$$

and where $y(0) = 0$, $y(a) = b$.

Note

The constraint is a very simple one, and in this easy example it is possible to eliminate one of the variables, say z, to give

$$J = \frac{1}{\sqrt{(2g)}}\int_0^a \frac{(1 + \dot{y}^2)^{\frac{1}{2}}}{x^{\frac{1}{2}}}\,\mathrm{d}x.$$

This is the cost functional in the brachistochrone problem, and we have already seen that the equation for the optimum path satisfies

$$\frac{\dot{y}}{[x(1+\dot{y}^2)]^{\frac{1}{2}}} = c.$$

We check this result by using the constraint with a Lagrange multiplier.

Solution

Introducing the Lagrange multiplier λ, the augmented functional is

$$J^* = \int_0^a \left\{ \frac{1}{\sqrt{(2g)}} \left[\frac{1+\dot{y}\dot{z}}{x} \right]^{\frac{1}{2}} + \lambda(y-z-1) \right\} \mathrm{d}x,$$

since the constraint is $y - z - 1 = 0$.

The end point conditions are automatically satisfied, since y and hence z are fixed at the end points. Euler's equations, (9.20), with $n = 2$ and x_1, x_2 replaced by y, z respectively (and t by x) give

$$y: \lambda - \frac{\mathrm{d}}{\mathrm{d}x}\left[\frac{z}{2\sqrt{(2g)}[x(1+\dot{y}\dot{z})]^{\frac{1}{2}}} \right] = 0$$

$$z: -\lambda - \frac{\mathrm{d}}{\mathrm{d}x}\left[\frac{\dot{y}}{2\sqrt{(2g)}[x(1+\dot{y}\dot{z})]^{\frac{1}{2}}} \right] = 0.$$

Eliminating λ, we obtain

$$\frac{\mathrm{d}}{\mathrm{d}x}\left[\frac{\dot{y}+\dot{z}}{2\sqrt{(2g)}[x(1+\dot{y}\dot{z})]^{\frac{1}{2}}} \right] = 0$$

Integrating,

$$\frac{\dot{y}+\dot{z}}{[x(1+\dot{y}\dot{z})]^{\frac{1}{2}}} = A;$$

and using the constraint, $z = y - 1$, we obtain for the optimum path

$$\frac{\dot{y}}{[x(1+\dot{y}^2)]^{\frac{1}{2}}} = \frac{A}{2},$$

as expected. In this example, the Lagrange multiplier method has been shown to yield the expected result.

In the final example of this chapter we illustrate one method of dealing with functionals containing derivatives of order greater than one.

Example 9

Minimise the functional $J = \frac{1}{2}\int_0^2 (\ddot{x})^2\,\mathrm{d}t$ where $x(t)$ satisfies

$$x(0) = 1, \quad \dot{x}(0) = 1;$$

$$x(2) = 0, \quad \dot{x}(2) = 0.$$

Solution

The integrand contains the second derivative $\ddot{x}$, and our theory only allows the integrand to be a function of x, $\dot{x}$ and t. We overcome this by introducing two variables, x_1 and x_2 say, where

$$x_1 = x, \quad x_2 = \dot{x}_1.$$

The functional is now

$$J = \frac{1}{2}\int_0^2 \dot{x}_2^2\,\mathrm{d}t$$

and we have the constraint

$$x_2 - \dot{x}_1 = 0.$$

The augmented functional is

$$J^* = \int_0^2 [\tfrac{1}{2}\dot{x}_2^2 + \lambda(x_2 - \dot{x}_1)]\,\mathrm{d}t,$$

where λ is the Lagrange multiplier.

Forming Euler's equations (9.20) for x_1 and x_2 gives

$$x_1 : 0 - \frac{\mathrm{d}}{\mathrm{d}t}(-\lambda) = 0, \quad \text{that is,} \quad \dot{\lambda} = 0;$$

$$x_2 : \lambda - \frac{\mathrm{d}}{\mathrm{d}t}(\dot{x}) = 0, \quad \text{that is,} \quad \lambda - \ddot{x}_2 = 0.$$

Eliminating λ we have $\frac{\mathrm{d}^3x_2}{\mathrm{d}t^3} = 0$, and integrating, $x_2 = At^2 + Bt + C$, where A, B and C are constants of integration. Since $\dot{x}_1 = x_2$, we obtain

$$x_1 = \frac{At^3}{3} + \frac{Bt^2}{2} + Ct + D,$$

D being a further constant of integration. Applying the end conditions implies

$$x_1 = \tfrac{1}{2}t^3 - \tfrac{7}{4}t^2 + t + 1,$$

$$x_2 = \tfrac{3}{2}t^2 - \tfrac{7}{2}t + 1.$$

We have determined the optimal path, and substitution into J yields

$$J = \tfrac{1}{2}\int_0^2 (3t - \tfrac{7}{2})^2\,\mathrm{d}t = 13/4.$$

[This value of J is in fact a minimum.]

PROBLEMS

(1) Find the path $y = y(x)$ which gives a stationary value for the functional

$$J = \int_0^1 (y'^2 + 1)^2\,\mathrm{d}x$$

when

(i) $y(0) = 1$, $y(1) = 2$

(ii) $y(0) = 1$ and $y(1)$ not specified.

(2) Find the curve $y = y(x)$ passing through $y(0) = y_0$, $y(X) = 0$ which minimises the cost functional

$$J = \int_0^X (y^2 + \alpha^2 y'^2)\,\mathrm{d}x$$

where α is a constant and X is fixed.
Minimise the same functional but without the restriction $y(X) = 0$. Determine the minimum cost in each case and compare the results.

(3) Find the stationary path for the functional

$$J = \int_0^2 (\tfrac{1}{2}y'^2 + yy' + y' + y)\,\mathrm{d}x$$

(4) The cost of a chemical process is described by the functional

$$J = \int_0^1 (\tfrac{1}{2}y'^2 + y)\,\mathrm{d}x,$$

where $y(0) = 1$. If also $y(1) = 0$, find the optimal trajectory for y, and evaluate the corresponding minimum value of J.
If $y(1)$ is not specified, use the transversality condition at $x = 1$ to determine the optimal trajectory. Evaluate the corresponding value of J and compare with the value found above.

(5) Find the extremals of the fixed end point problem corresponding to the functional

$$J = \int_0^{\pi/2} (2x_1x_2 - 2x_1^2 + \dot{x}_1^2 - \dot{x}_2^2)\mathrm{d}t$$

where x_1, x_2 are functions of t.

(6) Find the minimum of $x^2 + y^2 + z^2$ subject to the constraint

$$x + 3y - 2z = 4$$

by

(i) using the constraint to eliminate one of the variables;

(ii) using a Lagrange multiplier.

(7) Find the dimensions of the rectangular box which encloses a volume V with minimum surface area.

(8) Find the curve joining the points (0,0) and (1,0) for which the integral

$$\int_0^1 \left(\frac{\mathrm{d}^2y}{\mathrm{d}x^2}\right)^2 \mathrm{d}x$$

is a minimum if

(i) $\dfrac{\mathrm{d}y}{\mathrm{d}x}(0) = a, \; \dfrac{\mathrm{d}y}{\mathrm{d}x}(1) = b,$

(ii) no other conditions are prescribed.

(9) Find stationary paths $x = x(t)$ for the functional

$$J = \int_0^1 \left[1 + \left(\frac{\mathrm{d}^2x}{\mathrm{d}t^2}\right)^2\right]\mathrm{d}t$$

subject to the boundary conditions

$$x(0) = 0, \frac{\mathrm{d}x}{\mathrm{d}t}(0) = 1, x(1) = 1, \frac{\mathrm{d}x}{\mathrm{d}t}(1) = 1.$$

(10) Find the extremals of the functional

$$J = \int_0^{\pi/2} \left[\left(\frac{\mathrm{d}^2y}{\mathrm{d}x^2}\right)^2 - y^2 + x^2\right]\mathrm{d}x$$

subject to the boundary conditions

$$y(0) = 1, \frac{\mathrm{d}y}{\mathrm{d}x}(0) = 0, y(\pi/2) = 0, \frac{\mathrm{d}y}{\mathrm{d}x}(\pi/2) = 1.$$

CHAPTER 10

Optimal control with unbounded continuous controls

10.1 INTRODUCTION

The methods appropriate for variational calculus can readily be extended to optimal control problems, where the control vector $\mathbf{u}$ is unbounded. Techniques for dealing with the general case $\mathbf{u} \in U$, a bounded admissible control region, will be discussed in the next chapter.

We start with a simple example. Suppose we wish to find *extremum* values of the functional

$$J = \int_0^T (x^2 + u^2)\mathrm{d}t \tag{10.1}$$

where x, the state variable, and u, the control variable, satisfy the differential equation

$$\dot{x} + x = u \tag{10.2}$$

and $x(0) = x_0$, $x(T) = 0$. The direct method of solving this problem would be to eliminate u from (10.2) in (10.1), which reduces it to a straightforward calculus of variations problem. In general, though, we will not be able to eliminate the control, so we will develop another method based on the Lagrange multiplier technique for dealing with constraints.

We introduce the Lagrange multiplier, λ, and form the augmented functional

$$J^* = \int_0^T [x^2 + u^2 + \lambda(\dot{x} + x - u)]\mathrm{d}t. \tag{10.3}$$

The integrand $F = x^2 + u^2 + \lambda(\dot{x} + x - u)$ is a function of the two variables x and u. We can evaluate Euler's equation for x and u; that is,

$$\frac{\partial F}{\partial x} - \frac{\mathrm{d}}{\mathrm{d}t}\left(\frac{\partial F}{\partial \dot{x}}\right) = 0, \quad \text{which gives} \quad 2x + \lambda - \dot{\lambda} = 0; \tag{10.4}$$

$$\frac{\partial F}{\partial u} - \frac{\mathrm{d}}{\mathrm{d}t}\left(\frac{\partial F}{\partial \dot{u}}\right) = 0, \quad \text{which gives} \quad 2u - \lambda = 0. \tag{10.5}$$

Substituting λ from (10.5) into (10.4) gives

$$x + u - \dot{u} = 0$$

and substituting u from (10.2) gives

$$x + (\dot{x} + x) - (\ddot{x} + \dot{x}) = 0,$$

that is, $\quad \ddot{x} - 2x = 0.$

Solving for x gives

$$x = A\mathrm{e}^{\sqrt{2}\,t} + B\mathrm{e}^{-\sqrt{2}\,t}$$

or, alternatively,

$$x = a \cosh \sqrt{2}\,t + b \sinh \sqrt{2}\,t$$

Using the alternative expression for x and substituting for the initial conditions, that is, $x(0) = x_o$, gives $x_o = a$, and $x(T) = 0$ gives

$$0 = x_o \cosh \sqrt{2}\,T + b \sinh \sqrt{2}\,T.$$

Hence

$$x = x_o \cosh \sqrt{2}\,t - x_o \frac{\cosh \sqrt{2}\,T}{\sinh \sqrt{2}\,T} \sinh \sqrt{2}\,t,$$

that is, $$x = \frac{x_o \sinh \sqrt{2}\,(T - t)}{\sinh \sqrt{2}\,T}.$$

The corresponding optimal control is recovered from (10.2), giving

$$u = x_o \frac{(\sinh \sqrt{2}\,(T-t) - \sqrt{2} \cosh \sqrt{2}\,(T-t))}{\sinh \sqrt{2}\,T}.$$

The control and state variable are illustrated in Fig. 10.1.

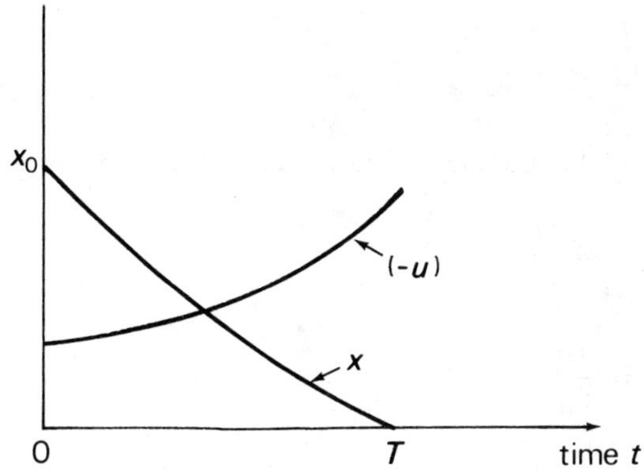

Fig. 10.1 Optimal Control and State Variable

10.2 THE HAMILTONIAN

Consider the general problem of minimising

$$J = \int_0^T f_0(x, u, t)\mathrm{d}t \tag{10.6}$$

subject to the differential equation

$$\dot{x} = f(x, u, t) \tag{10.7}$$

and boundary conditions $x(0) = a$. With Lagrange multiplier, λ, we form the augmented functional

$$J^* = \int_0^T [f_0(x, u, t) + \lambda(f(x, u, t) - \dot{x})]\mathrm{d}t.$$

The integrand $F = f_0 + \lambda(f - \dot{x})$ is a function of the two variables x and u, and so we have two Euler equations:

$$\frac{\partial F}{\partial x} - \frac{\mathrm{d}}{\mathrm{d}t}\left(\frac{\partial F}{\partial \dot{x}}\right) = 0 \text{ which gives } \frac{\partial f_0}{\partial x} + \lambda\frac{\partial f}{\partial x} + \dot{\lambda} = 0; \tag{10.8}$$

$$\frac{\partial F}{\partial u} - \frac{\mathrm{d}}{\mathrm{d}t}\left(\frac{\partial F}{\partial \dot{u}}\right) = 0 \text{ which gives } \frac{\partial f_0}{\partial u} + \lambda\frac{\partial f}{\partial u} = 0. \tag{10.9}$$

Now defining

$$H = f_0 + \lambda f, \tag{10.10}$$

which we call the **Hamiltonian**, (10.8) and (10.9) can be written as

$$\dot{\lambda} = -\frac{\partial H}{\partial x} \tag{10.11}$$

$$0 = \frac{\partial H}{\partial u}. \tag{10.12}$$

These are the equations, together with (10.2), which govern the optimal paths. Since we have two first order differential equations, on integrating, we need two conditions to complete the solution. One is given by $x(0) = a$, and the other either by a condition $x(T) = b$, or, if x is not fixed at $t = T$, by the 'free-end' condition (see (9.16)), $\partial F/\partial \dot{x} = 0$, that is, $\lambda = 0$ at $t = T$.

Example 1

Using the Hamiltonian method, find the optimal control u which yields a stationary value for the functional

$$J = \int_0^1 (x^2 + u^2)\mathrm{d}t$$

where

$$\dot{x} = u,$$

and $x(0) = 1$ but x is not fixed at $t = 1$.

Solution

[Note that the control can be eliminated, giving a variational calculus problem. Readers might like to check that the same result is obtained.]

The Hamiltonian is

$$H = x^2 + u^2 + \lambda u$$

and so (10.12) becomes

$$2u + \lambda = 0$$

so that $2\dot{u} + \dot{\lambda} = 0$.

Also (10.11) becomes

$$\dot{\lambda} = -2x$$

and on eliminating λ,

$$\dot{u} = x.$$

Hence, $\ddot{x} = \dot{u} = x$; and on solving the equation $\ddot{x} - x = 0$, we obtain

$$x = a \sinh t + b \cosh t$$

The condition $x = 1$ at $t = 0$ gives $b = 1$. The other condition is $\lambda = 0$ at $t = 1$ (since x is not fixed at $t = 1$); so that since $2u + \lambda = 0$, $u = 0$ at $t = 1$. But

$$u = \dot{x} = a \cosh t + \sinh t,$$

so that

$$0 = a \cosh 1 + \sinh 1.$$

Thus the optimal control is given by

$$u = -\frac{\sinh(1-t)}{\cosh 1}$$

(using the relation $\sinh(x-y) = \sinh x \cosh y - \cosh x \sinh y$) and the corresponding state trajectory is

$$x = \frac{\cosh(1-t)}{\cosh 1}.$$

10.3 EXTENSION TO HIGHER ORDER SYSTEMS

The method described in the last section can be readily extended to higher order systems. For example, suppose we wish to minimise

$$J = \tfrac{1}{2}\int_0^\infty (x^2 + \alpha u^2)\mathrm{d}t \qquad (10.13)$$

where,

$$\ddot{x} = -\dot{x} + u, \ \alpha \text{ constant}, \qquad (10.14)$$

and initially $x = a$, $\dot{x} = b$, and as $t \to \infty$, both x and $\dot{x} \to 0$. Following the usual technique, we define

$$x_1 = x$$

$$x_2 = \dot{x}$$

so that we have two constraints

$$x_2 - \dot{x}_1 = 0 \tag{10.15}$$

$$-x_2 + u - \dot{x}_2 = 0. \tag{10.16}$$

The augmented function is

$$J^* = \int_0^\infty [\tfrac{1}{2}(x_1^2 + \alpha u^2) + \lambda_1(x_2 - \dot{x}_1) + \lambda_2(-x_2 + u - \dot{x}_2)]\mathrm{d}t.$$

We now have three Euler equations, namely,

$$\frac{\partial F}{\partial x_1} - \frac{\mathrm{d}}{\mathrm{d}t}\left(\frac{\partial F}{\partial \dot{x}_1}\right) = 0, \quad \text{that is,} \quad x_1 + \dot{\lambda}_1 = 0; \tag{10.17}$$

$$\frac{\partial F}{\partial x_2} - \frac{\mathrm{d}}{\mathrm{d}t}\left(\frac{\partial F}{\partial \dot{x}_2}\right) = 0, \quad \text{that is,} \quad \lambda_1 - \lambda_2 + \dot{\lambda}_2 = 0; \tag{10.18}$$

$$\frac{\partial F}{\partial u} - \frac{\mathrm{d}}{\mathrm{d}t}\left(\frac{\partial F}{\partial \dot{u}}\right) = 0, \quad \text{that is,} \quad \alpha u + \lambda_2 = 0; \tag{1019}$$

(F is the integrand). Eliminating λ_2 fron (10.18) and (10.19) gives

$$\lambda_1 = -\alpha u + \alpha \dot{u},$$

where we have used $\dot{\lambda}_2 = -\alpha \dot{u}$ from (10.19). Substituting for $\dot{\lambda}_1$ in (10.17) yields

$$x_1 - \alpha \dot{u} + \alpha \ddot{u} = 0.$$

We can now eliminate u from (10.16) and x_2 from (10.15) to obtain

$$x_1 - \alpha(\ddot{x}_2 + \dot{x}_2) + \alpha(\dddot{x}_2 + \ddot{x}_2) = 0$$

that is, $$\frac{\mathrm{d}^4 x_1}{\mathrm{d}t^4} - \frac{\mathrm{d}^2 x_1}{\mathrm{d}t^2} + \frac{x_1}{\alpha} = 0. \tag{10.20}$$

This is the equation which determines the optimal path, and its solution will depend on the value of the parameter α. For instance, if $\alpha = 4$, (10.20) will have solutions of the form $x_1 = e^{mt}$, where m satisfies

$$m^4 - m^2 + \tfrac{1}{4} = 0;$$

that is, $$(m^2 - \tfrac{1}{2})^2 = 0.$$

Hence $m = \pm 1/\sqrt{2}$, repeated, and we can write the solution as

$$x_1 = (A + Bt)e^{t/\sqrt{2}} + (C + Dt)e^{-t/\sqrt{2}} \tag{10.21}$$

and from (10.15)

$$x_2 = \left(\frac{A}{\sqrt{2}} + B + \frac{Bt}{\sqrt{2}}\right)e^{t/\sqrt{2}} + \left(D - \frac{C}{\sqrt{2}} - \frac{Dt}{\sqrt{2}}\right)e^{-t/\sqrt{2}}. \tag{10.22}$$

The conditions on x_1 and x_2 are

$$x_1 = a,\ x_2 = b \quad \text{at} \quad t = 0$$

$$x_1, x_2 \to 0 \quad \text{as} \quad t \to \infty.$$

The conditions as $t \to \infty$ immediately give $A = B = 0$, whereas the conditions at $t = 0$ give

$$a = C$$

$$b = D - C/\sqrt{2},$$

that is, $C = a, D = b + a/\sqrt{2}$ giving

$$x_1 = [a + (b + a/\sqrt{2})t]e^{-t/\sqrt{2}}. \tag{10.23}$$

We can similarly determine x_2 from (10.22) and consequently the optimal control u from (10.16); that is,

$$x_2 = [b - (b + a/\sqrt{2})\,t/\sqrt{2}\,]e^{-t/\sqrt{2}} \tag{10.24}$$

$$u = [-a/2 - (\sqrt{2} - 1)\,b - (b + a/\sqrt{2})(1 - 1/\sqrt{2})\,t/\sqrt{2}\,]e^{-t/\sqrt{2}}. \tag{10.25}$$

It is of interest to note that the optimal control, u, can be expressed as a linear function of x_1 and x_2; for, if we assume that

$$u = Ax_1 + Bx_2$$

then

$$u = \{Aa + Bb + [A(b + a/\sqrt{2})t - B(b + a/\sqrt{2})t/\sqrt{2}]\}e^{-t\sqrt{2}}$$

and to agree with (10.25) we must have

$$Aa + Bb = -a/2 - (\sqrt{2} - 1)b$$

$$(A - B/\sqrt{2})(b + a/\sqrt{2}) = (b + a/\sqrt{2})(1 - 1/\sqrt{2})/\sqrt{2}.$$

These equations have solutions

$$A = -\tfrac{1}{2}, \quad B = -(\sqrt{2} - 1)$$

and so the optimal control is given by

$$u = -\tfrac{1}{2}x_1 - (\sqrt{2} - 1)x_2 \tag{10.26}$$

which does not depend on the initial conditions on x_1 and x_2. We have calculated the optimal feedback control for this system, which minimises the cost functional (10.13), when $\alpha = 4$.

The method described above can be generalised for linear systems with interesting mathematical results. We refer readers to Barnett (1975) for details.

GENERAL PROBLEM

We are now in a position to solve the general optimal control problem described in section 8.4 in the case when the controls are continuous and unbounded. The basic problem is to find the optimal controls u which yield extremal values of

$$J = \int_0^T f_0(\mathbf{x}, \mathbf{u}, t)\mathrm{d}t \tag{10.27}$$

subject to the differential constraints

$$\dot{x}_i = f_i(\mathbf{x}, \mathbf{u}, t) \qquad (i = 1, 2, \ldots, n), \tag{10.28}$$

where $\mathbf{x} = [x_1 x_2 \ldots x_n]'$, $\mathbf{u} = [u_1 u_2 \ldots u_m]'$

We introduce the Lagrange multipliers, say p_i $(i = 1, 2, \ldots, n)$ usually called the **adjoint variables** and form the augmented functional

$$J^* = \int_0^T \{f_0 + \sum_{i=1}^{n} p_i(f_i - \dot{x}_i)\} \mathrm{d}t.$$

We also define the Hamiltonian, H, as

$$H = f_0 + \sum_{i=1}^{n} p_i f_i \tag{10.29}$$

so that

$$J^* = \int_0^T (H - \sum_{i=1}^{n} p_i \dot{x}_i) \mathrm{d}t.$$

The integrand, $F = H - \sum_{i=1}^{n} p_i \dot{x}_i$ depends on $\mathbf{x}$, $\mathbf{u}$ and t, and we form $(n+m)$ Euler equations, namely

$$\frac{\partial F}{\partial x_i} - \frac{\mathrm{d}}{\mathrm{d}t}\left(\frac{\partial F}{\partial \dot{x}_i}\right) = 0 \qquad (i = 1, 2, \ldots, n)$$

that is,

$$\dot{p}_i = -\frac{\partial H}{\partial x_i} \tag{10.30}$$

which are known as the **adjoint equations**; and

$$\frac{\partial F}{\partial u_j} - \frac{\mathrm{d}}{\mathrm{d}t}\left(\frac{\partial F}{\partial \dot{u}_j}\right) = 0 \qquad (j = 1, 2, \ldots, m)$$

that is,

$$\frac{\partial H}{\partial u_j} = 0. \tag{10.31}$$

The optimal solutions for $\mathbf{x}$, $\mathbf{u}$ and $\mathbf{p} = [p_1 p_2 \ldots p_n]'$ are determined from the $(2n+m)$ equations (10.28), (10.30) and (10.31). Using the boundary conditions given in section 8.4, we have $x_i(0)$ given $(i = 1, 2, \ldots, n)$ and $x_l(T)$ $(l = 1, 2, \ldots, q)$ are given. The remaining values $x_{q+1}(T), \ldots, x_n(T)$ are free, and so we must apply the free end point condition (9.23)

$$\frac{\partial F}{\partial \dot{x}_k} = 0 \qquad (k = q+1, \ldots, n) \text{ at } t = T.$$

This gives

$$p_k(T) = 0 \qquad (k = q+1, \ldots, n) \tag{10.32}$$

which is known as the **transversality condition.** In general

$$p_i(T) = 0 \quad \text{for any } x_i \text{ not specified at } t = T.$$

Example 2

Find the control $u(t)$ which minimises

$$J = \int_0^1 u^2 \, dt$$

where $\dot{x} = u + ax$, a is a constant, and

(i) $x(0) = 1$

(ii) $x(0) = 1, \quad x(1) = 0.$

Solution

As in earlier examples in this chapter, we could eliminate the control u. But using the method outlined above, we form the Hamiltonian

$$H = u^2 + p(u + ax)$$

where p is the adjoint variable. (10.31) yields

$$2u + p = 0$$

and the adjoint equation, (10.30), becomes

$$\dot{p} = -ap.$$

On solving, we obtain $p = Ae^{-at}$, A being a constant. Thus $u = -\frac{A}{2}e^{-at}$ and

$$\dot{x} - ax = -\frac{A}{2}e^{-at}.$$

This equation has solution

$$x = Be^{at} + \frac{A}{4a}e^{-at}.$$

To find the constants A and B, we must look at each case separately.

(i) $x(0) = 1, x(1)$ not specified

Here

$$B + A/4a = 1$$

and since $x(1)$ is not specified, we apply the transversality condition (10.32) at $t = 1$, which is $p(1) = 0$,

that is, $\quad A = 0.$

Thus $A = 0$ and $B = 1$, and the optimal control is $u = 0$, $0 \leqslant t \leqslant 1$. This gives $J = 0$, which is clearly a minimum as $J \geqslant 0$.

(ii) $x(0) = 1, \quad x(1) = 0$

We now have

$$B + A/4a = 1$$

$$Be^{a} + Ae^{-a}/4a = 0.$$

Hence, $A = 4a/(1 - e^{-2a})$ and the optimal control is

$$u = -2ae^{-at}/(1 - e^{-2a}), \quad 0 \leqslant t \leqslant 1.$$

Example 3

Find the optimal control which gives an extremum value ot the functional

$$J = \int_0^1 u^2 \, dt$$

where

$$\dot{x}_1 = x_2$$

$$\dot{x}_2 = u$$

and $x_1(0) = 1$, $x_1(1) = 0$, $x_2(0) = 1$ but $x_2(1)$ is not specified.

Solution

We define the Hamiltonian

$$H = u^2 + p_1 x_2 + p_2 u$$

where p_1, p_2 are the adjoint variables. (10.31) gives

$$\frac{\partial H}{\partial u} = 0 = 2u + p_2, \qquad \text{that is,} \qquad u = -\tfrac{1}{2}p_2.$$

The adjoint equations (10.30) become

$$\dot{p}_1 = 0, \qquad \text{that is,} \quad p_1 = A$$

$$\dot{p}_2 = -p_1 = -A, \qquad \text{that is,} \quad p_2 = -At + B.$$

Hence, $u = \dfrac{A}{2}t - \dfrac{B}{2}$ and

$$\dot{x}_2 = \frac{A}{2}t - \frac{B}{2}.$$

Integrating $x_2 = \dfrac{A}{4}t^2 - \dfrac{B}{2}t + C,$ and integrating again

$$x_1 = \frac{A}{12}t^3 - \frac{B}{4}t^2 + Ct + D.$$

The boundary conditions give

$$D = 1, \frac{A}{12} - \frac{B}{4} + C + D = 0, \qquad C = 1.$$

We also have the transversality condition, $p_2(1) = 0$, since $x_2(1)$ is not specified; that is,

$$-A + B = 0.$$

Solving, we obtain $A = B = 12$ and the optimal control is

$$u = 6(t - 1), \qquad 0 \leqslant t \leqslant 1.$$

PROBLEMS

(1) Find extremum values of the functional

$$J = \int_0^T (x^2 + u^2)\,\mathrm{d}t$$

subject to $\dot{x} + x = u$, $x(0) = x_0$, $x(T) = 0$ by direct elimination of u. Compare your result with that obtained in section 10.1.

(2) A control system is described by the differential equation

$$\dot{x} + ax = u \quad (a \text{ constant})$$

where $x = x(t)$ is the state variable, $u = u(t)$ is the control variable. If $x = 0$ at $t = 0$, and $x = Q$ at $t = T$, determine the optimal control u which minimises

$$I = \int_0^T [(Q - x)^2 + u^2]\mathrm{d}t.$$

(3) The system

$$\dot{x}_1 = x_2$$
$$\dot{x}_2 = -x_2 + u$$

is to be controlled so that its control effort is conserved; that is, the performance measure

$$J = \int_0^T \tfrac{1}{2}u^2 \mathrm{d}t$$

is to be minimised.
Use the Hamiltonian method to find the governing optimal path. Solve the problem completely when

$$x_1(0) = 1, \quad x_2(0) = 1$$
$$x_1(1) = 0, \quad x_2(1) = 0.$$

(4) A system is characterised by the equations

$$\dot{x}_1 = x_2$$
$$\dot{x}_2 = -x_2 + u$$

where u is the control variable, chosen so as to minimise the cost functional

$$\int_0^\infty (x_1^2 + \tfrac{16}{3}u^2)\mathrm{d}t.$$

Find the optimum control u if $x_1(0) = a$, $x_2(0) = b$, $x_1, x_2 \to 0$ as $t \to \infty$.

(5) A first order system is described by the state equation

$$\dot{x} = x + u.$$

(i) Find the unconstrained control which minimises the functional

$$J = \int_0^T \tfrac{1}{2}(3x^2 + u^2)\mathrm{d}t$$

where T is fixed, $x(0)$ is given and $x(T)$ is free.

(ii) Show that for $T \to \infty$, the optimal control law is of the form

$$u = Fx$$

where F is a constant. Find F.

(6) A system is described by the differential equation

$$\dot{x} = -ax + u.$$

It is desired to bring the system from some arbitrary fixed initial state $x(0)$ to the origin in T seconds whilst minimising the performance measure

$$J = \int_0^T u^2 \, dt.$$

(i) Find the expression for the optimal trajectory in terms of $x(0)$, a, t and T;

(ii) Find the expression for the optimal control in terms of $x(0)$, a, t and T;

(iii) The optimal control can be expressed in terms of x in the form

$$u = F(t, T, a)x.$$

Find $F(t, T, a)$.

CHAPTER 11

Bang-bang control

11.1 INTRODUCTION

In the previous chapter we dealt with a number of optimal control problems, but in all cases we assumed that the controls were continuous and no restrictions were put on the range of possible values of the controls. For many practical applications, the control will be bounded, and it might well be possible to have discontinuities in the control values. For example, a light switch is a control which is either on or off and in fact does not take intermediate values. The steering wheel of a car is a control which can take any intermediate values between two end values.

To illustrate some of the basic concepts involved when controls are bounded and allowed to have discontinuities we start with a simple physical problem. Suppose we wish to drive a car from a stationary position on a driveway into a stationary position in a garage, having moved a total distance a. The available controls for the driver are the accelerator and the brake (for simplicity, we assume no gear changes), and we take the equation of motion of the car as

$$\frac{\mathrm{d}^2x}{\mathrm{d}t^2} = u \tag{11.1}$$

where $u = u(t)$ represents the applied acceleration or deceleration (braking) and x the distance travelled. The control u is subject to both a lower bound (maximum braking) and upper bound (maximum acceleration), that is,

$$-\alpha \leqslant u \leqslant \beta \tag{11.2}$$

where α, β are positive constants. The problem can now be stated: minimise the time taken, T, that is,

$$T = \int_0^T 1.\,\mathrm{d}t \tag{11.3}$$

subject to (11.1), (11.2) and boundary conditions

$$x(0) = 0, \quad \dot{x}(0) = 0, \quad x(T) = a, \quad \dot{x}(T) = 0. \tag{11.4}$$

The methods we developed in the last chapter would be appropriate for this problem except that they cannot cope with *inequality constraints* of the form (11.2). But we can change this constraint into an equality constraint by introducing another control variable, say v, where

$$v^2 = (u + \alpha)(\beta - u). \tag{11.5}$$

Since v is real, u must satisfy (11.2). We introduce the usual state variable notation $x_1 = x$ so that

$$\dot{x}_1 = x_2 \tag{11.6}$$

$$\dot{x}_2 = u \tag{11.7}$$

replaces the second order differential equation (11.1), with boundary conditions $x_1(0) = x_2(0) = 0$, $x_1(T) = a$, $x_2(T) = 0$.

We now form the augmented functional

$$T^* = \int_0^T \{1 + p_1(x_2 - \dot{x}_1) + p_2(u - \dot{x}_2) + \mu[v^2 - (u + \alpha)(\beta - u)]\}\,dt \tag{11.8}$$

where p_1, p_2 and μ are Lagrange multipliers associated with the constraints (11.6), (11.7) and (11.5) respectively. The Euler equations for the state variables x_1 and x_2 and control variables u and v are (F is the integrand in (11.8))

$$\frac{\partial F}{\partial x_1} - \frac{d}{dt}\left(\frac{\partial F}{\partial \dot{x}_1}\right) = 0, \quad \text{that is,} \quad \dot{p}_1 = 0 \tag{11.9}$$

$$\frac{\partial F}{\partial x_2} - \frac{d}{dt}\left(\frac{\partial F}{\partial \dot{x}_2}\right) = 0, \quad \text{that is,} \quad \dot{p}_2 = -p_1 \tag{11.10}$$

$$\frac{\partial F}{\partial u} - \frac{d}{dt}\left(\frac{\partial F}{\partial \dot{u}}\right) = 0, \quad \text{that is,} \quad p_2 = \mu(\beta - \alpha - 2u) \tag{11.11}$$

$$\frac{\partial F}{\partial v} - \frac{d}{dt}\left(\frac{\partial F}{\partial \dot{v}}\right) = 0, \quad \text{that is,} \quad 2v\mu = 0. \tag{11.12}$$

Now from (11.12), either $v = 0$ or $\mu = 0$ (or both). We first consider $\mu = 0$.

(11.11) gives $p_2 = 0$ and (11.10) $p_1 = 0$. (11.9) is satisfied and we have not determined any solution to the problem. So we can discard this possibility (in this simple problem) and we are left with the alternative $\nu = 0$. But from (11.5), if $\nu = 0$, we must have either $u = \beta$ or $u = -\alpha$, its two extreme values. Hence on the optimal path, the acceleration takes either its maximum or its minimum value, but *no* intermediate values. From the physical situation, it is clear that we must first have $u = \beta$ followed by $u = -\alpha$; so that, assuming there is only *one switch* in the control, we have

$$\dot{x}_2 = \begin{cases} \beta & 0 \leqslant t < \tau \\ -\alpha & \tau < t \leqslant T, \end{cases}$$

the switch taking place at time τ. Integrating

$$x_2 = \dot{x}_1 = \begin{cases} \beta t & 0 \leqslant t < \tau \\ -\alpha(t - T) & \tau < t \leqslant T \end{cases} \tag{11.13}$$

using the boundary conditions on x_2. Integrating again, using the boundary conditions on x_1,

$$x_1 = \begin{cases} \frac{1}{2}\beta t^2 & 0 \leqslant t < \tau \\ -\frac{1}{2}\alpha(t - T)^2 + a & \tau < t \leqslant T. \end{cases} \tag{11.14}$$

Since both x_1 (distance) and x_2 (speed) are continuous at $t = \tau$, we must have

$$\beta\tau = \alpha(T - \tau)$$

$$\tfrac{1}{2}\beta\tau^2 = a - \tfrac{1}{2}\alpha(t - \tau)^2.$$

Eliminating T gives the switching time as

$$\tau = [2a\alpha/\beta(\alpha + \beta)]^{\frac{1}{2}} \tag{11.15}$$

and the final time is

$$T = [2a(\alpha + \beta)/\alpha\beta]^{\frac{1}{2}}. \tag{11.16}$$

The problem has now been completely solved, and we note that the optimal control is specified by

$$u = \begin{cases} \beta & 0 \leqslant t < \tau \\ -\alpha & \tau < t \leqslant T \end{cases} \tag{11.17}$$

This is illustrated in Fig. 11.1.

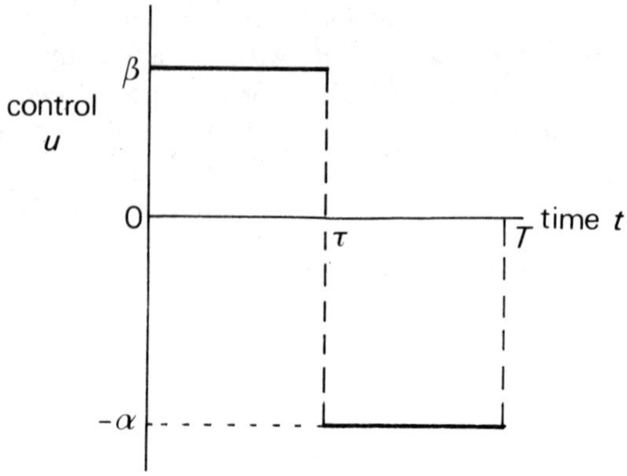

Fig. 11.1 Optimal Control

This figure brings out clearly that the control has a switch (discontinuity) at time $t = \tau$, and that the control only takes its maximum and minimum values. This type of control is called **bang-bang control**. Of course the solution to this rather trivial problem was just about obvious. On the other hand it has given some indications of how to deal with bounded controls. Before moving onto the general theory it is of value to note the form of the state-space trajectory. From (11.13) and (11.14), we see that for $0 \leqslant t < \tau$,

$$x_2^2 = 2\beta x_1,$$

and for $\tau < t \leqslant T$,

$$x_2^2 = 2\alpha(a - x_1).$$

These trajectories are illustrated in the $x_1 - x_2$ plane in Fig. 11.2.

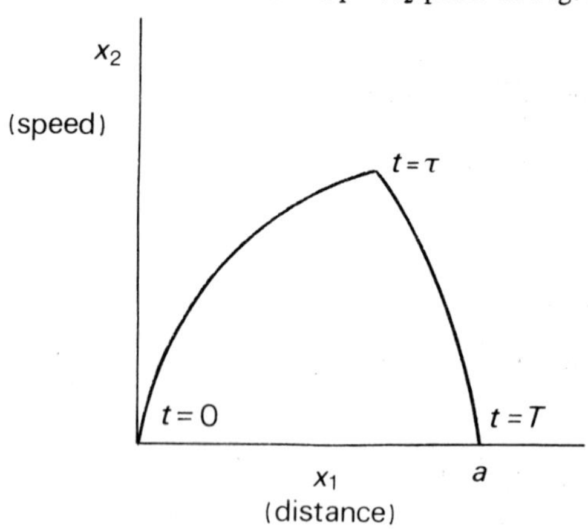

Fig. 11.2 Phase Plane Trajectory

11.2 PONTRYAGIN'S PRINCIPLE

Early in the 1960s, Pontryagin and his Russian colleagues published a general principle (referred to as either a 'maximum' or 'minimum' principle) which deals, not only with continuous controls (as in Chapter 10) but also with bounded and possibly discontinuous controls. A rigorous proof of the principle is far beyond the scope of this book, but we will present a non-rigorous verification of the principle.

We first restate our general problem. We are seeking extremum values of the functional

$$J = \int_0^T f_0(\mathbf{x}, \mathbf{u}, t)\,\mathrm{d}t, \tag{11.18}$$

subject to state equations

$$\dot{x}_i = f_i(\mathbf{x}, \mathbf{u}, t) \qquad (i = 1, 2, \ldots, n), \tag{11.19}$$

initial conditions $\mathbf{x} = \mathbf{x}_0$ and final conditions on $x_1, x_2, \ldots, x_q$ $(q \leqslant n)$; and subject to $\mathbf{u} \in U$, the **admissible control region**. For example, in the problem in the section above, the admissible control region is defined by

$$U = \{u : -\alpha \leqslant u \leqslant \beta\}.$$

As in section 10.4, we form the augmented functional

$$J^* = \int_0^T \{f_0 + \sum_{i=1}^{n} p_i(f_i - \dot{x}_i)\}\,\mathrm{d}t \tag{11.20}$$

and define the Hamiltonian

$$H = f_0 + \sum_{i=1}^{n} p_i f_i. \tag{11.21}$$

For simplicity we will drop the dependence of f_0 and f_i explicitly on time t, so that H is a function of the state vector $\mathbf{x}$, control vector $\mathbf{u}$, and adjoint vector $\mathbf{p}$, that is, $H = H(\mathbf{x}, \mathbf{u}, \mathbf{p})$.

We can express J^* as

$$J^* = \int_0^T (H - \sum_{i=1}^{n} p_i \dot{x}_i)\,\mathrm{d}t \tag{11.22}$$

and evaluating the Euler equations for x_i, we obtain as in section 10.4 the adjoint equations

$$\dot{p}_i = -\frac{\partial H}{\partial x_i} \qquad (i = 1, 2, \ldots, n). \qquad (11.23)$$

The Euler equations for the control variables, u_j, do not follow as in section 10.4 as it is possible that there are discontinuities in u_j, and so we cannot assume that the partial derivatives $\partial H/\partial u_j$ exist. On the other hand, we can apply the free end point condition (9.23) to obtain (as in section 10.4)

$$p_k(T) = 0, \qquad k = q+1, \ldots, n, \qquad (11.24)$$

that is, the adjoint variable is zero at every end point where the corresponding state variable is not specified. As before, we refer to (11.24) as the *transversality conditions*.

Our difficulty now lies in obtaining the analogous equation to $\partial H/\partial u_j = 0$ for continuous controls. For the moment, let us assume that we can differentiate H with respect to u, and consider a small variation $\delta\mathbf{u}$ in the control $\mathbf{u}$ such that $\mathbf{u} + \delta\mathbf{u}$ still belongs to U, the admissible control region. Corresponding to the small change in $\mathbf{u}$, there will be small changes in $\mathbf{x}$, say $\delta\mathbf{x}$, and in $\mathbf{p}$, say $\delta\mathbf{p}$. The change in the value of J^* will be δJ^*, where

$$\delta J^* = \delta \int_0^T \{H - \sum_{i=1}^{n} p_i \dot{x}_i\} \mathrm{d}t \ .$$

The small change operator, δ, obeys the same sort of properties as the differential operator $\mathrm{d}/\mathrm{d}x$. For example, if $f = f(x)$, then

$$\begin{aligned} \delta f &= f(x + \delta x) - f(x) \\ &= \delta x f'(x) + 0(\delta x^2). \end{aligned}$$

Similarly
$$\begin{aligned} \delta(fg) &= f(x + \delta x)g(x + \delta x) - f(x)g(x) \\ &= (f(x+\delta x) - f(x))g(x+\delta x) + f(x)(g(x+\delta x) - g(x)) \\ &= (\delta f)g(x+\delta x) + f(x)(\delta g) \\ &= (\delta f)g + f(\delta g) + 0(\delta x^2); \end{aligned}$$

so to the first order in small quantities, $\delta(fg) = (\delta f)g + f(\delta g)$. Assuming we can interchange the small change operator, δ, and integral sign, we obtain

$$\delta J^* = \int_0^T \delta(H - \sum_{i=1}^{n} p_i \dot{x}_i)\mathrm{d}t$$

$$= \int_0^T [\delta H - \sum_{i=1}^{n} \delta(p_i \dot{x}_i)]\mathrm{d}t$$

$$= \int_0^T [\delta H - \sum_{i=1}^{n} \dot{x}_i \delta p_i - \sum_{i=1}^{n} p_i \delta \dot{x}_i]\mathrm{d}t.$$

But, using the chain rule for partial differentiation,

$$\delta H = \sum_{j=1}^{m} \frac{\partial H}{\partial u_j}\delta u_j + \sum_{i=1}^{n} \frac{\partial H}{\partial x_i}\delta x_i + \frac{\partial H}{\partial p_i}\delta p_i$$

so that

$$\delta J^* = \int_0^T \left[\sum_{j=1}^{m} \frac{\partial H}{\partial u_j}\delta u_j + \sum_{i=1}^{n}\left(\frac{\partial H}{\partial p_i}\delta p_i - \dot{p}_i \delta x_i - \dot{x}_i \delta p_i - \dot{x}_i \delta p_i - p_i \delta \dot{x}_i\right)\right]\mathrm{d}t$$

since $\dot{p}_i = -\partial H/\partial x_i$. Also, from (11.21)

$$\frac{\partial H}{\partial p_i} = f_i = \dot{x}_i,$$

using (11.19). Thus

$$\delta J^* = \int_0^T \left[\sum_{j=1}^{m} \frac{\partial H}{\partial u_j}\delta u_j - \sum_{i=1}^{n} (\dot{p}_i \delta x_i + p_i \delta \dot{x}_i)\right]\mathrm{d}t$$

$$= \int_0^T \left[\sum_{j=1}^{m} \frac{\partial H}{\partial u_j}\delta u_j - \sum_{i=1}^{n} \frac{\mathrm{d}}{\mathrm{d}t}(p_i \delta x_i)\right]\mathrm{d}t.$$

We can now integrate the second part of the integrand to yield

$$\delta J^* = -\sum_{i=1}^{n} p_i \delta x_i \Big|_0^T + \int_0^T \sum_{j=1}^{m} \frac{\partial H}{\partial u_j}\delta u_j \ \mathrm{d}t. \tag{11.25}$$

Now at $t = 0$, all the values of x_i $(i = 1, 2, \ldots, n)$ are specified and so $\delta x_i(0) = 0$. Similarly at $t = T$, $\delta x_i(T) = 0$ for $i = 1, 2, \ldots, q$, since those values of x_i are fixed at $t = T$. For $i = q + 1, \ldots, n$, from the transversality conditions, (11.24), we have $p_i(T) = 0$. Hence

$$p_i(T)\delta x_i(T) = 0$$

for $i = 1, 2, \ldots, q, q+1, \ldots, n$. We now have

$$\delta J^* = \int_0^T \sum_{j=1}^{m} \frac{\partial H}{\partial u_j} \delta u_j \ dt$$

where δu_j is the small variation in the jth component of the control vector $\mathbf{u}$. Since all these variations are independent, and we require $\delta J^* = 0$ for a turning point when the controls are continuous, we conclude that

$$\frac{\partial H}{\partial u_j} = 0 \qquad (j = 1, 2, \ldots, m), \qquad (11.26)$$

that is, equation (10.31) again. But this is only valid when the controls *are continuous* and *not* constrained. In our present case when $\mathbf{u} \in U$, the admissible control region and discontinuities in $\mathbf{u}$ are allowed. The arguments presented above follow through in the same way, except that $(\partial H/\partial u_j)\delta u_j$ must be replaced by

$$H(\mathbf{x}; u_1, u_2, \ldots, u_j + \delta u_j, \ldots, u_m; \mathbf{p}) - H(\mathbf{x}, \mathbf{u}, \mathbf{p}).$$

We thus obtain

$$\delta J^* = \int_0^T \sum_{j=1}^{m} [H(\mathbf{x}; u_1, \ldots, u_j + \delta u_j, \ldots, u_m; \mathbf{p}) - H(\mathbf{x}, \mathbf{u}, \mathbf{p})]\, dt.$$

In order for $\mathbf{u}$ to be a minimising control, we must have $\delta J^* \geqslant 0$ for all admissible controls $\mathbf{u} + \delta\mathbf{u}$. This implies that

$$H(\mathbf{x}; u_1, \ldots, u_j + \delta u_j, \ldots, u_m; \mathbf{p}) \geqslant H(\mathbf{x}, \mathbf{u}, \mathbf{p}) \qquad (11.27)$$

for all admissible δu_j and for $j = 1, \ldots, m$. So we have established that on the optimal control H is minimised with respect to the control variables, $u_1, u_2, \ldots, u_m$. This is known as **Pontryagin's minimum principle**. Of course in the case of continuous unbounded controls, this result immediately gives (11.26), and so the principle holds for all cases.

We first illustrate its use by returning to the simple problem posed in the first section. We required to minimise

$$J = \int_0^T 1 \, dt$$

subject to

$$\dot{x}_1 = x_2$$

$$\dot{x}_2 = u$$

where

$$-\alpha \leqslant u \leqslant \beta$$

and $x_1(0) = x_2(0) = 0$, $x_1(T) = a$, $x_2(T) = 0$. Introducing adjoint variables p_1 and p_2, the Hamiltonian is given by

$$H = 1 + p_1 x_2 + p_2 u.$$

We must minimise H with respect to u, and where $u \in U = [-\alpha, \beta]$, the admissible control region. Since H is linear in u, it clearly attains its minimum on the boundary of the control region, that is, either at $u = -\alpha$ or $u = \beta$. This is illustrated in Fig. 11.3. In fact we can write the optimal control as

$$u = \begin{cases} -\alpha \text{ if } p_2 > 0 \\ \beta \text{ if } p_2 < 0. \end{cases}$$

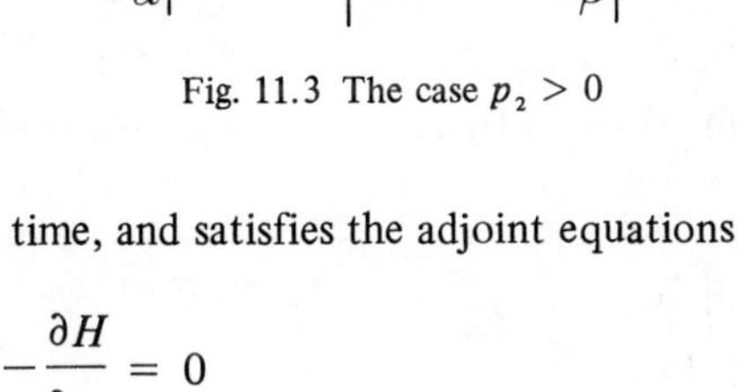

Fig. 11.3 The case $p_2 > 0$

But p_2 will vary in time, and satisfies the adjoint equations,

$$\dot{p}_1 = -\frac{\partial H}{\partial x_1} = 0$$

$$\dot{p}_2 = -\frac{\partial H}{\partial x_2} = -p_1.$$

Thus $p_1 = A$, a constant, and $p_2 = -At + B$, where B is a constant. Since p_2 is a linear function of t, there will *at most* be one switch in the control, since

p_2 has at most one zero, and from the physical situation there must be at least one switch. So we conclude that

(i) the control $u = -\alpha$ or $u = \beta$, that is, bang-bang control;

(ii) there is one and only one switch in the control.

Again, it is clear from the basic problem that initially $u = \beta$, followed by $u = -\alpha$ at the appropriate time (see section 11.1).

11.3 SWITCHING CURVES

In the last section we met the idea of a switch in a control. The time (and position) of switching from one extremum value of the control to another does of course depend on the initial starting point in the phase plane. By considering a specific example we shall show how these switching positions define a *switching curve*.

Suppose a system is described by the state variables x_1, x_2 where

$$\dot{x}_1 = -x_1 + u \tag{11.28}$$

$$\dot{x}_2 = u. \tag{11.29}$$

Here u is the control variable which is subject to the constraints $-1 \leqslant u \leqslant 1$, that is, $|u| \leqslant 1$. Given that at $t = 0$, $x_1 = a$, $x_2 = b$, we wish to find the optimal control which takes the system to $x_1 = x_2 = 0$ in minimum time; that is, we wish to minimise

$$J = \int_0^T 1 \, dt \tag{11.30}$$

whilst moving from (a, b) to $(0, 0)$ in the $x_1 - x_2$ phase plane and subject to (11.28), (11.29) and

$$-1 \leqslant u \leqslant 1. \tag{11.31}$$

Following the procedure outlined in section 11.2, we introduce the adjoint variables p_1 and p_2 and the Hamiltonian

$$H = 1 + p_1(-x_1 + u) + p_2 u$$

$$= u(p_1 + p_2) + 1 - p_1 x_1.$$

Since H is linear in u, and $|u| \leqslant 1$, H is minimised with respect to u by taking

$$u = \begin{cases} +1 & \text{if } p_1 + p_2 < 0 \\ -1 & \text{if } p_1 + p_2 > 0. \end{cases}$$

So the control is bang–bang and the number of switches will depend on the sign changes in $p_1 + p_2$. As the adjoint equations, (11.23), are

$$\dot{p}_1 = -\frac{\partial H}{\partial x_1} = p_1,$$

$$\dot{p}_2 = -\frac{\partial H}{\partial x_2} = 0.$$

Hence $p_1 = Ae^t$ and $p_2 = B$, where A and B are constants. Thus $p_1 + p_2 = Ae^t + B$, and this function has *at most* one sign change.

So we know that from any initial point (a, b), the optimal control will be bang–bang, that is, $u = \pm 1$, with at most one switch in the control. Now suppose $u = k$, when $k = \pm 1$, then the state equations for the system are

$$\dot{x}_1 = -x_1 + k$$

$$\dot{x}_2 = k.$$

We can integrate each equation to give

$$x_1 = Ae^{-t} + k$$

$$x_2 = kt + B$$

where A and B are constants. The $x_1 - x_2$ plane trajectories are found by eliminating t, giving

$$x_2 - B = -k \log|(x_1 - k)/A|.$$

Now if $u = 1$, that is, $k = 1$, then the trajectories are of the form

$$x_2 - B = -\log|(x_1 - 1)/A|$$

that is, $$x_2 = -\log|x_1 - 1| + C \qquad (11.32)$$

where C is a constant. The curves for different values of C are illustrated in Fig. 11.4.

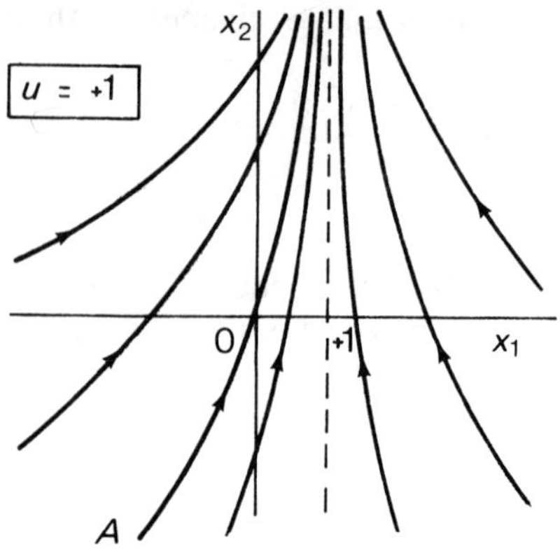

Fig. 11.4 Trajectories for $u = 1$

The directions of the trajectories can be seen by noting that $\dot{x}_2 = 1$, that is, x_2 is an increasing function; and it can also be checked from $\dot{x}_1 = 1 - x_1$, that is, x_1 is increasing for $x_1 < 1$ and decreasing for $x_1 > 1$.

We can follow the same procedure for $u = -1$, that is, $k = -1$, giving

$$x_2 = \log|x_1 + 1| + C. \tag{11.33}$$

C being constant; and these curves are illustrated in Fig. 11.5. The directions indicated on the trajectories are determined from $\dot{x}_2 = -1$ which shows that x_2 is a decreasing function.

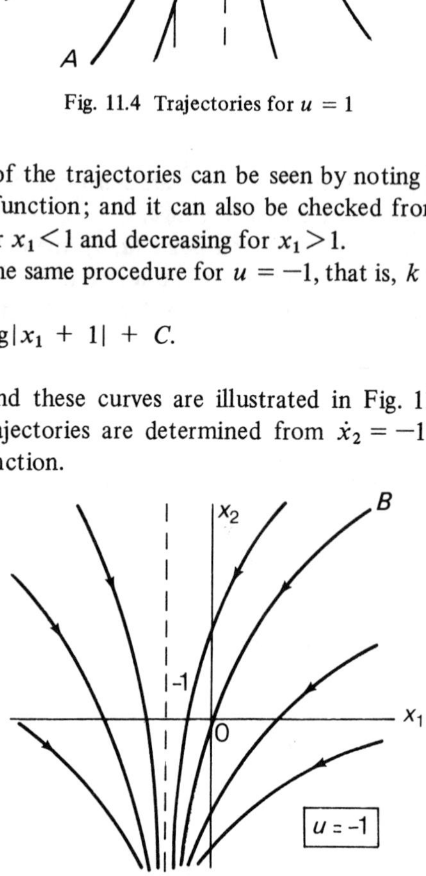

Fig. 11.5 Trajectories for $u = -1$

The basic problem is to reach the origin from an arbitrary initial point. All the possible trajectories are illustrated in Figs 11.4 and 11.5, and we can see that there are only *two* possible paths which reach the origin, namely AO in Fig. 11.4

and BO in Fig. 11.5. So the optimal trajectory must either reach the origin on AO or BO. On AO, we have $u = +1$, the previous part of the trajectory must be on $u = -1$, and since there is at most only one switch, there are *no* other switches. Similarly for BO. So the optimal trajectories are made up of moving either on a trajectory corresponding to $u = -1$ until AO is reached and thence along AO with $u = +1$, or on a trajectory corresponding to $u = +1$ until BO is reached and thence along BO with $u = -1$. Combining the two diagrams we develop the picture shown in Fig. 11.6.

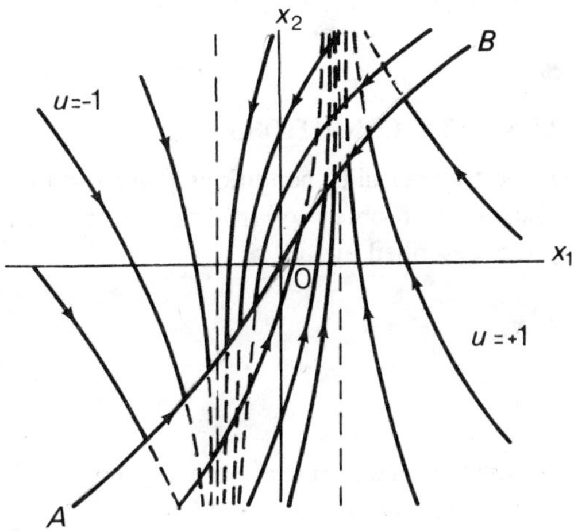

AOB is called the **switching curve**. For initial points below AOB, we take $u = +1$ until the switching curve is reached, followed by $u = -1$ until the origin is reached. Similarly for points above AOB, $u = -1$ until the switching curve is reached, followed by $u = +1$ until the origin is reached. Of course, for points on the switching curve, there is no switch at all; on AO, we take $u = +1$ and on BO, $u = -1$. So we have solved our problem, which was to find the optimal trajectory from an arbitrary starting point.

The vital criterion is whether the initial point is above, on, or below the switching curve; so we finish this section by determining the equation of this curve. For AO, we need the curve (11.32) which passes through the origin; that is,

$$0 = -\log|0 - 1| + C.$$

Thus $C = 0$, and AO has equation

$$x_2 = -\log(1 - x_1), x_1 \leqslant 0.$$

Similarly BO is determined from (11.33) given $C = 0$ and

$$x_2 = \log(1 + x_1), x_1 \geqslant 0.$$

Thus the switching curve has equation

$$x_2 = \begin{cases} \log(1 + x_1) & \text{for} \quad x_1 > 0 \\ -\log(1 - x_1) & \text{for} \quad x_1 < 0. \end{cases} \tag{11.34}$$

11.4 TRANSVERSALITY CONDITIONS

To illustrate how the transversality conditions ($p_i(T) = 0$ if x_i is not specified) are used, we consider the problem of finding the optimum control u where $|u| \leqslant 1$ for the system described by

$$\dot{x}_1 = x_2 \tag{11.35}$$

$$\dot{x}_2 = u \tag{11.36}$$

which takes the system from an arbitrary initial point $x_1(0) = a$, $x_2(0) = b$ to any point on the x_2 axis, that is, $x_1(T) = 0$ but $x_2(T)$ not given, in minimum time. This means minimising

$$J = \int_0^T 1.\mathrm{d}t \tag{11.37}$$

subject to (11.35), (11.36), the above boundary conditions on x_1 and x_2, and such that

$$-1 \leqslant u \leqslant 1. \tag{11.38}$$

Following the usual procedure, we form the Hamiltonian

$$H = 1 + p_1 x_2 + p_2 u \tag{11.39}$$

where p_1 and p_2 are the adjoint variables associated with x_1 and x_2. We minimise H with respect to u, where $-1 \leqslant u \leqslant 1$, which gives

$$u = \begin{cases} +1 & \text{if} \quad p_2 < 0 \\ -1 & \text{if} \quad p_2 > 0. \end{cases}$$

The adjoint variables satisfy

$$\dot{p}_1 = -\frac{\partial H}{\partial x_1} = 0$$

$$\dot{p}_2 = -\frac{\partial H}{\partial x_2} = -p_1$$

On integration, we obtain $p_1 = A$ and $p_2 = -At + B$. Since $x_2(T)$ is not specified the transversality condition becomes $p_2(T) = 0$. Hence $0 = -AT + B$ and $p_2 = A(T - t)$. For $0 < t < T$, there is no change in the sign of p_2, and hence no switch in u. Thus either $u = +1$, or $u = -1$, but with no switches.

Now for $u = k$ $(= \pm 1)$, we have from (11.35) and (11.36),

$$x_2 = kt + A$$

and $$x_1 = (kt + A)^2/2k + B,$$

that is, $$x_2^2 = 2k(x_1 - B).$$

When $u = 1$, we have

$$x_2^2 = 2(x_1 - B) \tag{11.40}$$

and when $u = -1$,

$$x_2^2 = -2(x_1 - B). \tag{11.41}$$

These trajectories are illustrated in Fig. 11.7, the direction of the arrows being determined from $\dot{x}_2 = u$.

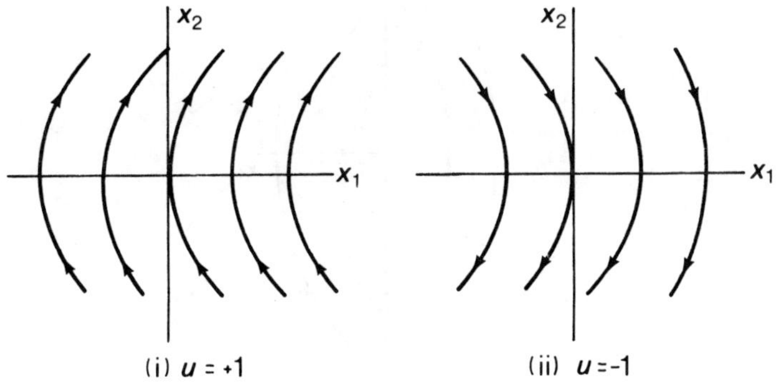

Fig. 11.7 Possible Optimal Trajectories

We first consider initial points (a, b) for which $a > 0$. For points above the curve OA, there is only one trajectory, $u = -1$ which reaches the x_2-axis, and this must be the optimal curve. For points below the curve OA, there are two possible curves, as shown in Fig. 11.8.

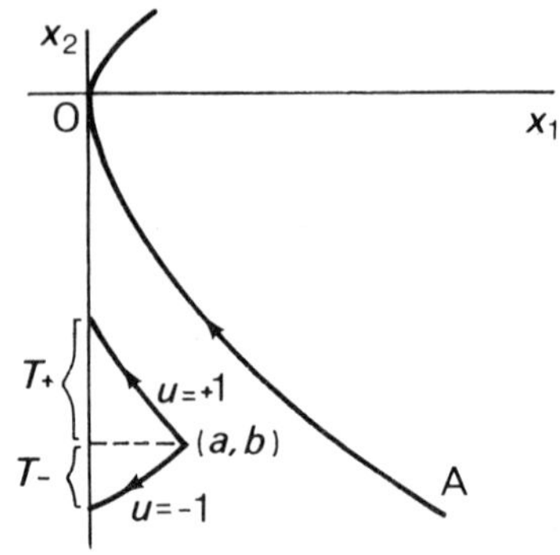

Fig. 11.8 Initial Point below OA

We note though, from (11.36), that $\dot{x}_2 = \pm 1$, and integrating between 0 and T gives

$$x_2(T) - x_2(0) = \pm T$$

that is, $$T = |x_2(T) - x_2(0)|. \tag{11.42}$$

Hence the modulus of the difference in final and initial values of x_2 gives the time taken. This is shown in the diagram as T_+ for $u = +1$ and T_- for $u = -1$.

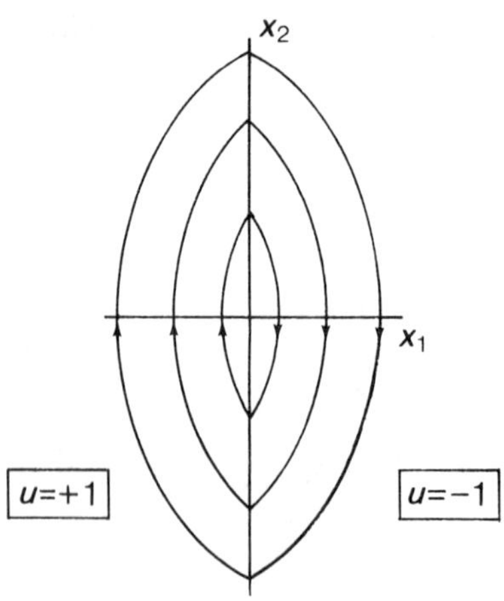

Fig. 11.9 Optimal Trajectories to reach x_2-axis in minimum time

The diagram indicates that $T_- < T_+$ (and this can be checked analytically) so that even for points $a > 0$ under OA, we still take $u = -1$ for the optimal trajectory. Hence for all points initially to the right of the x_2-axis the optimal control is $u = -1$, and similarly for all points to the left of the x_2-axis, we take $u = +1$. The complete set of optimal trajectories is illustrated in Fig. 11.9.

11.5 EXTENSION TO THE BOLTZA PROBLEM

There are many generalisations and extensions which can be made to the basic theory outlined in section 11.2. We refer interested readers to Kirk [31], in which many of the possible extensions are developed. In this text we will just consider the extension to the case of Boltza problems, in which we wish to minimise

$$J = \phi(\mathbf{x}(T)) + \int_0^T f_0(\mathbf{x}, \mathbf{u})\mathrm{d}t \tag{11.43}$$

subject to the differential equation constraints

$$\dot{x}_i = f_i(\mathbf{x}, \mathbf{u}) \qquad (i = 1, 2, \ldots, n) \tag{11.44}$$

and initial boundary conditions $\mathbf{x} = \mathbf{x}_0$, but no conditions on $\mathbf{x}$ at the final time T.

In the usual way we form the augmented functional

$$\begin{aligned} J^* &= \phi(\mathbf{x}(T)) + \int_0^T [f_0 + \sum_{i=1}^n p_i(f_i - \dot{x}_i)]\mathrm{d}t \\ &= \phi(\mathbf{x}(T)) + \int_0^T (H - \sum_{i=1}^n p_i\dot{x}_i)\mathrm{d}t, \end{aligned}$$

where, as usual, we define the Hamiltonian as

$$H = f_0 + \sum_{i=1}^n p_i f_i. \tag{11.45}$$

If, as in section 11.2, we consider small changes in the control $\mathbf{u}$ which yield small changes in the state variables, say δx_i, and

$$\delta J^* = \sum_{i=1}^n \frac{\partial \phi}{\partial x_i}\delta x_i(T) - \sum_{i=1}^n p_i \delta x_i\Big|_0^T + \int_0^T \left(\sum_{j=1}^m \frac{\partial H}{\partial u_j}\delta u_j\right)\mathrm{d}t,$$

using (11.25) and assuming $\mathbf{u}$ is continuous and unbounded. Since $\mathbf{x}(0)$ is specified $\delta x_i(0) = 0$ $(i = 1, 2, \ldots, n)$ and we have

$$\delta J^* = \sum_{i=1}^{n} \left(\frac{\partial \phi}{\partial x_i} - p_i \right) \delta x_i \Big|_{t=T} + \int_0^T \left(\sum_{j=1}^{m} \frac{\partial H}{\partial u_j} \delta u_j \right) \mathrm{d}t.$$

As usual the adjoint equations hold, and the integral term is taken care of by the minimum principle, that is, H is minimised with respect to the control variables. The only difference is the transversality condition which gives

$$\frac{\partial \phi}{\partial x_i} = p_i \quad \text{at} \quad t = T \tag{11.46}$$

for any value of i where $x_i(t)$ is not given.

Example

Minimise

$$\tfrac{1}{2}x^2(T) + \tfrac{1}{2}\int_0^T u^2\,\mathrm{d}t$$

subject to

$$\dot{x} = -x + u$$

and $$x(0) = x_0.$$

Solution

We form the Hamiltonian

$$H = \tfrac{1}{2}u^2 + p(-x + u)$$

where p is the adjoint variable. Since u has no constraint, Pontryagin's principle becomes

$$\frac{\partial H}{\partial u} = 0, \quad \text{that is,} \quad u + p = 0.$$

Hence $p = -u$. The adjoint equation is

$$\dot{p} = -\frac{\partial H}{\partial x} = p.$$

from which we obtain $p = Ae^t$. This gives $u = -Ae^t$ and

$$\dot{x} = -x - Ae^t,$$

which has solution

$$x = Be^{-t} - \frac{A}{2}e^{t}.$$

The two constants A and B are determined from the initial condition on x and the transversality condition at T. Now $x(0) = x_0$ gives

$$x_0 = B - \frac{A}{2},$$

and the transversality condition, (11.46), at $t = T$, gives

$$x(T) = p(T) = Ae^{T},$$

that is, $$Be^{-T} - \frac{A}{2}e^{T} = Ae^{T}.$$

This gives $\frac{3}{2}Ae^{T} = Be^{-T}$ and $x_0 = \frac{3}{2}Ae^{2T} - \frac{A}{2}$. Thus $A = 2x_0/(3e^{2T} - 1)$ and the optimal control is

$$u = -2x_0e^{t}/(3e^{2T} - 1).$$

Note: An alternative method for dealing with problems of the type which does not require the use of Boltza transversality condition (11.46) is to write

$$J = \tfrac{1}{2}x^2(T) + \int_0^T u^2\,dt$$

as

$$\begin{aligned} J &= \int_0^T \frac{d}{dt}(\tfrac{1}{2}x^2)\,dt + \tfrac{1}{2}\int_0^T u^2\,dt \\ &= \int_0^T (x\dot{x} + \tfrac{1}{2}u^2)\,dt \\ &= \int_0^T [x(-x + u) + \tfrac{1}{2}u^2]\,dt \\ &= \int_0^T (\tfrac{1}{2}u^2 - x^2 + xu)\,dt. \end{aligned}$$

We now wish to minimise J subject to $\dot{x} = -x + u$, $x(0) = x_0$ and $x(T)$ not given. We form the Hamiltonian

$$H = \tfrac{1}{2}u^2 - x^2 + xu + p(-x + u)$$

and $\partial H/\partial u = 0$ gives

$$u + x + p = 0.$$

Here the adjoint variable satisfies

$$\dot{p} = -\frac{\partial H}{\partial x} = 2x - u + p.$$

Eliminating u,

$$\dot{p} = 2p + 3x$$

and we also have the state equation

$$\dot{x} = -2x - p.$$

These coupled differential equations have solution

$$x = a\mathrm{e}^t + b\mathrm{e}^{-t}$$

$$p = -3a\mathrm{e}^t - b\mathrm{e}^{-t}$$

(a, b constants). The initial condition $x(0) = x_0$ and transversality condition, now just $p(T) = 0$, determine a and b, which are

$$a = x_0/(1 - 3\mathrm{e}^{2T}), \quad b = -3\mathrm{e}^{2T}x_0/(1 - 3\mathrm{e}^{2T})$$

and so the optimal control is

$$u = -(x + p) = 2a\mathrm{e}^t = -2x_0\mathrm{e}^t/(3\mathrm{e}^{2T} - 1),$$

which agrees with the earlier result.

PROBLEMS

(1) The system of equations

$$\dot{x}_1 = -2x_1 + u$$
$$\dot{x}_2 = u$$

is controlled by the variable u. If u is bounded so that $|u| \leqslant 1$, find the switching curve for minimum time trajectories of the system from an arbitrary point (ξ_1, ξ_2) at $t = 0$ to the origin.

(2) A dynamical system is described by the state variables $x_1(t)_1 x_2(t)$ which are subject to the equations

$$\frac{dx_1}{dt} = -x_1 - u$$
$$\frac{dx_2}{dt} = -2x_2 - 2u$$

where $u = u(t)$ is a control variable such that $|u| \leqslant 1$. Use the maximum principle to find the admissible control which will force the system from an arbitrary state (x_1^0, x_2^0) to the origin $(0,0)$ in minimum time. Show that there is at most one switch for u, and find the switching curve. Indicate by a sketch the optimal trajectory from any arbitrary initial point.

(3) Determine the optimal control law for transferring the system

$$\dot{x}_1 = x_2$$
$$\dot{x}_2 = u$$

from an arbitrary initial state to $x_1 = x_2 = 2$ in minimum time with $|u| \leqslant 1$.

(4) It is required to find the optimal trajectory which minimises the transfer time for driving the state vector $\mathbf{x} = (x_1, x_2)$ from an initial fixed point $\mathbf{x}_0$ to a terminal point anywhere on the x_2-axis, where

$$\frac{dx_1}{dt} = -x_1 + u$$
$$\frac{dx_2}{dt} = u$$

and the one-dimensional control vector u belongs to the control region $|u| \leqslant 1$. Use the maximum principle and transversality condition at the terminal point to show that $u = 1$ or $u = -1$ on the optimal trajectory with *no* switches. Deduce that optimal trajectories are of the form

$$x_2 - C = \mp \log|x_1 \mp 1|$$

where C is a constant and the minus sign is taken for $u = 1$ and the plus sign for $u = -1$. Hence sketch the optimal trajectory from any arbitrary initial point.

(5) Consider the first order system

$$\dot{x} = 2x + u, \quad |u| \leqslant 1.$$

It is desired to transfer the system from an arbitrary initial state to the origin in minimum time; however there are some states that *cannot* be transferred to the origin no matter how much time is allowed. Find the initial states for which there is no time-optimal control to the origin.

CHAPTER 12

Applications of optimal control

12.1 INTRODUCTION

In the previous two chapters we have developed the basic optimal control theory, and we are now in a position to apply it to the problems outlined in Chapter 8. We should stress, at this stage, that the problems posed in Chapter 8 are necessarily of a simple type. You might well feel that some of the important aspects of the problem have been badly modelled or even neglected, and you might well be right. In presenting these models we are only aiming to show the varied 'use' to which optimal control has been applied in the last two decades. We do not expect you to be experts in modelling situations in economics, biology, engineering, etc., and we have restricted our attention to models which can be readily understood. Even so, we will find that it is often difficult to obtain solutions to these simple models, particularly analytical ones. In order to cope with more complex models, there has been much research and development of numerical techniques for solving optimal control problems. Interested readers should consult Sage [1967] or Lootsma [1972] for details.

We hope that you will enjoy seeing the theory being applied to problems, motivated from practical situations. References for further details are given at the end of each section.

12.2 ECONOMIC GROWTH

The problem, as outlined in section 8.2.1, is to choose the consumption path $c(t)$, $0 \leqslant t \leqslant T$ which maximises the welfare integral

$$W = \int_0^T e^{-\delta t} u(c) \mathrm{d}t$$

whilst satisfying the growth equation

$$\frac{\mathrm{d}k}{\mathrm{d}t} = f(k) - (\lambda + \mu)k - c, \quad k(0) = k_0; \tag{12.1}$$

k being the capital, $f(k)$ the production function, λ, μ positive constants and δ the discount factor, (c is the control variable and k the state variable).

We first note that we can convert the maximisation to a minimisation of

$$-W = \int_0^T -e^{-\delta t} U(c) dt \tag{12.2}$$

and so form the Hamiltonian

$$H = -e^{-\delta t} U(c) + p[f(k) - (\lambda + \mu)k - c] \tag{12.3}$$

p being the adjoint variable. Using the minimum principle, and assuming no bounds on c, we have

$$\frac{\partial H}{\partial c} = 0, \quad \text{that is,} \quad -e^{-\delta t} U' - p = 0 \tag{12.4}$$

where $U' = dU/dc$. The adjoint equation is

$$\dot{p} = -\partial H/\partial k = -p(f' - (\lambda + \mu)) \tag{12.5}$$

where $f' = df/dk$. Using p from (12.5) in (12.4) gives

$$\frac{d}{dt}(e^{-\delta t} U') = -e^{-\delta t} U'(f' - (\lambda + \mu))$$

that is, $$U'' \dot{c} e^{-\delta t} - \delta e^{-\delta t} U' = -e^{-\delta t} U'(f' - (\lambda + \mu))$$

[since $\frac{d}{dt}(U') = \frac{d}{dc}(U') \dot{c} = U'' \dot{c}$]. Hence

$$\dot{c} = \frac{U'}{U''}(\lambda + \mu + \delta - f').$$

An important quantity in economic terms is

$$\eta = -cU''/U' \tag{12.7}$$

which is called the **elasticity of marginal utility**. It is often assumed constant; for if η is constant, we can integrate (12.7), with respect to c, to give

$$\eta \log c = -\log U' + \log A$$

that is, $U' = Ac^{-\eta}$,

and integrating again,

$$U(c) = \frac{Ac^{-\eta+1}}{(1-\eta)} + B.$$

Assuming $U(0) = 0$, we obtain $B = 0$, so that

$$U(c) = Ec^{1-\eta} \qquad (E \text{ constant}) \qquad (12.8)$$

which, if $0 < \eta < 1$, has the required form for the utility function as illustrated in Fig. 8.2.

For simplicity we assume therefore that η is a constant, and so (12.6) becomes

$$\dot{c} = \eta[f'(k) - \beta]c \qquad (12.9)$$

where $\beta = \lambda + \mu + \delta$. So the equations governing the optimal path are (12.9) and (12.1) which are two coupled non-linear first order differential equations.

In general they cannot be solved analytically. There is, though, one special case, namely $f(k) = bk$, b constant, which can be readily solved although the linear assumption for the production function is not very realistic. If $f(k) = bk$, $f'(k) = b$ and (12.8) becomes

$$\dot{c} = \eta(b-\beta)c$$

and solving, we have

$$c = c_0 e^{\eta(b-\beta)t} = c_0 e^{\alpha_1 t}, \; \alpha_1 = \eta(b-\beta). \qquad (12.10)$$

Returning to (12.1) we obtain

$$\frac{dk}{dt} = (b-\lambda-\mu)k - c_0 e^{\alpha_1 t}$$

$$= \alpha_2 k - c_0 e^{\alpha_1 t}$$

when $\alpha_2 = b - \lambda - \mu$. This has solution

$$k = Ae^{\alpha_2 t} - c_0 e^{\alpha_1 t}/(\alpha_2 - \alpha_1).$$

The constants A and c_0 are determined from the initial condition $k(0) = k_0$ and, say, a final condition $k(T) = k_T$.

The important quantity is the control, in this case, consumption c, given by (12.10). If $b > \beta$, consumption is increasing exponentially. In practical cases, there is often an upper limit to the value of c, say $\bar{c}$, so that when reached by the optimal path (12.10), the value of c will remain constant at $\bar{c}$ for the rest of the time. Of course the initial value of c, c_0, must be chosen so that the conditions on capital are met. A typical optimal path for c is illustrated in Fig. 12.1.

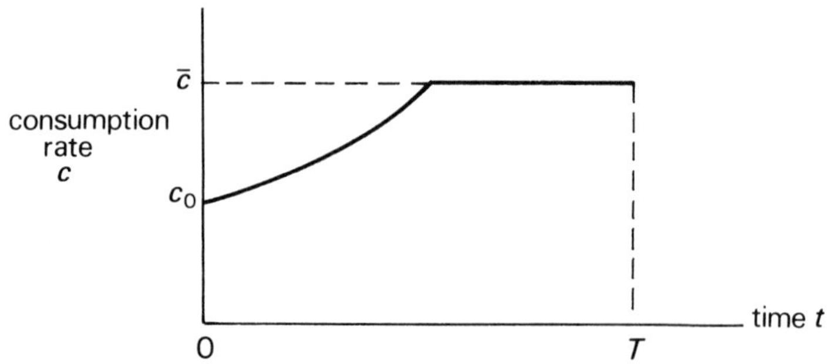

Fig. 12.1 Typical Optimal Path

In this solution, we have both the possibility of a combination of an interior solution, (that is, not at the boundary of the admissible control region) and extremum solution for c.

This problem and extensions has been dealt with extensively by Dasgupta and Heal (1974) and accounts can be found in Hadley and Kemp (1971) and Intrilligator (1971).

12.3 RESOURCE DEPLETION

In section 8.2.2 we presented a simple model which led to the problem of maximising

$$W = \int_0^T e^{-\delta t} U(c) \mathrm{d}t \tag{12.11}$$

subject to

$$\frac{\mathrm{d}K}{\mathrm{d}t} = F(K, R) - C \tag{12.12}$$

$$\frac{\mathrm{d}S}{\mathrm{d}t} = -R \tag{12.13}$$

where the capital, K, and resource stock, S, are the state variables, the consumption rate, C, and depletion rate, R, are the controls, and $F = F(K, R)$ is the production function.

In the usual way we form the Hamiltonian, with adjoint variables P_1 and P_2

$$H = -e^{-\delta t}U(c) + P_1(F - C) + P_2(-R) \tag{12.14}$$

and assuming there are no bounds on the controls C and R, from the minimum principle, we minimise H with respect to C and R. We obtain

$$\frac{\partial H}{\partial C} = 0, \quad \text{that is,} \quad -e^{-\delta t}U' - P_1 = 0 \quad (U' = \mathrm{d}U/\mathrm{d}C) \tag{12.15}$$

$$\frac{\partial H}{\partial R} = 0, \quad \text{that is,} \quad P_1\frac{\partial F}{\partial R} - P_2 = 0. \tag{12.16}$$

We also have the adjoint equations,

$$\dot{P}_1 = -\frac{\partial H}{\partial K} = -P_1\frac{\partial F}{\partial K} \tag{12.17}$$

$$\dot{P}_2 = -\frac{\partial H}{\partial S} = 0. \tag{12.18}$$

Now using P_2 from (12.18) gives

$$\frac{\mathrm{d}}{\mathrm{d}t}\, P_1\frac{\partial F}{\partial R} = 0$$

that is, $$\dot{p}_1\frac{\partial F}{\partial R} + P_1\frac{\mathrm{d}}{\mathrm{d}t}\left(\frac{\partial F}{\partial R}\right) = 0;$$

and using (12.17),

$$-P_1\frac{\partial F}{\partial R}\frac{\partial F}{\partial K} + P_1\frac{\mathrm{d}}{\mathrm{d}t}\left(\frac{\partial F}{\partial R}\right) = 0$$

that is, $$\frac{\mathrm{d}}{\mathrm{d}t}\left(\frac{\partial F}{\partial R}\right) = \frac{\partial F}{\partial K}\frac{\partial F}{\partial R}. \tag{12.19}$$

Also, using P_1 from (12.11) in (12.17) yields

$$\frac{\mathrm{d}}{\mathrm{d}t}(e^{-\delta t}U') = e^{-\delta t}U'\frac{\partial F}{\partial K}$$

that is, $$\delta e^{-\delta t}U' - e^{-\delta t}U''\dot{C} = e^{-\delta t}U'\frac{\partial F}{\partial K}.$$

As in section 12.2 we define $\eta = -CU''/U'$, so that

$$\dot{C} = C\left(\frac{\partial F}{\partial K} - \delta\right)/\eta \tag{12.20}$$

which is analogous to equation (12.8) in the last section. Equations (12.19) and (12.20) are the basic equations governing the optimal path, together with the state equations. The equations are non-linear, and coupled, and to proceed further we must specify the production function and (as in section 12.2) assume η to be constant. For the production function, we consider the function

$$F = K^{\alpha}R^{1-\alpha} \tag{12.21}$$

where α is a positive constant such that $0 < \alpha < 1$. This means that

$$\frac{\partial F}{\partial K} = K^{\alpha-1}R^{1-\alpha} \qquad \frac{\partial F}{\partial R} = (1-\alpha)\,K^{\alpha}R^{-\alpha}$$

$$= \alpha x^{\alpha-1} \qquad = (1-\alpha)x^{\alpha}$$

where we have defined a new variable $x = K/R$, the capital/resource depletion rate ratio. Hence

$$\frac{\mathrm{d}}{\mathrm{d}t}\left(\frac{\partial F}{\partial R}\right) = (1-\alpha)\alpha x^{\alpha-1}\,\dot{x},$$

and (12.19) reduces to

$$\dot{x} = x^{\alpha}. \tag{12.22}$$

This can be integrated to give

$$x = [(1-\alpha)\,t + x_0^{1-\alpha}]^{1/(1-\alpha)} \tag{12.23}$$

where x_0 is a constant. We can now deal with (12.20) which gives

$$\frac{\dot{C}}{C} = \frac{1}{\eta}(\alpha x^{\alpha-1} - \delta) = \frac{\alpha}{\eta}\frac{\dot{x}}{x} - \frac{\delta}{\eta}$$

and so

$$\log C = \frac{\alpha}{\eta}\log x - \frac{\delta t}{\eta} + \log A,$$

that is, $$C = Ax^{\alpha/\eta}\,\mathrm{e}^{-\delta t/\eta}$$

and if at $t = 0$, $C = C_0$, we have

$$C = C_0 (x/x_0)^{\alpha/\eta} e^{-\delta t/\eta}$$

that is,
$$C = C_0 e^{-\delta t/\eta} [1 + (1-\alpha)t/x_0^{1-\alpha}]^{\alpha/\eta(1-\alpha)} \tag{12.24}$$

We can also determine the other control, the depletion rate R, from

$$\frac{\mathrm{d}}{\mathrm{d}t}\left(\frac{K}{R}\right) = \frac{\dot{K}}{R} - K\frac{\dot{R}}{R^2} = \frac{1}{R}(F - C) - x\frac{\dot{R}}{R}, \quad \text{using (12.12)}$$

that is,
$$\dot{x} = x^\alpha - \frac{C}{R} - x\frac{\dot{R}}{R},$$

using (12.21) and $x = K/R$. But from (12.22), $\dot{x} = x^\alpha$, so

$$\dot{R} = -C/x$$

that is,
$$\dot{R} = C_0 e^{-\delta t/\eta} [1 + (1-\alpha)t/x_0^{1-\alpha}]^{(\alpha-\eta)/\eta(1-\alpha)}/x_0.$$

This can only be integrated numerically, except in the rather special case when $\alpha = \eta$, when

$$\dot{R} = -C_0 e^{-\delta t/\eta}/x.$$

On integration this becomes

$$R = R_0 - \frac{\eta C_0}{\delta x_0}(e^{-\delta t/\eta} - 1) \tag{12.25}$$

where the constant of integration has been chosen so that $R = R_0$, at $t = 0$. We can also find K from $x = K/R$. Since we know both x and R, finally we can determine S from (12.25) with $\mathrm{d}S/\mathrm{d}t = -R$.

We still have to determine the appropriate values for the constants introduced, using, say, boundary conditions

$$K(0) = K_0, \quad K(T) = K_T$$

$$S(0) = S_0, \quad S(T) = 0.$$

A typical solution is illustrated in Fig. 12.2. These problems have received much attention in the last decade, for example see Dasgupta and Heal (1974), Pearce (1975), and Davison (1978).

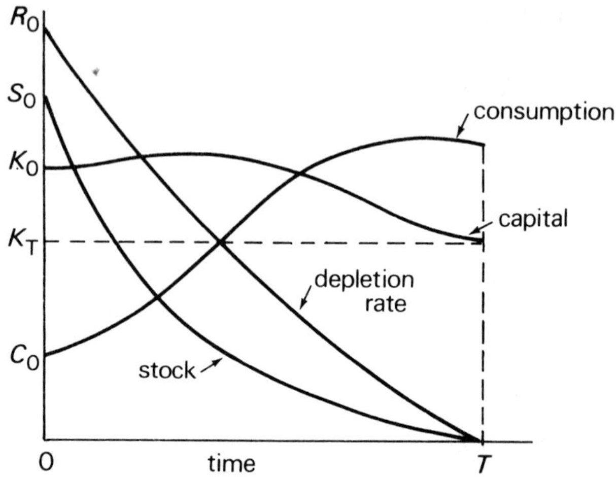

Fig. 12.2 Typical Optimal Trajectories

12.4 EXPLOITED POPULATIONS

In section 8.2.3, we formulated the problem as maximise the 'present value' of the economic rent,

$$P.V. = \int_0^\infty e^{-\delta t}(pEx - cE)\,dt \tag{12.26}$$

where the population level obeys

$$\frac{dx}{dt} = F(x) - h(t) \tag{12.27}$$

and the harvesting rate, h, is related to the fishing effort, E, by

$$h = Ex. \tag{12.28}$$

The effort, E, is an appropriate control, so we write (12.27) as

$$\frac{dx}{dt} = F(x) - Ex. \tag{12.29}$$

Initially we have a given population level, say x_0. To find the optimum fishing effort, we form the Hamiltonian

$$H = -e^{-\delta t}(pEx - cE) + \lambda(F(x) - Ex) \tag{12.30}$$

where λ is the adjoint variable, and we minimise H with respect to E, that is

$$\frac{\partial H}{\partial E} = 0 \quad \text{yields} \quad -\mathrm{e}^{-\delta t}(px - c) - \lambda x = 0. \tag{12.31}$$

Now λ satisfies

$$\dot{\lambda} = -\frac{\partial H}{\partial x} = -\mathrm{e}^{-\delta t}pE - \lambda(F' - E)$$

where $F' = \mathrm{d}F/\mathrm{d}x$. From (12.31)

$$\lambda = \mathrm{e}^{-\delta t}(c/x - p)$$

and substituting into the above equation gives

$$-\delta\mathrm{e}^{-\delta t}(c/x - p) + \mathrm{e}^{-\delta t}(-c/x^2)\dot{x}$$

$$= -\mathrm{e}^{-\delta t}pE - \mathrm{e}^{-\delta t}(c/x - p)(F' - E).$$

Eliminating $\dot{x}$ from (12.30) reduces the equation to

$$F'(x) + \frac{cF(x)}{x(px - c)} = \delta. \tag{12.32}$$

Equation (12.32) is an implicit equation for the optimum population size, say x^*, which will be an *equilibrium* solution. To see if (12.32) does in fact have a solution, we must specify the form of the growth function $F(x)$. One well used model is the Verhulst model

$$F(x) = rx(1 - x/k) \tag{12.33}$$

where r is a positive constant and k is the saturation level of population. In this case, (12.32) becomes

$$r - \frac{2rx}{k} + \frac{c(rx - rx^2/k)}{x(px - c)} = \delta$$

which yields a quadratic equation with one positive real root, and so one appropriate solution x^* for the optimun population.

What happens, though, if $x(0) \neq x^*$? We must return to (12.30) and note that H is linear in E, and in practice there must be some bounds to the fishing effort, E, that is,

$$0 \leqslant E \leqslant \bar{E}. \tag{12.34}$$

So if $x(0) > x^*$ it is clear the $E = \bar{E}$, the maximum level until x^* is reached. The population must now remain at x^*, E being given from (12.29) and $E = F(x^*)/x^*$. Similarly if $x(0) < x^*$ we must take $E = 0$ until the population level increases to x^*, whence $E = F(x^*)/x^*$ and the population remains at x^*. We can summarise the optimal control as

$$E = \begin{cases} \bar{E} & \text{whenever } x > x^* \\ F(x^*)/x^* & \text{whenever } x = x^* \\ 0 & \text{whenever } x < x^* \end{cases} \tag{12.35}$$

and the population level is illustrated in Fig. 12.3. An excellent introduction to this subject is given by Clarke (1976).

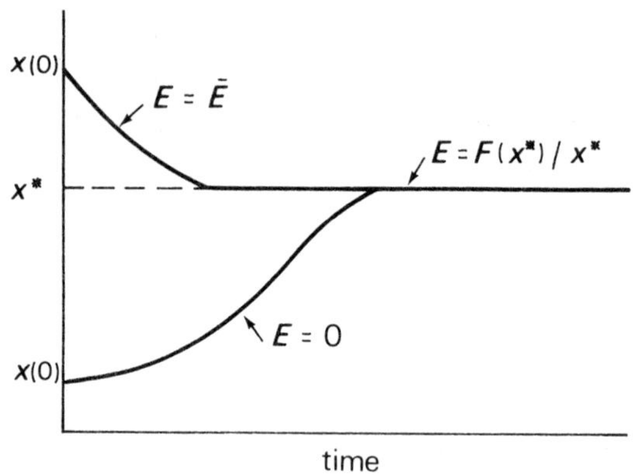

Fig. 12.3 Population Level

12.5 ADVERTISING POLICIES

In section 8.2.4 we formulated the problem as that of maximising the total sales

$$S = \int_0^T s(t)\mathrm{d}t \tag{12.36}$$

subject to the constraints

$$\frac{\mathrm{d}s}{\mathrm{d}t} = rA(1 - s/M) - \lambda s \tag{12.37}$$

where the advertising rate satisfies

$$B = \int_0^T A(t)\,\mathrm{d}t, \tag{12.38}$$

B being the total advertising budget. We can deal with the differential constraint (12.31), and the integral constraint can be converted to a differential constraint by introducing the total amount spent on advertising up to time t by

$$b(t) = \int_0^t A(\tau)\,\mathrm{d}\tau,$$

then

$$\frac{\mathrm{d}b}{\mathrm{d}t} = A \tag{12.39}$$

and we have the end point conditions $b(0) = 0$ $b(T) = B$, together with the initial sales condition $s(0) = s_0$. The optimal control problem is now to maximise the functional (12.36) subject to the constraints (12.37) and (12.39), and the end-point conditions. s and b are the state variables and A the control. As usual we form the Hamiltonian

$$H = -s + p_1(rA(1 - s/M) - \lambda s) + p_2 A \tag{12.40}$$

and this we minimise with respect to the control, A.

As in the previous example, we see that H is linear in the control A, and so will only achieve a minimum with respect to A if there are bounds on A. So suppose

$$0 \leqslant A \leqslant \bar{A}$$

then the optimum control is given by

$$A = \begin{cases} 0 & \text{if} \quad (p_1 r + p_2) > 0 \\ \bar{A} & \text{if} \quad (p_1 r + p_2) < 0. \end{cases}$$

Now the adjoint equations are

$$\dot{p}_1 = -\frac{\partial H}{\partial s} = 1 + p_1(rA/M + \lambda)$$

$$\dot{p}_2 = -\frac{\partial H}{\partial b} = 0.$$

Unfortunately, although we can readily solve for p_2, the equation for p_1 involves the unknown control A, and so cannot be immediately solved. So our techniques do not give an answer to this problem, and more advanced theory concerning the number of switches is needed. Another possibility is that $p_1 r + p_2 = 0$ for a non-zero interval of time (this is called **singular control**).

Assuming, though, that there is just one switch between the bounds of the control, there are two possibilities, namely

$$\text{(i) } A = \begin{cases} \bar{A} & \text{for } 0 < t < t \\ 0 & \text{for } t_1 < t < T \end{cases} \qquad \text{(ii) } A = \begin{cases} 0 & \text{for } 0 < t < t_2 \\ \bar{A} & \text{for } t_2 < t < T. \end{cases}$$

The switchover time, t_1 or t_2, is determined from the condition on the total amount spent on advertising. Fig. 12.4 illustrates the two possibilities.

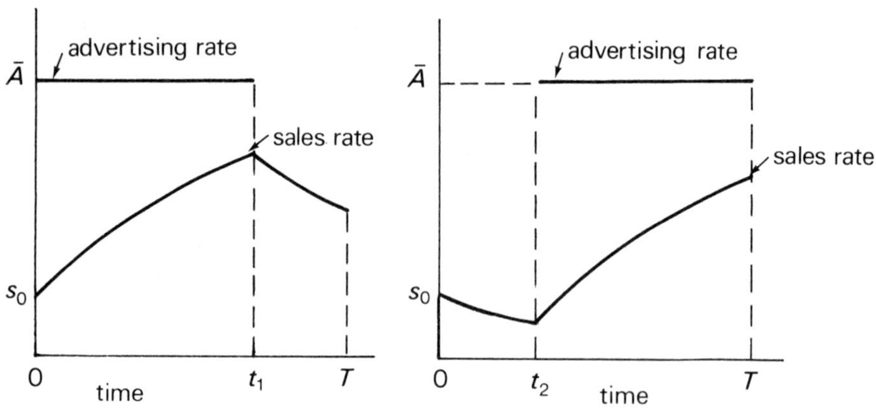

Fig. 12.4 Possible advertising strategies and corresponding response in sales

Now the total sales are given by the area under the sales rate curve, and since the area in the first case for $(0, t_1)$ is always greater than the area in the second case for (t_2, T), and similarly for (t_1, T) and $(0, t_2)$ it is clear that the optimum strategy is given by (i).

Unfortunately we cannot answer the original question posed in section 8.2.4, but by now we might well be questioning the whole formulation of the model. For example, what happens after time T; why not look at the net revenue produced rather than the total sales; shouldn't future sales be time discounted and so on! Readers interested in these problems can refer to Sethi (1973), although many of the difficulties have yet to be resolved.

12.6 ROCKET TRAJECTORIES

The problem described in section 8.2.5 involves minimising the fuel

$$J = \int_0^T \beta \mathrm{d}t \tag{12.41}$$

subject to the differential constraints

$$\frac{\mathrm{d}v_1}{\mathrm{d}t} = \frac{c\beta}{m}\cos\phi \tag{12.42}$$

$$\frac{\mathrm{d}v_2}{\mathrm{d}t} = \frac{c\beta}{m}\sin\phi - g \tag{12.43}$$

and also

$$\frac{\mathrm{d}x}{\mathrm{d}t} = v_1 \tag{12.44}$$

$$\frac{\mathrm{d}y}{\mathrm{d}t} = v_2. \tag{12.45}$$

The last two equations must also be included since there are end conditions on x, y, v_1 and v_2.

$$t = 0 : x = 0,\ y = 0,\ v_1 = 0,\ v_2 = 0$$

$$t = T : x \text{ not specified},\ y = \bar{y},\ v_1 = \bar{v},\ v_2 = 0.$$

Thus we have four state variables, namely x, y, v_1 and v_2 and two controls β, the rate at which the exhaust gases are emitted, and ϕ, the thrust attitude angle. In practice we must have bounds on β, that is,

$$0 \leqslant \beta \leqslant \bar{\beta}; \tag{12.46}$$

so that $\beta = 0$ corresponds to the rocket motor being shut down and $\beta = \bar{\beta}$ corresponds to the motor at full power.

The Hamiltonian for the system is

$$H = \beta + p_1 v_1 + p_2 v_2 + p_3 \frac{c\beta}{m}\cos\phi + p_4\left(\frac{c\beta}{m}\sin\phi - g\right) \tag{12.47}$$

where p_1, p_2, p_3 and p_4 are the adjoint variables associated with x, y, v_1 and v_2 respectively. We first note that H is linear in the control β, so that assuming

$(1+\frac{p_3c\beta}{m}\cos\phi+\frac{p_4c\beta}{m}\sin\phi)\neq 0$, we see that the control β is bang-bang, that is $\beta=0$ or $\beta=\bar{\beta}$. We must clearly start with $\beta=\bar{\beta}$ so that, with one switch in β, we have

$$\beta = \begin{cases} \bar{\beta} & \text{for} \quad 0 \leqslant t < t_1 \\ 0 & \text{for} \quad t_1 < t \leqslant T. \end{cases} \tag{12.48}$$

Now H must also be maximised with respect to the second control, ϕ, that is, $\partial H/\partial\phi = 0$ giving

$$-p_3\frac{c\beta}{m}\sin\phi + p_4\frac{c\beta}{m}\cos\phi = 0$$

which yields

$$\tan\phi = p_4/p_3.$$

The adjoint variables satisfy the equations

$$\dot{p}_1 = -\frac{\partial H}{\partial x} = 0, \quad \text{that is,} \quad p_1 = A, \text{ constant:}$$

$$\dot{p}_2 = -\frac{\partial H}{\partial y} = 0, \quad \text{that is,} \quad p_2 = B, \text{ constant;}$$

$$\dot{p}_3 = -\frac{\partial H}{\partial v_1} = -p_1 = -A, \quad \text{that is,} \quad p_3 = C - At, \; C \text{ constant;}$$

$$\dot{p}_4 = -\frac{\partial H}{\partial v_2} = -p_2 = -B, \quad \text{that is,} \quad p_4 = D - Bt, \; D \text{ constant.}$$

Thus

$$\tan\phi = \frac{D-Bt}{C-At}.$$

Since x is not specified at $t=T$, the transversality condition is $p_1(T)=0$, that is, $A = 0$, and so

$$\tan\phi = a - bt \tag{12.49}$$

where $a=D/C$, $b=B/C$.

The problem, in principle, is now solved. To complete it requires just integration and algebraic manipulation. With ϕ given by (12.49) and β by

(12.48), we can integrate (12.42) to (12.45). This will bring four further constants of integration; together with a, b and the switchover time t_1 we have seven unknown constants.

These are determined from the seven end-point conditions at $t = 0$ and $t = T$. A typical trajectory is shown in Fig. 12.5.

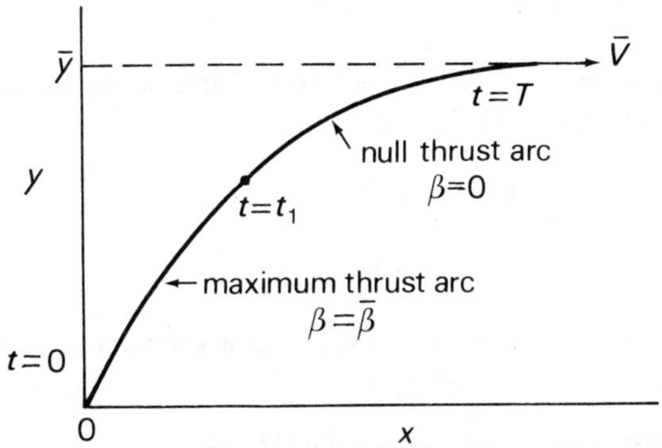

Fig. 12.5 Typical rocket trajectory

12.7 SERVO PROBLEM

The problem here is to minimise

$$T = \int_0^T 1\,\mathrm{d}t \tag{12.50}$$

subject to

$$\frac{\mathrm{d}^2\theta}{\mathrm{d}t^2} + \alpha\frac{\mathrm{d}\theta}{\mathrm{d}t} + \omega^2\theta = u \tag{12.51}$$

with, at $t = 0$, $\theta = \theta_0$, $\mathrm{d}\theta/\mathrm{d}t = \theta_0'$, and at $t = T$, $\theta = \mathrm{d}\theta/\mathrm{d}t = 0$ and $|u| \leqslant 1$. We must first introduce the state variables x_1 and x_2 where $x_1 = \theta$, and $x_2 = \mathrm{d}\theta/\mathrm{d}t$. Thus

$$\dot{x}_1 = x_2$$

$$\dot{x}_2 = ax_2 - \omega^2 x_1 + u. \tag{12.52}$$

As usual, we form the Hamiltonian

$$H = 1 + p_1 y_2 + p_2(-ax_2 - \omega^2 x_1 + u) \tag{12.53}$$

where p_1, p_2 are the adjoint variables satisfying

$$\dot{p}_1 = -\frac{\partial H}{\partial x_1} = \omega^2 p_2 \tag{12.54}$$

$$\dot{p}_2 = -\frac{\partial H}{\partial x_2} = -p_1 + ap_2. \tag{12.55}$$

Since H is linear in u, and $|u| \leqslant 1$, we again immediately see that the control u is bang-bang, that is, $u = \pm 1$; in fact

$$u = \begin{cases} +1 & \text{if } p_2 < 0 \\ -1 & \text{if } p_2 > 0. \end{cases}$$

To ascertain the number of switches, we must solve for p_2. From (12.55), we see that

$$\ddot{p}_2 = -\dot{p}_1 + a\dot{p}_2 = -\omega^2 p_2 + a\dot{p}_2$$

that is,
$$\ddot{p}_2 - a\dot{p}_2 + \omega^2 p_2 = 0. \tag{12.56}$$

We have already met this differential equation (see section 4.2) and the type of solution depends on the sign of $(a^2 - 4\omega^2)$. The parameter a represents the damping in the system, and if this is small, oscillatory solutions for p_2 are obtained. This means that as time increases p_2 will regularly pass through 0 which will correspond to a switch in u. To be more precise, we will just consider the case with *no* damping, that is, $a = 0$. In this special case, (12.56) is

$$\ddot{p}_2 + \omega^2 p_2 = 0$$

and so $p_2 = A \sin \omega t + B \cos \omega t$. A typical solution is illustrated in Fig. 12.6 on a time interval $[0, T]$. We note that $p_2 = 0$ at time $t = \tau$ and then at equal intervals π/ω.

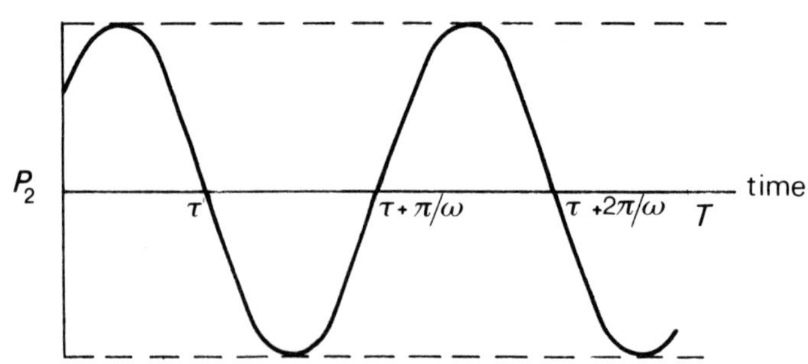

Fig. 12.6 Adjoint Variable p_2

The corresponding optimal control u is illustrated in Fig. 12.7.

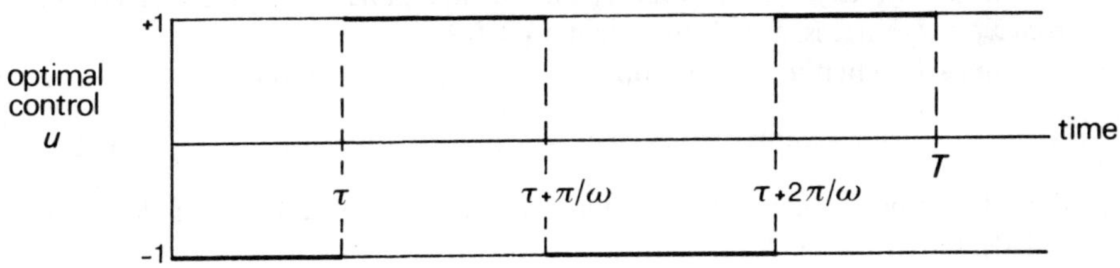

Fig. 12.7 Corresponding Optimal Control

If $u = 1$, the state equations are

$$\dot{x}_1 = y_2$$

$$\dot{x}_2 = -\omega^2 x_1 + 1$$

that is, $\ddot{x}_1 + \omega^2 x_1 = 1.$

This has the solution

$$x_1 = \frac{1}{\omega^2} + A \sin(\omega x + \alpha)$$

where A and α are the constants of integration; this gives

$$x_2 = A\omega \cos(\omega x + \alpha)$$

so that

$$(x_1 - 1/\omega^2)^2 + (x_2/\omega)^2 = A^2. \tag{12.57}$$

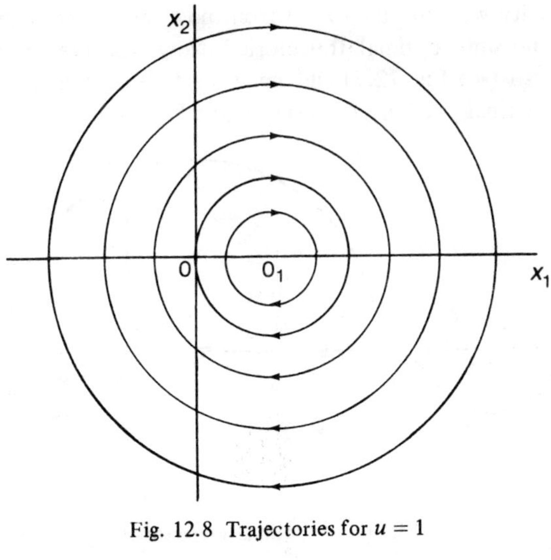

Fig. 12.8 Trajectories for $u = 1$

The solutions are illustrated, when $u = 1$, in Fig. 12.8. The trajectories are circles, centre (1,0), that is, point O_1 and the directions indicated are determined from $\dot{x}_1 = x_2$, that is, x_1 is increasing if $x_2 > 0$ etc.

Similarly, when $u = -1$ we obtain the trajectory equation

$$(x_1 + 1/\omega^2)^2 + (x_2/\omega)^2 = B^2. \tag{12.58}$$

These trajectories when $u = 1$ are illustrated in Fig. 12.9; they are circles, centre $(-1,0)$, that is, point O_{-1}.

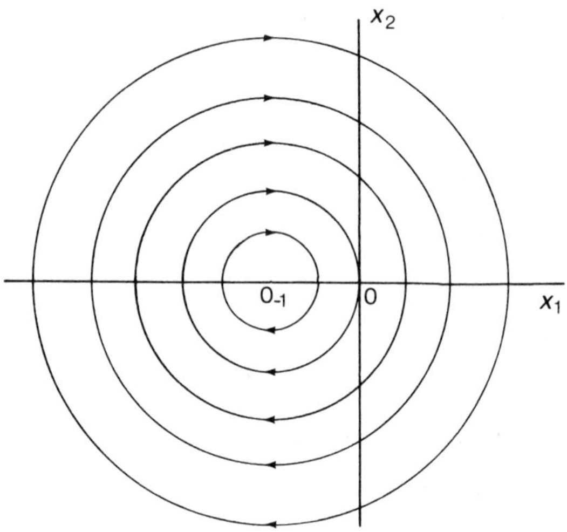

Fig. 12.9 Trajectories for $u = -1$

For simplicity we will take $\omega = 1$ from now on; so that Figs 12.8 and 12.9 represent the possible optimal trajectories. Now, for the final time interval, $[\tau + 2\pi, \tau]$, $u = 1$ (see Fig. 12.7) and since the trajectory must end at (0,0), we can illustrate the final portion of the trajectory, Fig. 12.10(a).

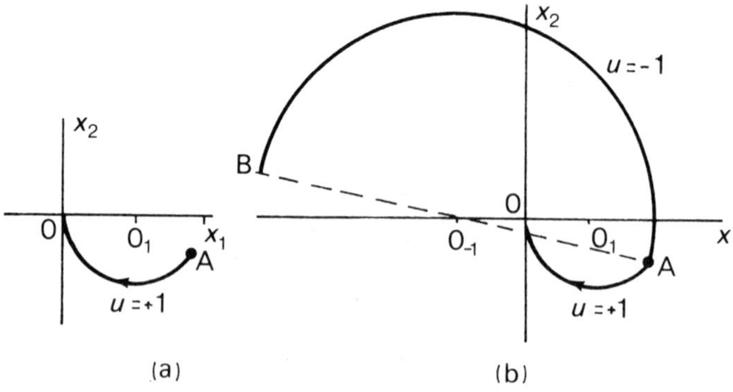

Fig. 12.10 Optimal Trajectories

As the period for a complete circular trajectory is 2π, so for intervals of time of length π, a semicircle is traversed. Hence, moving backwards through time on the interval $[\tau + \pi, \tau + 2\pi]$, $u = -1$, and the semicircle (centre O_{-1}) is traversed. Similarly on $[\tau, \tau + \pi]$, $u = +1$, and again a semicircle (centre O_1) is traversed. This is shown in Fig. 12.11.

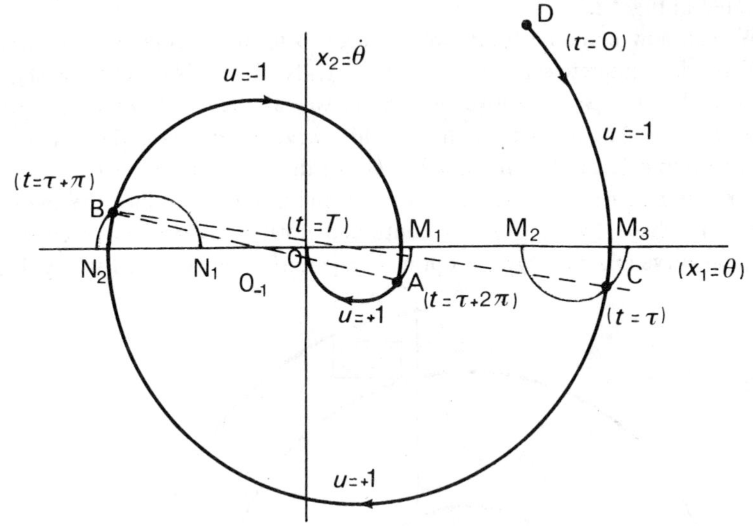

Fig. 12.11 Optimal Trajectory

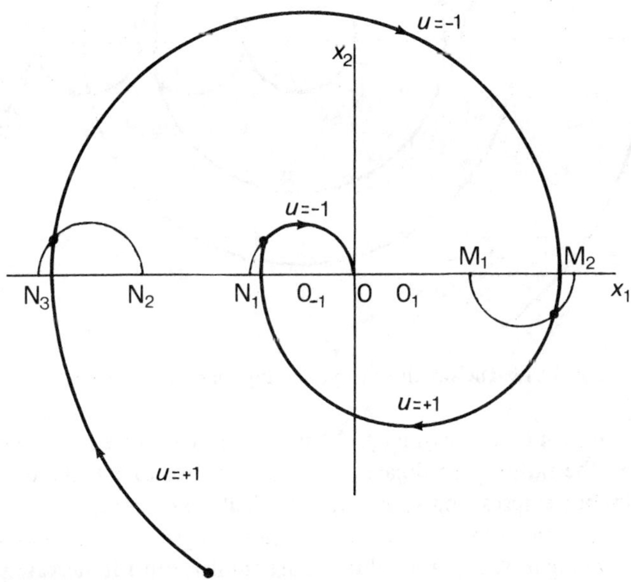

Fig. 12.12 Optimal Trajectory, finishing with $u = -1$

It should be noted that the point B is on the semicircle N_1N_2, and C on the semicircle M_2M_3, both of which have radius 1. We can now complete the trajectory by sketching the portion of the curve for $[0, T]$, centre O_{-1}, from the initial point (θ_0, θ_0') to C.

There is also a second type of trajectory, ending with $u = -1$, which is illustrated in Fig. 12.12.

We can now combine these two pictures to form the complete picture. We first form the complete set of semicircles ... N_3N_2, N_2N_1, N_1O, OM_1, M_1M_2, ..., and note that for points above this curve, we take $u = -1$, whilst for points below $u = +1$. The switch occurs on this curve, so we have determined the *switching curve*. If the initial point lies above this curve, we take $u = -1$, until the curve is reached. We switch to $u = +1$ and again move until the switching curve is reached. This process is continued until the origin is reached. The switching curve and two typical optimal trajectories are sketched in Fig. 12.13.

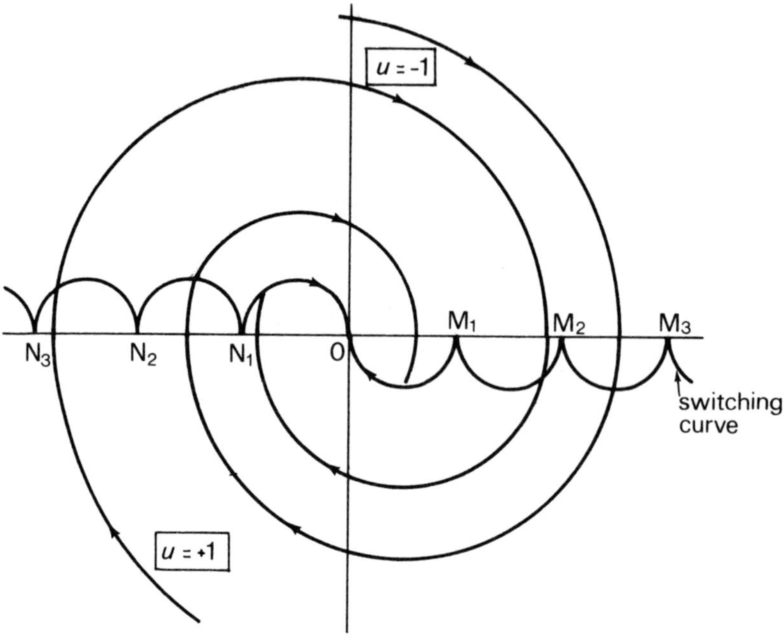

Fig. 12.13 The switching curve and typical optimal trajectories

We now return to our original problem, and note that we have only achieved a solution in the rather special case $\omega = 1$, $a = 0$. For $\omega \neq 1$, the curves are no longer circles but ellipses, and so are more difficult to construct, but the solution is essentially the same. Non-zero damping ($a \neq 0$) presents far more problems. For small damping ($a < 2\omega$) the solutions are similar, but for increased damping, the characteristics change.

PROBLEMS

(1) From equations (12.1) and (12.9), assuming η constant, determine the second order non-linear differential equation which the optimal capital per worker satisfies.

(2) Generalise the renewable resource problem to the case when harvesting and costs are modelled by

$$h = x^{\alpha}E^{\beta}, \quad 0 < \alpha, \beta < 1$$

$$\text{costs} = cE^{\gamma}, \quad 0 < \gamma < 1.$$

Show that, when $\beta = \gamma$, an equilibrium optimal solution is given by

$$\delta = \frac{\alpha c\gamma x^{-\alpha-1}F(x)}{(p\beta - c\gamma x^{-\alpha})} + F'(x).$$

(3) The system of equations

$$\dot{x}_1 = x_2$$

$$\dot{x}_2 = -x_1 + u$$

describes the motion of an harmonic oscillator controlled by the function u. If $|u| \leqslant 1$, find the control which forces the system from any state (ξ_1, ξ_2) to the origin in minimum time. Define the switching curve.
By considering the motion from the point (1,1), verify that the minimum time optimal control to the origin is not necessarily the control with the least number of switchings.

(4) A rocket ascends vertically above the earth, which may be assumed flat, under constant gravitational attraction and subject to negligible aerodynamic forces. The thrust from the rocket engine acts vertically downwards and the equations governing the motion are

$$\dot{h} = v,$$

$$\dot{h} = v,$$

$$\dot{v} = -g + \frac{c\beta}{m},$$

$$\dot{m} = -\beta,$$

where h represents height, v vertical velocity, m rocket mass, c is a positive constant and β is the propellant mass flow which is subject to the constraint

$$0 \leqslant \beta \leqslant \bar{\beta},$$

Assuming that at time $t = 0$ the mass $m = m_0$, height $h = 0$ and velocity $v = 0$, demonstrate by the application of Pontryagin's Minimum Principle that the optimal control to maximise the height achieved is of bang-bang type. Determine the complete solution to the problem, assuming only one switch in the control.

(5) The angular motion of a ship is described in terms of the variable θ by the equation

$$\frac{d^2\theta}{dt^2} + \frac{d\theta}{dt} = p,$$

where p is the rudder setting which is subject to the constraint $|p| \leqslant 1$. A change in course from $\theta = \alpha$, $d\theta/dt = 0$, to $\theta = d\theta/dt = 0$ is required. Find p to minimise the time taken for the correction.

CHAPTER 13

Dynamic programming

13.1 INTRODUCTION

Dynamic programming is concerned with the solution of optimisation problems which can be formulated as a sequence of decisions. The method was first developed by Bellman and co-workers in the 1950s, and has since been widely used in such areas as planning, stock control, maintenance and replacement, resource allocation and transportation.

In many operational problems, the 'system' has a number of possible states, and a transition between them is required. For example, in problems of machine replacement, the machine is the system and a state could be defined as its age or condition.

In this chapter we will consider a system, which has a finite set of discrete states, through which we must move, following one of a number of possible paths. The movement is controlled by a decision-maker who, at each state, uses one of a set of feasible actions. As the movement proceeds, it generates a sequence of returns (such as distance, time, cost). The decision-maker wishes to find the sequence of actions which optimises some function of the returns generated by the movement through the system. We illustrate this process in the next section with a simple example.

We then introduce the D.P. (dynamic programming) notation and illustrate it use in a number of examples. The development and examples are based on the excellent text by Hastings (1972).

We finally show the connection of D.P. with variational calculus and optimal control by considering continuous rather than discrete problems.

13.2 ROUTING PROBLEM

In Fig. 13.1, a network of roads is shown between two towns A and H. The length of the roads is given on the lines, for example, 7 miles from A to B. The problem is to find the path from A to H which minimises the distance between A and H. We can only move along the roads indicated and in the directions marked.

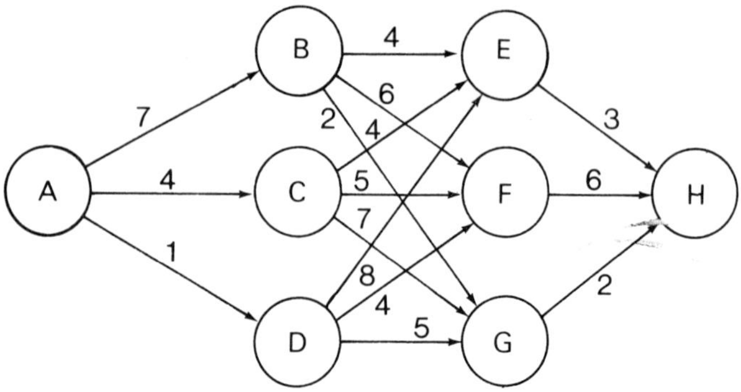

Fig. 13.1 Network of roads between A and H

We first define the D.P. terminology in the context of this problem.

STATE: A **state** is a configuration of the system; towns A,B,C,D, E,F,G, and H are the states in the example above.

STAGE: In D.P. we move through the system from state to state and so on. A **stage** is a single step in such a process; in the problem above there are three stages. The first is from A to B or C or D.

ACTION: At each state there is a set of possible **actions** from amongst which a choice must be made. At C, the three possible actions are to go to E or F or G.

PLAN: A **plan** is a set of actions, one for each of a number of states. An optimal plan is the best set of actions in accordance with a given objective.

RETURN: A **return** is a quantity which the system generates over one stage of a process. In the above, it is the distance between towns.

VALUE OF STATE: The **value of state** is a function (usually the sum) of the returns generated when the system starts in that state and a particular plan is followed.

STATE	Town
STAGE	Transition from town to adjacent town
ACTION	Taking a particular route from a town
RETURN	Distance from town to adjacent town
VALUE OF STATE	Distance from town to terminal town under given plan.

We next solve the problem, not using the D.P. terminology, but using the D.P. philosophy. Consider the last stage ending at H. There is no choice here from state E, F or G. Consider the previous stage. From B we have three possible paths BEH, BFH, BGH, with values 4 + 3, 6 + 6, 2 + 2 respectively. Clearly BGH with value 4 is the best choice here. For C we consider CEH, CFH, CGH with values 4 + 3, 5 + 6, 7 + 2 and so CEH is the best choice. For D, consider DEH, DFH, DGH with values 8 + 3, 4 + 6, 5 + 2, giving DGH as the best choice. Finally we look at A. We have three choices, to B, C or D and then on to the best choice from there. Thus for AB, we move on ABGH with value 4 + 7 = 11; For AC, we move on ACEH with value 7 + 4 = 11; and for AD we move on ADGH with value 1 + 7 = 8. Thus our optimum path is

A D G H

with value 8.

We note two important points. Firstly we have assumed that each part of an optimal path is optimal. That is, if the overall optimal path from A to H passes through D, say, it will necessarily move on the optimal path from D to the terminal town H. Secondly, the number of additions made at each stage is 0, 3 + 3 + 3, 3: a total of 12. If the value of each possible path (and there are 9) is evaluated, we would have 18 additions, that is, we have a 30% saving in additions. The saving in the number of additions becomes quite remarkable as the number of stages increases.

13.3 D.P. NOTATION

In order to be able to formulate D.P. problems in a precise way, suitable for computation, we introduce a stage and state variable. The stage variable n, indicates the number of stages which remain until a terminal state is reached. The state at stage n by (n, i). The routing problem from the previous section is illustrated in Fig. 13.2.

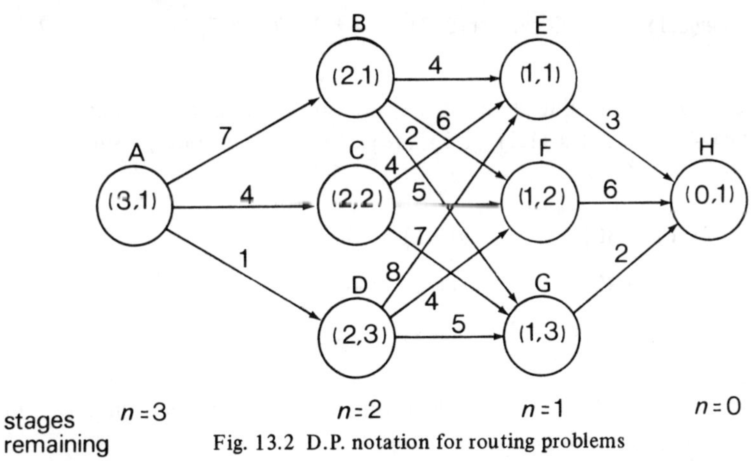

Fig. 13.2 D.P. notation for routing problems

In the state (n, i) there are a number of possible actions. Let K_{ni} denote the set of actions available at the state (n, i), and we number the actions $k = 1, 2, \ldots, k_{ni}$; that is,

$$K_{ni} = \{1, 2, \ldots, k_{ni}\}.$$

For example

$$K_{21} = \{1, 2, 3\};$$

here 1 denotes the move from (2,1) to (1,1), 2 denotes the move to (1,2), 3 denotes the move to (1,3).

In general, the successor state $(n-1, j)$ from state (n, i) is determined from the current stage, state and action variables, so that we can introduce a **transition function** t, and write

$$j = t(n, i, k).$$

In the example above, using the action numbering as indicated, we see that $j = k$. Thus at state (2,1) action $k = 1$ provides the move to (1,1); $k = 2$ provides the move to (1,2) etc.

We also define the return associated with state (n, i) and action k as $r(n, i, k)$, where r is called the **return function**. Thus, in the example,

$$r(2,1,1) = 4, \quad r(2,1,2) = 6 \quad r(2,1,3) = 2.$$

The value of state (n, i) under an optimal plan is denoted by $v(n, i)$, v being the optimal value function; so that

$$v(2,1) = 2+2=4, \quad v(2,2) = 4+3=7, \quad v(2,3) = 5+2=7.$$

Using the notation introduced above, we now solve the routing example again. Now $v(0,1) = 0$. In states (1,1), (1,2), (1,3), there is only one possible action, $k = 1$, that is,

$$\begin{aligned} &(1,1) \to (0,1) \quad \text{and} \quad && v(1,1) = 3 \\ &(1,2) \to (0,1) && v(1,2) = 6 \\ &(1,3) \to (0,1) && v(1,3) = 2. \end{aligned}$$

At stage 2, we consider each state in turn. For (2,1) there are three possible actions, $k = 1, 2, 3$, which give values

$$r(2,1,1) + v(1,1) = 4 + 3 = 7$$
$$r(2,1,2) + v(1,2) = 6 + 6 = 12$$
$$r(2,1,3) + v(1,3) = 2 + 2 = 4.$$

Thus the value under the optimal plan is 4 for $v(2,1)$. These results are indicated in the table, the optimal plans for each stage and state being indicated by underlining. We can summarise the method of obtaining the optimal actions and value for state (2,1) by

$$v(2,1) = \min_{k \in K_{21} = \{1,2,3\}} [r(2,1,k) + v(1,k)].$$

Stage n	State i	Action k	Value v
1	1	<u>1</u>	<u>3</u>
	2	<u>1</u>	<u>6</u>
	3	<u>1</u>	<u>2</u>
2	1	1	4 + 3 = 7
		2	6 + 6 = 12
		<u>3</u>	2 + 2 = <u>4</u>
	2	<u>1</u>	4 + 3 = <u>7</u>
		2	5 + 6 = 11
		3	7 + 2 = 9
	3	1	8 + 3 = 11
		2	4 + 6 = 10
		<u>3</u>	5 + 2 = <u>7</u>
3	1	1	7 + 4 = 11
		2	4 + 7 = 11
		<u>3</u>	1 + 7 = <u>8</u>

In general we can write

$$v(n, i) = \min_{k \in K_{ni}} [r(n, i, k) + v(n-1, k)].$$

We can now summarise the main computational algorithm by

$$\text{Recurrence relation: } v(n, i) = \min_{k \in K_{ni}} [r(n, i, k) + v(n-1, j)] \quad (13.1)$$

$$\text{Transition equation: } j = t(n, i, k) \quad (13.2)$$

and $v(0, i)$ is given for all terminal states.

Note that D.P. can equally well be used for maximisation problems. In the recurrence relation we just replace minimum by maximum. If we return to the routing problem, we can assume that the numbers represent the returns made by a carrier along the various routes from the initial to the terminal state. The calculations for the maximum are shown below:

Stage	State	Action	Value
1	1	1 (underlined)	3 (underlined)
	2	1 (underlined)	6 (underlined)
	3	1 (underlined)	2 (underlined)
2	1	1	4 + 2 = 7
		2 (underlined)	6 + 6 = 12 (underlined)
		3	2 + 2 = 4
	2	1	4 + 3 = 7
		2 (underlined)	5 + 6 = 11 (underlined)
		3	7 + 2 = 9
	3	1 (underlined)	8 + 3 = 11 (underlined)
		2	4 + 6 = 10
		3	5 + 2 = 7
3	1	1 (underlined)	7 + 12 = 19 (underlined)
		2	4 + 11 = 15
		3	1 + 11 = 12

13.4 EXAMPLES

13.4.1 Resources Allocation

A dairy manufacturer can make three products: cheese, butter and yoghurt from his incoming resource, milk. The profit which he can obtain by using 1, 2, 3 or 4 units of milk are given in the table:

Units of milk	1	2	3	4
Cheese	8	18	22	24
Butter	3	6	9	12
Yoghurt	6	7	8	10

The manufacturer wishes to find the resource allocation to the three products so as to maximise his profits when using a total of *four* units of milk.

This problem is quite simple, and it would not take long to find the solution by a trial and error method. We will use the problem to illustrate the techniques of D.P. again. We must first define the relevant terms.

STAGE n	Number of products available (label product cheese as 1 etc.)
STATE i	Number of units of milk available
ACTION k	Number of units of milk allocated to product n
RETURN r	Profit obtained on allocating k units of milk to product n
VALUE v	Total profit on allocation of i units.

The same procedure as in section 13.3 is followed. We first consider product 1, cheese, the first stage, and the allocation of up to four units of milk to this product. At the second stage, products 1 and 2 are utilised, and so on.

The optimal policies are underlined, and having reached stage 3 at state 4 the overall optimum policy is

1 resource unit to product 3 Yoghurt
0 ,, ,, ,, ,, 2 Butter
3 ,, ,, ,, ,, 1 Cheese

giving a maximum profit of 28 units.

Stage	State	Action	Value
1	0	0	0
	1	1	8
	2	2	18
	3	3	22
	4	4	24
2	0	0	0
	1	$\underline{0}$	$0 + 8 = \underline{8}$
		1	$3 + 0 = 3$
	2	$\underline{0}$	$0 + 18 = \underline{18}$
		1	$3 + 8 = 11$
		2	$6 + 0 = 6$
	3	$\underline{0}$	$0 + 22 = \underline{22}$
		1	$3 + 18 = 21$
		2	$6 + 8 = 14$
		3	$9 + 0 = 9$
	4	0	$0 + 24 = 24$
		$\underline{1}$	$3 + 22 = \underline{25}$
		2	$6 + 18 = 24$
		3	$9 + 8 = 17$
		4	$12 + 0 = 12$
3	4	0	$0 + 25 = 25$
		1	$6 + 22 = \underline{28}$
		2	$7 + 18 = 25$
		3	$8 + 8 = 16$
		4	$10 + 10 = 20$

If you are having difficulties following the calculations, let us consider the stage 2, state 2; which means we are considering products 1 and 2 (cheese and butter), and 2 units of milk are available. There are three possible actions:

Action 0 Allocate 0 units of milk to product 2 (butter); so that 2 units of milk are allocated to product 1 (cheese), and the value is 0 (for product 2) + 18 (product 1) = 18.

Action 1 1 unit of milk to product 2 and so 1 unit to product 1; value = 3 (product 2) + 8 (product 1) = 11.

Action 2 2 units of milk to product 2 and so 0 units to product 1; value = 6 (product 2) + 0 (product 1) = 6.

13.4.2 Production Planning

We take, as an example, the problem of a boat builder, who has the following orders for delivery in a particular year, at the end of the months indicated.

Mar.	Apr.	May	June	July	Aug.
1	2	5	3	2	1

He can build up to four boats in any month, and can hold up to three boats in stock. If boats are built in a particular month, there is an overhead cost per month of £4000, and a construction cost of £10 000 per boat. Stock holding costs are £1000 per boat per month. If boats are not built in a particular month, there are no overhead costs. In what months should boats be built, and in what quantities, if costs are to be minimised?

In D.P. terms, each month corresponds to a *stage*. Let n be the number of months or stages which remain until the planning horizon, that is, the end of August, is reached. Let *state* (n, i) define the number of boats, i, which are in stock at the start of the month corresponding to stage n. (For example, (3.2) corresponds to 2 boats in stock at the beginning of June.) Let *action* k at a given state be to build k boats in the current month. The ***return*** in any month is the cost incurred in that month, and is given by $r = r(i, k)$, not depending on the stage variable.

The recurrence relation where $v(n, i)$ is the total cost from state (n, i) under optimum plan, is

$$v(n,i) = \min_{k \in K} \{r(i,k) + v(n-1,j)\} \quad (= \text{value}). \tag{13.3}$$

The state variable j in the successor state $(n-1, j)$ is determined by the transition equation

$$j = i + k - d \quad (= \text{current stock} + \text{production} - \text{demand}) \tag{13.4}$$

where d is the demand at the current stage.

The total construction costs £140 000 must be met, however production is scheduled, and so need not be considered. If i boats are in stock at the start of the month then

$$r(i,0) = i$$

$$r(i,k) = 4 + i (k \neq 0)$$

working in units of £1000.

The calculations are given in the table below. To understand the calculations the reader must work through them, remembering that

(i) $0 \leqslant i \leqslant 3$, that is, only three boats at most can be stocked;
(ii) the stock, i, must be less than or equal to the sums of the remaining demands;
(iii) demands *must* be met at the end of each month;
(iv) $0 \leqslant k \leqslant 4$, that is, only four boats at most can be made in any one month.

The explanation for the calculations for state (2,1) are given at the end.

Stage	State	Action	Value	Demand
1	0	1	4	1
	1	0	1	
2	0	2	4 + 4 = 8	2
		3	4 + 1 = 5	
	1	1	5 + 4 = 9	
		2	5 + 1 = 6	
	2	0	2 + 4 = 6	
		1	6 + 1 = 7	
	3	0	3 + 1 = 4	
3	0	3	4 + 5 = 9	3
		4	4 + 6 = 10	
	1	2	5 + 5 = 10	
		3	5 + 6 = 11	
		4	5 + 6 = 11	
	2	1	6 + 5 = 11	
		2	6 + 6 = 12	
		3	6 + 6 = 12	
		4	6 + 4 = 10	
	3	0	3 + 5 = 8	
		1	7 + 6 = 13	
		2	7 + 6 = 13	
		3	7 + 4 = 11	

Stage	State	Action	Value	Demand
4	1	4	5 + 9 = 14	5
	2	3	6 + 9 = 15	
		4	6 + 10 = 16	
	3	2	7 + 9 = 16	
		3	7 + 10 = 17	
		4	7 + 10 = 17	
5	0	3	4 + 14 = 18	2
		4	4 + 15 = 19	
	1	2	5 + 14 = 19	
		3	5 + 15 = 20	
		4	5 + 16 = 21	
	2	1	6 + 14 = 20	
		2	6 + 15 = 21	
		3	6 + 16 = 22	
	3	0	3 + 14 = 17	
		1	7 + 15 = 22	
		2	7 + 16 = 23	
6	0	1	4 + 18 = 22	1
		2	4 + 19 = 23	
		3	4 + 20 = 24	
		4	4 + 17 = 21	

The minimum cost is £21 000, and the plan which gives this cost is given below.

Optimal plan:

	Mar.	Apr.	May	June	July	Aug.
Boats built in month	4	0	4	3	3	0
Boats stored during month	0	3	1	0	0	1

If we consider state (2,1) (that is, stage 2 and state 1) then we are at the start of July (2 months to go) with 1 boat in stock, having met the demand at the end of June. Since the demand at the end of the month is 2, we must build at most 1 boat, but since only 1 more boat is required (at the end of August), we must not build more than 2 boats; hence we have two possible actions, 1 and 2;

Action 1: Build 1 boat with
value = 4 (overhead cost) + 1 (stock holding cost) + 4 (August costs) = 9.

Action 2: Build 2 boats, with
value = 4 (overhead cost) + 1 (stock holding cost) + 1 (August costs) = 6.

13.5 BELLMAN'S EQUATION

So far we have developed D.P. for discrete decision problems. In this section we will modify the method for continuous processes, and in this way derive the Euler equation obtained in Chapter 9.

Consider the problem of minimising the functional

$$J[x] = \int_0^T f(x, \dot{x}, t)\,\mathrm{d}t \tag{13.5}$$

where $x(0) = \alpha$ and $x(T) = \beta$. We write $J[x]$ to denote the dependence of the functional on the path $x(t)$, $0 \leqslant t \leqslant T$ between $(\alpha, 0)$ and (β, T) in the x-t plane. The D.P. approach is to define the optimum value of the integral from time t to the final time T as

$$S(t, x) = \min_{y} \int_t^T f(y, \dot{y}, \tau)\,\mathrm{d}\tau \tag{13.6}$$

where $y(t) = x$ and $y(T) = \beta$. In other words, if we are at position x and time t, the minimum value of the integral

$$\int_t^T f(x, \dot{x}, t)\,\mathrm{d}t$$

ending with $x(T) = \beta$ is given by a certain path $y(\tau)$ and has value $S(t, x)$. This is illustrated in Fig. 13.3.

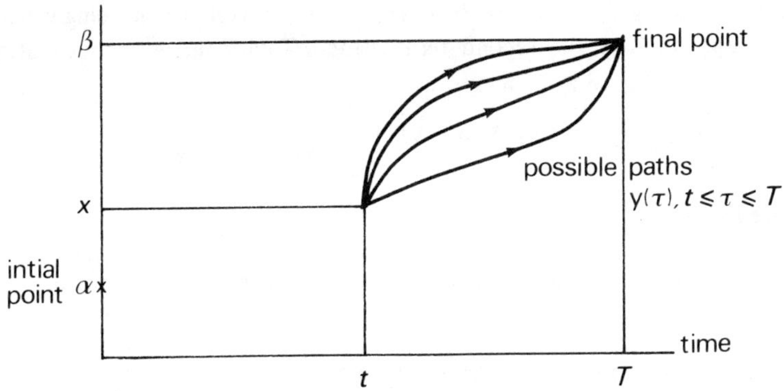

Fig. 13.3 Possible paths from (t, x) to (T, β)

Now we can split $S(t, x)$ up into two parts getting from x at time t to $x + \delta x$ at time $t + \delta t$ and then from $x + \delta x$ at $t + \delta t$ to β at the final time T. This is shown in Fig. 13.4.

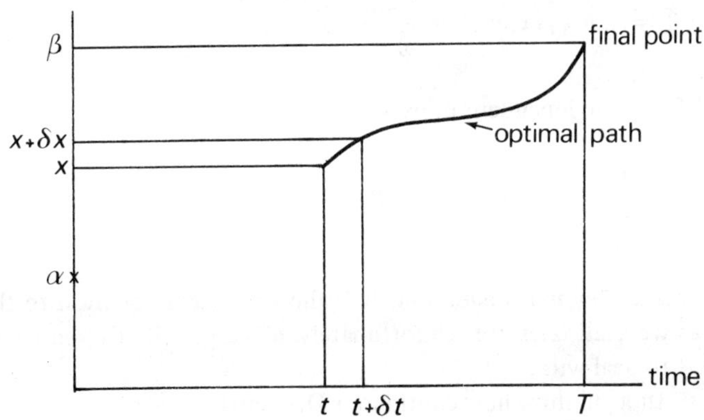

Fig. 13.4 Splitting path into two portions

In this way

$$S(t, x) = \int_t^{t+\delta t} f(y, \dot{y}, \tau)\,\mathrm{d}\tau + S(t + \delta t, x + \delta x)$$

since $S(t + \delta t, x + \delta x)$ represents the minimum value of the integral from $(t + \delta t, x + \delta x)$ to (T, β). We can approximate the first term by

$$\int_t^{t+\delta t} f(y, \dot{y}, \tau)\mathrm{d}\tau \approx \delta t f(y(t), \dot{y}(t), t) = \delta t f(x, \dot{x}, t).$$

(Here we are using $\int_a^b f(x)\mathrm{d}x \approx (b-a)f(a)$, the interval $b-a$ being small). We can also approximate the second term using a Taylor series approximation by

$$S(t+\delta t, x+\delta x) \approx S(t,x) + \delta t\frac{\partial S}{\partial t} + \delta x\frac{\partial S}{\partial x} + \ldots.$$

Thus we have

$$S(t,x) = \min_{\delta x}\{\delta t f(x,\dot{x},t) + S(t,x) + \delta t\frac{\partial S}{\partial t} + \delta x\frac{\partial S}{\partial x} + \ldots\} \tag{13.7}$$

the minimisation now being over δx which, since $\delta x \approx \dot{x}\delta t$, means $\dot{x}$; that is,

$$0 = \min_{\dot{x}}\{\delta t f(x,\dot{x},t) + \delta t\frac{\partial S}{\partial t} + \delta t\dot{x}\frac{\partial S}{\partial x} + 0(\delta t^2)\}.$$

We then obtain, since $\delta t \neq 0$,

$$0 = \min_{\dot{x}}\{f(x,\dot{x},t) + \frac{\partial S}{\partial t} + \dot{x}\frac{\partial S}{\partial x}\}$$

and since $\partial S/\partial t$ is independent of $\dot{x}$, we obtain

$$-\frac{\partial S}{\partial t} = \min_{\dot{x}}(\dot{x}\frac{\partial S}{\partial t} + f(x,\dot{x},t)). \tag{13.8}$$

This is known as **Bellman's equation**. It is the equivalent equation to the Euler equation, as we shall later see. Unfortunately, it is a partial differential equation and difficult to deal with.

We first, though, show how continuous D.P. can be used to solve, numerically, calculus of variations problems.

Example

Find the approximate minimum value of

$$\int_0^3 (x^2 + \dot{x}^2)\,\mathrm{d}t$$

where $x(0) = 0$, $x(3) = 3$.

Solution

Let $S(t,x) =$ minimum value of integral from (t,x) to $(3,3)$. From (13.7)

$$S(t,x) = \min_{y} \left\{ \int_{t}^{t+\delta t} (y^2 + \dot{y}^2)\,d\tau + S(t + \delta t, x + \delta x) \right\}.$$

In the interval δt, we can approximate $\delta x = \dot{y}\delta t$, and taking $y = x + \frac{1}{2}\delta x$ (halfway between x and $x + \delta x$) in this interval, we have

$$S(t,x) = \min_{\delta x} \{\delta t[(x + \tfrac{1}{2}\delta x)^2 + (\delta x/\delta t)^2] + S(t + \delta t, x + \delta x)\}$$

We now introduce a mesh, illustrated in Fig. 13.5, so that $\delta t = 1$ and δx can, in general, take the values $0, 1, 2, 3$. For example at $(0,0)$, in the time interval $\delta t = 1$, we can move to $(1,0)$ if $\delta x = 0$, $(1,1)$ if $\delta x = 1$, $(1,2)$ if $\delta x = 2$, and $(1,3)$ if $\delta x = 3$.

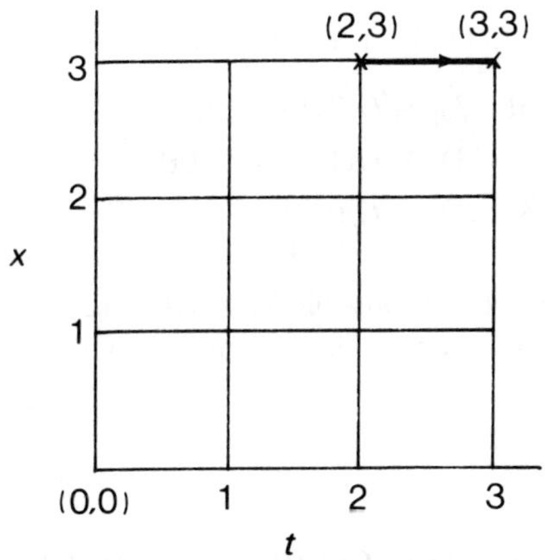

Fig. 13.5 Mesh size for calculations

Now with $\delta t = 1$, we have

$$S(t,x) = \min_{\delta x = 0,1,2,3} \{S(t + 1, x + \delta x) + (x + \tfrac{1}{2}\delta x)^2 + \delta x^2\}$$

and writing

$$G(x, \delta x) = (x + \tfrac{1}{2}\delta x)^2 + \delta x^2 \qquad (13.9)$$

we have

$$S(t,x) = \min_{\delta x = 0,1,2,3} \{S(t + 1, x + \delta x) + G(x, \delta x)\}. \qquad (13.10)$$

The values of the function $G(x, \delta x)$ for $x = 0, 1, 2$ or 3 and $\delta x = 0, 1, 2$ or 3 are given in the table below.

		δx			
		0	1	2	3
	0	0	1.25	5.00	11.25
	1	1.00	3.25	8.00	
x	2	4.00	7.25		
	3	9.00			

We now proceed backwards from the final point (3,3). Clearly $S(3,3) = 0$, and

$$\begin{aligned} S(2,3) &= \min_{\delta x=0} [S(3,3) + G(3,0)] = 9.00 \\ S(2,2) &= S(3,3) + G(2,1) = 7.25 \\ S(2,1) &= S(3,3) + G(1,2) = 8.00 \\ S(2,0) &= S(3,3) + G(0,3) = 11.25\,. \end{aligned}$$

For $S(2,3)$, there is only one possible path from (2,3) to (3,3), as shown in Fig. 13.3, and on this path $\delta x = 0$. Similarly for the points (2,2), (2,1) and (2,0).

We now take $t = 1$, and

$$S(1,3) = \min_{\delta x=0} [S(2,3) + G(3,0)] = 9.00 + 9.00 = 18.00$$

$$S(1,2) = \min \begin{Bmatrix} S(2,2) + G(2,0) \\ S(2,3) + G(2,1) \end{Bmatrix} = \min \begin{Bmatrix} 7.25 + 4.00 \\ 9.00 + 7.25 \end{Bmatrix} = 11.25$$

$$S(1,1) = \min \begin{Bmatrix} S(2,1) + G(1,0) \\ S(2,3) + G(1,1) \\ S(2,3) + G(1,2) \end{Bmatrix} = \min \begin{Bmatrix} 8.00 + 1.00 \\ 7.25 + 3.25 \\ 9.00 + 8.00 \end{Bmatrix} = 9.00$$

$$S(1,0) = \min \begin{Bmatrix} S(2,0) + G(0,0) \\ S(2,1) + G(0,1) \\ S(2,2) + G(0,2) \\ S(2,3) + G(0,3) \end{Bmatrix} = \min \begin{Bmatrix} 11.25 + 0 \\ 8.00 + 1.75 \\ 7.25 + 5.00 \\ 9.00 + 11.25 \end{Bmatrix} = 9.25$$

Finally, we see that

$$S(0,0) = \min \begin{Bmatrix} S(1,0) + G(0,0) \\ S(1,1) + G(0,1) \\ S(1,2) + G(0,2) \\ S(1,3) + G(0,3) \end{Bmatrix} = \min \begin{Bmatrix} 9.25 + 0 \\ 9.00 + 1.25 \\ 11.25 + 5.00 \\ 18.00 + 11.25 \end{Bmatrix} = 9.25$$

and we see that the optimum path is

$$(0,0) \rightarrow (1,0) \rightarrow (2,1) \rightarrow (3,3)$$

as shown in Fig. 13.6, and the integral has approximate value 9.25.

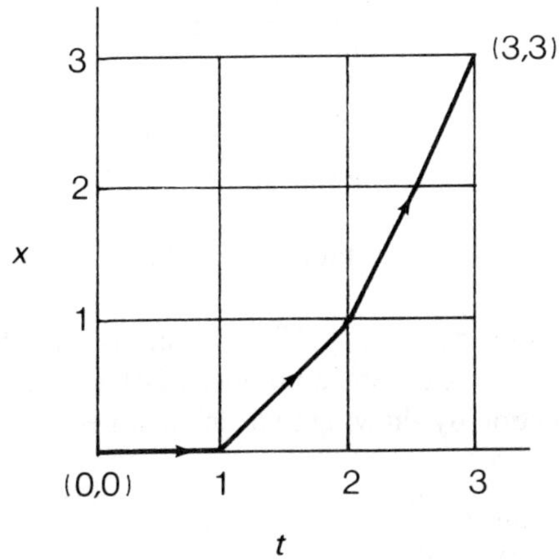

Fig. 13.6 Approximation to Optimum Curve

Note: In practice, to obtain a reasonable degree of accuracy, a much smaller mesh size must be used. This can of course be easily accomplished with a computer program. In fact, the exact solution to the problem is determined (as usual) by the Euler equation

$$\frac{\mathrm{d}}{\mathrm{d}t}\left(\frac{\partial F}{\partial \dot{x}}\right) - \frac{\partial F}{\partial x} = 0$$

where $F = x^2 + \dot{x}^2$. Thus

$$\frac{\mathrm{d}}{\mathrm{d}t}(2\dot{x}) = 2x$$

that is, $\ddot{x} - x = 0$

and $x = A\mathrm{e}^t + B\mathrm{e}^{-t}$. To pass through (0,0) and (3,3) we must have

$$0 = A + B$$

$$3 = A\mathrm{e}^3 + B\mathrm{e}^{-3}$$

which gives $x = 3 \sinh t/\sinh 3$, as the optimum path. On this path, the integral has value

$$\begin{aligned}\int_0^3 (x^2 + \dot{x}^2)\,\mathrm{d}t &= \int_0^3 [9(\sinh^2 t + \cosh^2 t)/\sinh^2 3]\,\mathrm{d}t \\ &= \frac{9}{\sinh^2 3}\int_0^3 \cosh 2t\,\mathrm{d}t \\ &= \frac{9}{2\sinh^2 3}\left[\sinh 2t\Big|_0^3\right] \\ &= 9 \tanh 3 = 8.96.\end{aligned}$$

The approximate minimum value, 9.25, is in fact remarkably close to the true minimum value, considering the small number of iterations made.

We finish this section by showing that Bellman's equation

$$-\frac{\partial S}{\partial t} = \min_{\dot{x}}\left(\dot{x}\frac{\partial S}{\partial x} + f(x, \dot{x}, \tau)\right) \tag{13.11}$$

is equivalent to the Euler equation in variational calculus.

To minimise the left-hand side with respect to $\dot{x}$, we take the partial derivations with respect to $\dot{x}$ and equate to zero, that is,

$$0 = \frac{\partial S}{\partial x} + \frac{\partial f}{\partial \dot{x}}. \tag{13.12}$$

Differentiating (13.12) with respect to t gives;

$$\begin{aligned}0 &= \frac{\mathrm{d}}{\mathrm{d}t}\left(\frac{\partial S}{\partial x}\right) + \frac{\mathrm{d}}{\mathrm{d}t}\left(\frac{\partial f}{\partial \dot{x}}\right) \\ &= \frac{\partial}{\partial t}\left(\frac{\partial S}{\partial x}\right) + \frac{\partial}{\partial x}\left(\frac{\partial S}{\partial x}\right)\frac{\mathrm{d}x}{\mathrm{d}t} + \frac{\mathrm{d}}{\mathrm{d}t}\left(\frac{\partial f}{\partial \dot{x}}\right)\end{aligned}$$

(using the chain rule for partial differentiation),

that is, $$0 = \frac{\partial}{\partial x}\left(\frac{\partial S}{\partial t}\right) + \frac{\partial}{\partial x}\left(\frac{\partial S}{\partial x}\right)\dot{x} + \frac{\mathrm{d}}{\mathrm{d}t}\left(\frac{\partial f}{\partial \dot{x}}\right)$$

$$\left(\text{since } \frac{\partial^2 S}{\partial x \partial t} = \frac{\partial^2 S}{\partial t \partial x}\right).$$

Thus

$$0 = \frac{\partial}{\partial x}\left(\frac{\partial S}{\partial t} + \dot{x}\frac{\partial S}{\partial x}\right) + \frac{d}{dt}\left(\frac{\partial f}{\partial \dot{x}}\right) \tag{13.13}$$

the $\dot{x}$ being inside the $\partial/\partial x$ bracket, since $\partial/\partial x$ means differentiating with respect to x, keeping $\dot{x}$ and t constant. But on the optimum path (13.11) becomes

$$-\frac{\partial S}{\partial t} = \dot{x}\frac{\partial S}{\partial x} + f$$

and so (13.13) can be written as

$$\frac{d}{dt}\left(\frac{\partial f}{\partial \dot{x}}\right) - \frac{\partial f}{\partial x} = 0 \tag{13.14}$$

which is Euler's equation.

13.6 THE MAXIMUM PRINCIPLE

We end this chapter by indicating how the maximum principle for optimal control (Chapter 11) can be derived from Bellman's equation (13.8). The case of scalar x in section 13.6 can be easily extended to the problem of finding the minimum value of

$$J = \int_0^T f(\mathbf{x},\dot{\mathbf{x}})\,dt \tag{13.15}$$

where $\mathbf{x}(0) = \mathbf{x}_0$, $\mathbf{x}(T) = \mathbf{x}_T$ and $\mathbf{x} = (x_1, x_2, \ldots, x_n)$.

Defining

$$S(t,\mathbf{x}) = \min_{\mathbf{y}}\left\{\int_t^T f(\mathbf{y},\dot{\mathbf{y}})\,d\tau\right\},\ \mathbf{y}(t) = \mathbf{x}(t),\ \mathbf{y}(T) = \mathbf{x}_T, \tag{13.16}$$

we can easily derive the equivalent Bellman equation

$$-\frac{\partial S}{\partial t} = \min_{\dot{\mathbf{x}}}\left\{\sum_{i=1}^{n} \dot{x}_i \frac{\partial S}{\partial x_i} + f(\mathbf{x},\dot{\mathbf{x}})\right\}. \tag{13.17}$$

Now we turn to the optimal control problem. Here we wish to find the control $\mathbf{u}$ which minimises

$$J = \int_0^T f_0(\mathbf{x},\mathbf{u})\,dt \tag{13.18}$$

subject to

$$\dot{x}_i = f_i(\mathbf{x},\mathbf{u}) \qquad (i = 1,2,\ldots,n) \tag{13.19}$$

and $\mathbf{x}(0) = \mathbf{x}_0$, $\mathbf{x}(T) = \mathbf{x}_T$. Again we define

$$S(t,\mathbf{x}) = \min_{\mathbf{u}} \left\{ \int_t^T f_0(\mathbf{y},\mathbf{u})\,d\tau \right\}, \; \mathbf{y}(t) = \mathbf{x}(t), \; \mathbf{y}(T) = \mathbf{x}_T, \tag{13.20}$$

and $\dot{y}_i = f_i(\mathbf{y},\mathbf{u})$ the minimisation now being over $\mathbf{u}$, instead of $\mathbf{y}$, since specifying the control $\mathbf{u}$, determines $\mathbf{y}$ and $\dot{\mathbf{y}}$.

The Bellman equation now takes the form

$$-\frac{\partial S}{\partial t} = \min_{\mathbf{u}} \left\{ \sum_{i=1}^{n} \frac{\partial S}{\partial x_i} f_i + f_0 \right\}. \tag{13.21}$$

Defining new variables $\mathbf{p} = (p_1, p_2, \ldots, p_n)$ by

$$p_i = -\partial S/\partial x_i \tag{13.22}$$

and the Hamiltonian

$$H = -f_0 + \sum_{i=1}^{n} p_i f_i \tag{13.23}$$

we see that (13.21) takes the form

$$-\frac{\partial S}{\partial t} = \min_{\mathbf{u}} (-H)$$

that is, $$\max_{\mathbf{u}}(H) = \partial S/\partial t. \tag{13.24}$$

So on the optimal path, the Hamiltonian H is minimised with respect to the control vector $\mathbf{u}$. This is just Pontryagin's maximum principle (derived in section 11.2).

We have still to derive the adjoint equations, which $\mathbf{p}$ satisfies. Now, from (11.22), we have

$$\frac{dp_i}{dt} = -\frac{d}{dt}\left(\frac{\partial S}{\partial x_i}\right) = -\frac{\partial}{\partial t}\left(\frac{\partial S}{\partial x_i}\right) - \sum_{k=1}^{n} \frac{\partial}{\partial x_k}\left(\frac{\partial S}{\partial x_i}\right)\dot{x}_k \qquad \text{(chain rule)}$$

$$= -\frac{\partial}{\partial x_i}\left(\frac{\partial S}{\partial t}\right) - \sum_{k=1}^{n} \frac{\partial}{\partial x_i}\left(\frac{\partial S}{\partial x_k}\right) f_k .$$

On the optimal path, (13.21) gives

$$-\frac{\partial S}{\partial t} = \sum_{k=1}^{n} \frac{\partial S}{\partial x_k} f_k + f_0 .$$

So we obtain,

$$\frac{dp_i}{dt} = \frac{\partial}{\partial x_i}\left(\sum_{k=1}^{n} \frac{\partial S}{\partial x_k} f_k + f_0\right) - \sum_{k=1}^{n} \frac{\partial}{\partial x_i}\left(\frac{\partial S}{\partial x_k}\right) f_k$$

$$= \sum_{k=1}^{n} \frac{\partial S}{\partial x_k}\frac{\partial f_k}{\partial x_i} + \frac{\partial f_0}{\partial x_i}$$

$$= -\sum_{k=1}^{n} p_k \frac{\partial f_k}{\partial x_i} + \frac{\partial f_0}{\partial x_i}, \quad \text{using (13.22);}$$

that is, $$p_i = -\frac{\partial H}{\partial x_i} \qquad (13.25)$$

from (13.23) regarding H as a function of **x**, **u** and **p**. These are the adjoint equations, which must be satisfied (see equations (11.23). We have now completed the derivation of the maximum principle method for optimal control problems.

PROBLEMS

(1) A Newcastle family wish to have a holiday in one of the following towns:

Clacton, Southend, London, Oxford, Gloucester.

To minimise their costs they wish to select the town which minimises the distance from Newcastle. The route network with distances in miles is shown in the figure below.

Find the optimal plan. Using dynamic programming, how many additions are made? How many possible paths are there, and how many additions would be made if the total distance for each path was evaluated separately?

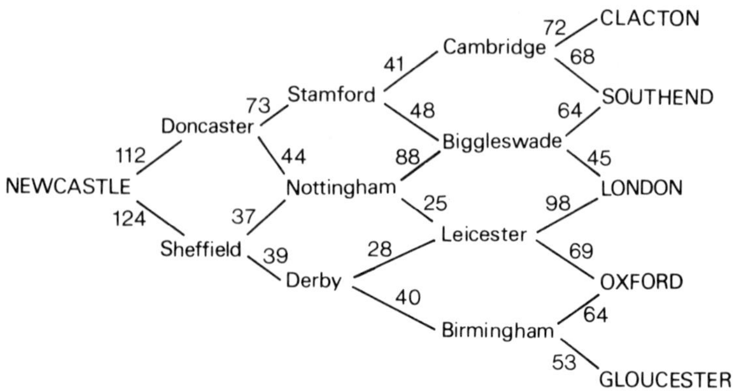

(2) The cost of taking an admissible step between nodes from left to right is indicated below. Find, using D.P. techniques, the path from A to B which gives minimum cost.

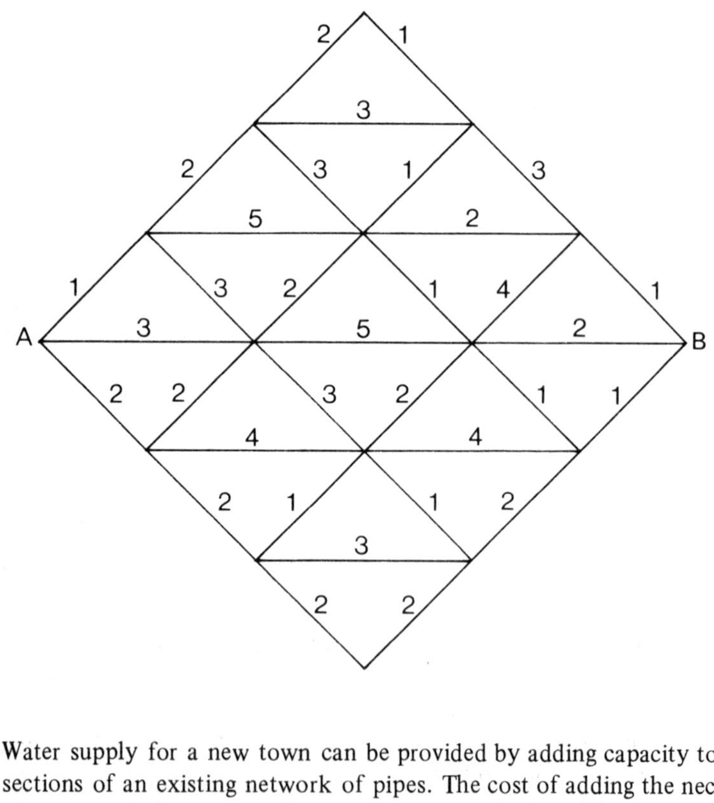

(3) Water supply for a new town can be provided by adding capacity to some sections of an existing network of pipes. The cost of adding the necessary capacity to each section is shown in the figure below.

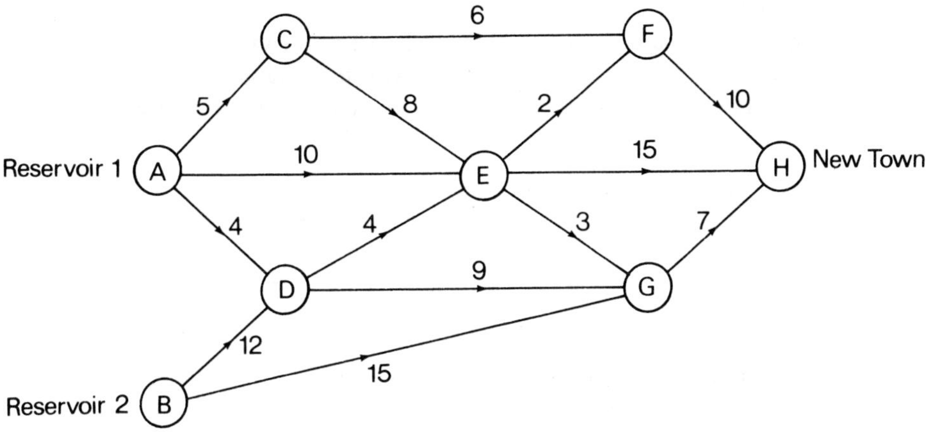

Determine the cheapest way to arrange for the new supply. Arrows indicate directions of flow. The whole supply must come from either reservoir 1 or reservoir 2.

(4) A firm produces three products 1, 2 and 3. The profits obtainable using 1, 2, 3 or 4 units of resource are given below:

x	1	2	3	4
$g_1(x)$	8	18	22	24
$g_2(x)$	3	6	9	12
$g_3(x)$	6	7	8	10

Here x denotes the number of units of resource, and $g_1(x)$ the profit obtained on using x units of resource in production of product 1, etc. Define the stage variable, n, as the number of products to be considered, the state variable, i, as the number of units of resource available and action, k, as the number of units of resource allocated to product labelled k.

Use D.P. to determine the optimal allocation of 4 units of resource.

(5) A company can buy t tons of a raw material for £$c(t)$. It can use the raw material to make products A, B, C in various amounts. The return resulting when k tons of raw material are allocated to the production of A is $r(A, k)$, and similar relationships apply for B and C. Use dynamic programming to

formulate a method for determining how much raw material the company should buy and how much it should allocate to the various products. Using your formulation, solve this problem with the data shown in Tables 1 and 2. Note that the variables k and t can take only the values which appear in the tables, or zero.

Table 1 The cost of raw materials

t tons	2	4	6	8
£$c(t)$	15	28	40	50

Table 2 Returns from raw materials usage

Allocations in tons	Returns (£) $r(A, k)$	$r(B, k)$	$r(C, k)$
1	9	6	8
2	18	12	10
3	23	21	11

(6) A plant overhaul has to be completed in ten days. The overhaul has three stages, stripping, repairing, and rebuilding, and these stages must be completed successively. The time taken to complete each stage is related to the cost of the resources which are used at that stage in the way shown in the following table.

Time to complete in days	Cost of stage (£1000s) Stripping	Repairing	Rebuilding
2	18	11	20
4	17	8	15
6	8	7	9
8	6	5	8

Develop a general dynamic programming formulation for allocating the available time to the stages of the overhaul in such a way as to minimise costs, and use your formulation to obtain the optimal allocation with the given data.

(7) A fish farmer keeps fish in a single large tank. He has a contract to supply a fixed number of fish at the end of each month for the period January to May. The price he receives for the fish depends on the month and the size of the fish in the way shown in the table below.

Selling price, pence per fish

Month	Length of fish (in) 8	9	10	11	12
Jan	10	15	20	–	–
Feb	8	10	14	20	24
Mar	4	6	10	14	20
Apr	2	4	8	14	20
May	0	2	8	14	22

At the beginning of January all the fish are 8 in long. Subsequently they can be fed at rates 1, 2 or 3 in each month. Feeding at rate 1 maintains a fish at its current size, feeding at rate 2 increases its size by one inch in one month, and feeding at rate 3 increases its size by two inches in one month. All the remaining fish must be fed at the same rate. The cost of feeding at the various rates depends on the size of the fish in the way shown in the table below.

Feeding cost, pence per fish

Length (inches)	Rate 1	2	3
8	1	2	5
9	3	4	7
10	5	6	7
11	8	9	–
12	10	–	–

Determine the feeding policy which maximises the total return and find the total return under that policy.

(8) Use the D.P. method of section 13.5, with a suitable mesh, to find the minimum of the functional

$$\int_2^1 (\dot{x}^2 - x^2)\mathrm{d}t, \quad x(0) = 0, \quad x(1) = 1.$$

Compare your result with the exact answer, obtained from the Euler equation.

APPENDIX 1

Partial fractions

We are concerned with rational functions, that is, ratios of polynomial functions of the form

$$G(s) = \frac{K(s^m + b_1 s^{m-1} + \ldots + b_m)}{s^n + a_1 s^{n-1} + \ldots + a_n}.$$

If $m \geqslant n$, $G(s)$ is said to be **improper**.
If $m < n$, $G(s)$ is said to be **proper**.

When $G(s)$ is improper, the denominator is divided into the numerator until the remainder is a proper fraction. For example,

$$G(s) = \frac{s^2 + 2s + 6}{s - 2} = s + 4 + \frac{14}{s - 2}.$$

In what follows we consider that $m < n$. There are two cases to consider.

Case 1: The roots of the denominator polynomial of $G(s)$ are *distinct*, in which case $G(s)$ has the form

$$G(s) = \frac{K(s^m + b_1 s^{m-1} + \ldots + b_m)}{(s + p_1)(s + p_2)\ldots(s + p_n)}.$$

To illustrate one of the methods used for the expansion of a proper rational function into a sum of partial fractions we consider the following example

$$G(s) = \frac{s^2 - 2s + 2}{(s + 1)(s + 2)(s + 3)} = \frac{K_1}{s + 1} + \frac{K_2}{s + 2} + \frac{K_3}{s + 3}.$$

To determine K_1, K_2 and K_3 we proceed as follows.

Multiply through by $(s+1)$, to obtain

$$(s+1)G(s) = \frac{s^2-2s+2}{(s+2)(s+3)} = K_1 + (s+1)\left[\frac{K_2}{s+2} + \frac{K_3}{s+3}\right].$$

Now let $s \to -1$; the second term on the right-hand side tends to zero so that

$$K_1 = \lim_{s \to -1} \frac{s^2-2s+2}{(s+2)(s+3)} = \frac{5}{2}.$$

This is also written as

$$K_1 = \left[\frac{s^2-2s+2}{(s+2)(s+3)}\right]_{s=-1} = \frac{5}{2}.$$

Similarly, we obtain

$$K_2 = \left[\frac{s^2-2s+2}{(s+1)(s+3)}\right]_{s=-2} = -10 \quad \text{and}$$

$$K_3 = \left[\frac{s^2-2s+2}{(s+1)(s+2)}\right]_{s=-3} = \frac{17}{2}.$$

The general formula for the residues K_i, is

$$K_i = [(s+p_i)G(s)]_{s=-p_i} \qquad (i = 1, 2, \ldots, n)$$

Case 2: The roots of the denominator polynomial are *multiple*. To simplify the mathematical manipulation, we consider that $G(s)$ has the form

$$G(s) = \frac{N(s)}{(s+p)^n}.$$

By division, we can write

$$N(s) = (s+p)^{n-1}q_1 + r_1(s), \qquad \text{(see example below)}$$

where q_1 is zero or a constant and $r_1(s)$ is a polynomial of degree $<(n-1)$. $r_1(s)$ can of course be zero. Also

$$r_1(s) = (s+p)^{n-2}q_2 + r_2(s)$$

where the degree of $r_2(s) < (n-2)$.

Similarly

$$
\begin{aligned}
r_2(s) &= (s+p)^{n-3}q_3 + r_3(s)\\
&\vdots\\
r_{n-2}(s) &= (s+p)q_{n-1} + r_{n-1}(s)\\
r_{n-1}(s) &= 1.q_n .
\end{aligned}
$$

We can now write

$$N(s) = (s+p)^{n-1}q_1 + (s+p)^{n-2}q_2 + \ldots + (s+p)q_{n-1} + q_n$$

so that

$$G(s) = \frac{N(s)}{(s+p)^n} = \frac{q_1}{s+p} + \frac{q_2}{(s+p)^2} + \ldots + \frac{q_{n-1}}{(s+p)^{n-1}} + \frac{q_n}{(s+p)^n}.$$

To illustrate the above procedure we consider the following example:

$$\frac{N(s)}{(s+p)^n} = \frac{2s^2+s-3}{(s+2)^3}.$$

Since $\dfrac{2s^2+s-3}{(s+2)^2} = 2 + \dfrac{-7s-10}{(s+2)^2}$, we have

$$2s^2 + s - 3 = (s+2)^2 2 + (-7s - 10).$$

Similarly,

$$
\begin{aligned}
-7s - 10 &= (s+2).(-7) + 4\\
4 &= 4.1.
\end{aligned}
$$

Hence

$$2s^2 + s - 3 = (s+2)^2.2 + (s+2).(-7) + 4,$$

so that

$$\frac{2s^2+s-3}{(s+2)^3} = \frac{2}{s+2} - \frac{7}{(s+2)^2} + \frac{4}{(s+2)^3}.$$

To see how to compute $q_1, q_2, \ldots, q_n$ in the partial fraction expansion of $G(s)$ above, we proceed as follows.

On multiplying the above expression for $G(s)$ by $(s+p)^n$, we obtain

$$(s+p)^n G(s) = q_n + (s+p)q_{n-1} + (s+p)^2 q_{n-2} + \ldots + (s+p)^{n-1}q_1.$$

Let $s \to -p$, we obtain

$$q_n = [(s+p)^n G(s)]_{s=-p}.$$

On differentiating with respect to s,

$$\frac{\mathrm{d}}{\mathrm{d}s}\{(s+p)^n G(s)\} = q_{n-1} + 2(s+p)q_{n-2} + \ldots + (n-1)(s+p)^{n-2}q_1$$

hence

$$q_{n-1} = \left[\frac{\mathrm{d}}{\mathrm{d}s}\{(s+p)^n G(s)\}\right]_{s=-p}.$$

Similarly,

$$\boxed{q_{n-r} = \frac{1}{r!}\left[\frac{\mathrm{d}^r}{\mathrm{d}s^r}\{(s+p)^n G(s)\}\right]_{s=-p}} \quad (r = 1, 2, \ldots, n-1).$$

For example, let

$$\begin{aligned} G(s) &= \frac{2s^2 - 3s + 1}{(s+1)(s+2)^2} \\ &= \frac{K_1}{s+1} + \frac{K_{21}}{s+2} + \frac{K_{22}}{(s+2)^2}, \end{aligned}$$

then

$$K_1 = \left[\frac{2s^2 - 3s + 1}{(s+2)^2}\right]_{s=-1} = 6$$

$$K_{22} = \left[\frac{2s^2 - 3s + 1}{s+1}\right]_{s=-2} = -15$$

$$K_{21} = \left[\frac{\mathrm{d}}{\mathrm{d}s}\left(\frac{2s^2 - 3s + 1}{s+1}\right)\right]_{s=-2} = -4.$$

APPENDIX 2

Notes on determinants and matrices

The notes that follow are based on the book *Matrix Theory and Applications for Engineers and Mathematicians* by A. Graham (published by Ellis Horwood Ltd, 1979).

A2.1 DETERMINANTS

A determinant is associated with a *square* matrix A, that is one of order $n \times n$. It is a single number, frequently denoted by $|A|$.

The determinant of the matrix A having one element only is defined as the value of this element, that is, if $A = [a_{11}]$, then $|A| = a_{11}$.

$$\text{When } A = \begin{bmatrix} a_{11} & a_{12} \\ a_{21} & a_{22} \end{bmatrix}, \text{ then } |A| = a_{11}a_{22} - a_{12}a_{21}.$$

Similarly, when

$$A = \begin{bmatrix} a_{11} & a_{12} & a_{13} \\ a_{21} & a_{22} & a_{23} \\ a_{31} & a_{32} & a_{33} \end{bmatrix}$$

then

$$|A| = a_{11}\begin{vmatrix} a_{22} & a_{23} \\ a_{32} & a_{33} \end{vmatrix} - a_{12}\begin{vmatrix} a_{21} & a_{23} \\ a_{31} & a_{33} \end{vmatrix} + a_{13}\begin{vmatrix} a_{21} & a_{22} \\ a_{31} & a_{32} \end{vmatrix}.$$

We write the above as

$$|A| = a_{11}A_{11} + a_{12}A_{12} + a_{13}A_{13}$$

$$\text{where } A_{11} = \begin{vmatrix} a_{22} & a_{23} \\ a_{32} & a_{33} \end{vmatrix}, \; A_{12} = -\begin{vmatrix} a_{21} & a_{23} \\ a_{31} & a_{33} \end{vmatrix}, \; A_{13} = \begin{vmatrix} a_{21} & a_{22} \\ a_{31} & a_{32} \end{vmatrix}$$

are called the **cofactors** of the elements a_{11}, a_{12} and a_{13} respectively.

The signs of the cofactors of the various elements are allocated according to the position of the element as suggested by the following checkboard pattern:

$$\begin{bmatrix} + & - & + & \cdots \\ - & + & - & \cdots \\ + & - & + & \cdots \\ \vdots & \vdots & \vdots & \end{bmatrix}$$

The cofactor A_{ij} of the element a_{ij} of A is the determinant obtained by removing the ith row and the jth column of A, multiplied by $+1$ or -1 according to the above pattern.

For example, in the 3×3 matrix above, the cofactor A_{23} is obtained by deleting the second row and the third column from A and multiplying the corresponding determinant by (-1), so that

$$A_{23} = (-1)\begin{vmatrix} a_{11} & a_{12} \\ a_{31} & a_{32} \end{vmatrix} = -(a_{11}a_{32} - a_{12}a_{31}) = (a_{12}a_{31} - a_{11}a_{32}).$$

It is not difficult to prove that the determinant is the sum of the products of the elements of *any* row, by their corresponding cofactors, so that

$$\begin{aligned} |A| &= a_{11}A_{11} + a_{12}A_{12} + a_{13}A_{13} \\ &= a_{21}A_{21} + a_{22}A_{22} + a_{23}A_{23} \\ &= a_{31}A_{31} + a_{32}A_{32} + a_{33}A_{33}. \end{aligned}$$

For example

$$A = \begin{vmatrix} 1 & 2 & -2 \\ -1 & 1 & 3 \\ 2 & -1 & 2 \end{vmatrix}$$

$$= 1\begin{vmatrix} 1 & 3 \\ -1 & 2 \end{vmatrix} - 2\begin{vmatrix} -1 & 3 \\ 2 & 2 \end{vmatrix} - 2\begin{vmatrix} -1 & 1 \\ 2 & -1 \end{vmatrix}$$

(expanding by 1st row)

$$= 2\begin{vmatrix} 2 & -2 \\ 1 & 3 \end{vmatrix} + 1\begin{vmatrix} 1 & -2 \\ -1 & 3 \end{vmatrix} + 2\begin{vmatrix} 1 & 2 \\ -1 & 1 \end{vmatrix}$$

(expanding by 3rd row)

$$= 23.$$

On the other hand, the sum of the products of the elements of any row, by cofactors corresponding to the elements of another row is zero. For example

$$a_{11}A_{31} + a_{12}A_{32} + a_{13}A_{33} = 0$$

so that for the above determinant

$$1\begin{vmatrix} 2 & -2 \\ 1 & 3 \end{vmatrix} - 2\begin{vmatrix} 1 & -2 \\ -1 & 3 \end{vmatrix} - 2\begin{vmatrix} 1 & 2 \\ -1 & 1 \end{vmatrix} = 0.$$

This result leads directly to an important rule for inverting a matrix. This rule makes use of the **adjoint** of A, denoted by adj.A; it is the transpose of the matrix of cofactors of the elements of the square matrix $A = [a_{ij}]$.

So that if

$$A = \begin{bmatrix} a_{11} & a_{12} & a_{13} \\ a_{21} & a_{22} & a_{23} \\ a_{31} & a_{32} & a_{33} \end{bmatrix} \text{ then adj.}A = \begin{bmatrix} A_{11} & A_{21} & A_{31} \\ A_{12} & A_{22} & A_{32} \\ A_{13} & A_{23} & A_{33} \end{bmatrix}.$$

The rule is the following:

if A is a non-singular square matrix (see below for definition), then

$$A^{-1} = \frac{1}{|A|}\,\text{adj.}\,A.$$

It is left to the reader to investigate further important properties of determinants.

For example, if A and B are matrices, both of order $n \times n$, then

$$|AB| = |A|.|B|$$

and that

$$|A| = |A'| \quad (A' \text{ being the transpose of } A).$$

The matrix A is said to be **non-singular** if $|A| \neq 0$ and **singular** if $|A| = 0$.

When A is non-singular, it is **invertible**, so that the inverse A^{-1} can be found such that

$$AA^{-1} = A^{-1}A = I.$$

On taking the determinants of both sides, we obtain

$$|A||A^{-1}| = 1.$$

A2.2 PARTITIONED MATRICES

It is frequently useful to partition a matrix into a number of smaller arrays. Partitioning can be done in a number of different ways and is obviously non-unique. For example, the matrix

$$\begin{bmatrix} 1 & 0 & 0 \\ 0 & 1 & 1 \\ -1 & 0 & 1 \end{bmatrix}$$

can be partitioned as

$$\left[\begin{array}{cc|c} 1 & 0 & 0 \\ 0 & 1 & 1 \\ \hline -1 & 0 & 1 \end{array}\right] = \begin{bmatrix} A_{11} & A_{12} \\ A_{21} & A_{22} \end{bmatrix}$$

where

$$A_{11} = \begin{bmatrix} 1 & 0 \\ 0 & 1 \end{bmatrix}, \; A_{12} = \begin{bmatrix} 0 \\ 1 \end{bmatrix}, \; A_{21} = [-1 \quad 0] \text{ and } A_{22} = [1].$$

This particular partitioning has the merit of creating the unit submatrix A_{11} of order 2×2, which may be an advantage in a number of applications.

A2.3 EIGENVECTORS AND EIGENVALUES

We consider a square matrix A of order $n \times n$. If there exists a vector $\mathbf{x} \neq \mathbf{0}$ and a scalar λ such that

$$A\mathbf{x} = \lambda\mathbf{x}$$

then $\mathbf{x}$ is called an **eigenvector** of A. λ is called an **eigenvalue** of A.

The above equation can be written in the form

$$[\lambda I - A]\mathbf{x} = \mathbf{0}$$

where I is the unit matrix (of order $n \times n$). It is known that the above homogeneous equation has non-trivial (that is, non-zero) solution only if the matrix $[\lambda I - A]$ is singular, that is if

$$|\lambda I - A| = 0.$$

This is an equation in λ of great importance. We denoted it by $c(\lambda)$, so that

$$c(\lambda) = |\lambda I - A| = 0.$$

It is called the **characteristic equation** of A. Written out in full, this equation has the form

$$c(\lambda) = \begin{vmatrix} \lambda - a_{11} & a_{12} & \dots & a_{1n} \\ a_{21} & \lambda - a_{22} & \dots & a_{2n} \\ \vdots & & & \\ a_{n1} & a_{n2} & & \lambda - a_{nn} \end{vmatrix} = 0.$$

On expansion of the determinant, $c(\lambda)$ is seen to be a polynomial of degree n in λ, having the form

$$c(\lambda) = \lambda^n + b_1\lambda^{n-1} + \dots + b_{n-1}\lambda + b_n$$

$$= (\lambda - \lambda_1)(\lambda - \lambda_2)\dots(\lambda - \lambda_n) = 0.$$

$\lambda_1, \lambda_2, \dots, \lambda_n$, the roots of $c(\lambda) = 0$, are the eigenvalues of A.

Assuming that A has distinct eigenvalues $\lambda_1, \lambda_2, \dots, \lambda_n$, the corresponding eigenvectors $\mathbf{x}_1, \mathbf{x}_2, \dots, \mathbf{x}_n$ are linearly independent. The (partitioned) matrix

$$X = [\mathbf{x}_1 \quad \mathbf{x}_2 \quad \dots \quad \mathbf{x}_n]$$

is called the **modal matrix** of A. Since

$$A\mathbf{x}_i = \lambda\mathbf{x}_i \qquad (i = 1, 2, \ldots, n)$$

it follows that

$$AX = \Lambda X$$

where

$$\Lambda = \begin{bmatrix} \lambda_1 & 0 & 0 & \ldots & 0 \\ 0 & \lambda_2 & 0 & \ldots & 0 \\ \vdots & & & & \\ 0 & 0 & 0 & \ldots & \lambda_n \end{bmatrix} = \operatorname{diag}\{\lambda_1, \lambda_2, \ldots, \lambda_n\}$$

is called the **eigenvalue matrix** of A.

Hence

$$\Lambda = X^{-1}AX.$$

For example, if

$$A = \begin{bmatrix} 3 & -1 \\ 4 & -2 \end{bmatrix}$$

then $c(\lambda) = (\lambda + 1)(\lambda - 2) = 0$ so that $\lambda_1 = -1$ and $\lambda_2 = 2$. The corresponding eigenvectors are

$$\mathbf{x}_1 = [1 \quad 4]' \quad \text{and} \quad \mathbf{x}_2 = [1 \quad 1]'.$$

It follows that the modal matrix is

$$X = \begin{bmatrix} 1 & 1 \\ 4 & 1 \end{bmatrix} \quad \text{so that} \quad X^{-1} = \frac{1}{3}\begin{bmatrix} -1 & 1 \\ 4 & -1 \end{bmatrix}$$

and $X^{-1}AX = \operatorname{diag}\{-1, 2\}$.

The eigenvalue matrix Λ and the matrix A are related by the equation

$$\Lambda = X^{-1}AX.$$

More generally, any two matrices A and B related by an equation of the form

$$B = P^{-1}AP \quad \text{(for any non-singular matrix } P\text{)}$$

are said to be **similar**.

Similar matrices have the same characteristic equation. This important result follows since

$$\begin{aligned} |\lambda I - B| &= |\lambda I - P^{-1}AP| = |\lambda P^{-1}IP - P^{-1}AP| \\ &= |P^{-1}||\lambda I - A||P| = |P^{-1}||P||\lambda I - A| \\ &= |\lambda I - A| \quad \text{since} \quad |P^{-1}||P| = 1. \end{aligned}$$

A2.4 THE COMPANION MATRIX

In section 3.4 it was stated that the characteristic equation of the matrix A in companion form

$$A = \begin{bmatrix} 0 & 1 & 0 & \dots & 0 \\ 0 & 0 & 1 & \dots & 0 \\ \vdots & & & & \\ 0 & 0 & 0 & \dots & 1 \\ -a_0 & -a_1 & -a_2 & \dots & -a_{n-1} \end{bmatrix}$$

is

$$c(\lambda) = \lambda^n + a_{n-1}\lambda^{n-1} + \dots + a_1\lambda + a_0 = 0.$$

We now prove this statement. The proof relies on a property of determinants, namely that the addition of a multiple of the elements of any column to the corresponding elements of another column does not alter the value of the determinant.

The characteristic equation of A is

$$|\lambda I - A| = \begin{vmatrix} \lambda & -1 & 0 & \dots & 0 \\ 0 & \lambda & -1 & \dots & 0 \\ \vdots & & & & \\ 0 & 0 & 0 & \dots & -1 \\ a_0 & a_1 & a_2 & \dots & \lambda + a_{n-1} \end{vmatrix}$$

Add λ times the second column, λ^2 times the third column, ... λ^{n-1} times the nth column, to the first column of the determinant; we obtain

$$|\lambda I - A| = \begin{vmatrix} 0 & -1 & 0 & \dots & 0 \\ 0 & \lambda & -1 & \dots & 0 \\ \vdots & & & & \\ 0 & 0 & 0 & \dots & -1 \\ c(\lambda) & a_1 & a_2 & \dots & \lambda + a_{n-1} \end{vmatrix}.$$

On expanding the determinant by the first column, we obtain

$$|\lambda I - A| = \pm c(\lambda).$$

The result follows.

A2.5 THE CAYLEY-HAMILTON THEOREM

Although, in general, operations with scalars are different from operations with matrices, there are a number of notable exceptions.

For example, given a square matrix A (of order $n \times n$) and integers r and s, then

$$A^r A^s = A^{r+s} = A^s A^r.$$

This property leads to the following definition.

Definition

Corresponding to a polynomial in a scalar variable λ

$$f(\lambda) = \lambda^k + b_1 \lambda^{k-1} + \dots + b_{k-1}\lambda + b_k$$

we define the (square) matrix $f(A)$, called a **matrix polynomial**, by

$$f(A) = A^k + b_1 A^{k-1} + \dots + b_{k-1} A + b_k I$$

where I is the unit matrix of order $n \times n$.

For example, corresponding to

$$f(\lambda) = \lambda^2 - 2\lambda + 3 \quad \text{and} \quad A = \begin{bmatrix} 1 & 1 \\ -1 & 3 \end{bmatrix},$$

we have

$$f(A) = \begin{bmatrix} 0 & 4 \\ -4 & 8 \end{bmatrix} - 2\begin{bmatrix} 1 & 1 \\ -1 & 3 \end{bmatrix} + 3\begin{bmatrix} 1 & 0 \\ 0 & 1 \end{bmatrix} = \begin{bmatrix} 1 & 2 \\ -2 & 5 \end{bmatrix}.$$

Of particular interest to us are polynomials f having the property that

$$f(A) = 0.$$

For every matrix A one such polynomial can be found by the **Cayley–Hamilton theorem** which states that:

Every square matrix A satisfies its own characteristic equations.

For the above example, the characteristic equation of A is

$$c(\lambda) = \lambda^2 - 4\lambda + 4$$

so that $$c(A) = A^2 - 4A + 4I$$

$$= \begin{bmatrix} 0 & 4 \\ -4 & 8 \end{bmatrix} - \begin{bmatrix} 4 & 4 \\ -4 & 12 \end{bmatrix} + \begin{bmatrix} 4 & 0 \\ 0 & 4 \end{bmatrix} = \begin{bmatrix} 0 & 0 \\ 0 & 0 \end{bmatrix}.$$

We have found a polynomial $c(\lambda)$ of degree 2 such that $c(A) = 0$ where A is of order 2×2. In fact the Cayley–Hamilton theorem guarantees the existence of a polynomial $c(\lambda)$ of degree n such that

$$c(A) = 0$$

whenever A is of order $n \times n$.

But it is sometimes possible to find a polynomial of degree smaller than n which has this property.

For example, when

$$A = \begin{bmatrix} 2 & 0 \\ 0 & 2 \end{bmatrix}$$

both

$$c(\lambda) = \lambda^2 - 4\lambda + 4 \quad \text{and} \quad m(\lambda) = \lambda - \tfrac{1}{2}$$

have this property.

Definition

The polynomial $m(\lambda)$ of lowest degree whose coefficient of the highest power of λ is unity (that is, a **monic polynomial**) and such that $m(A) = 0$, where A is a matrix of order $n \times n$, is called the **minimum** (or **minimal**) **polynomial** of A.

Definition

Any polynomial f such that $f(A) = 0$ is called an **annihilating polynomial** for A.

A2.6 LINEAR DEPENDENCE AND THE RANK OF A MATRIX

Let $\mathbf{x}_1, \mathbf{x}_2, \ldots, \mathbf{x}_r$ be vectors and $a_1, a_2, \ldots, a_r$ be scalars. We consider the equation

$$a_1\mathbf{x}_1 + a_2\mathbf{x}_2 + \ldots + a_n\mathbf{x}_n = \mathbf{0}.$$

If at least one $a_i \neq 0$ in the equation, the vectors $\mathbf{x}_1, \mathbf{x}_2, \ldots, \mathbf{x}_r$ are said to be **linearly dependent** and the equation is said to have a **non-trivial** solution. If, on the other hand, the above equation is satisfied only when

$$a_1 = a_2 = \ldots = a_r = 0$$

the vectors are said to be **linearly independent**, and this solution is said to be **trivial**.

For example, the vectors

$$\mathbf{x}_1 = \begin{bmatrix} 2 \\ 1 \\ 3 \end{bmatrix}, \ \mathbf{x}_2 = \begin{bmatrix} 1 \\ 0 \\ 2 \end{bmatrix} \text{ and } \mathbf{x}_3 = \begin{bmatrix} 1 \\ 2 \\ 0 \end{bmatrix}$$

are linearly dependent, since

$$a_1\mathbf{x}_1 + a_2\mathbf{x}_2 + a_3\mathbf{x}_3 = \mathbf{0}$$

when $a_1 = 2$, $a_2 = -3$ and $a_3 = -1$ (one of many possible solutions).

On the other hand, the vectors

$$\mathbf{x}_1 = \begin{bmatrix} 1 \\ 2 \\ 3 \end{bmatrix}, \ \mathbf{x}_2 = \begin{bmatrix} 1 \\ 1 \\ 0 \end{bmatrix} \text{ and } \mathbf{x}_3 = \begin{bmatrix} 1 \\ 0 \\ 1 \end{bmatrix}$$

are linearly independent, since the only solution in this case is the trivial one.

The concept of linear dependence is an extremely important one in linear algebra. It is closely associated with another concept, that of the *rank* of a matrix.

We consider a matrix A of order $m \times n$, say

$$A = \begin{bmatrix} a_{11} & a_{12} & \cdots & a_{1n} \\ a_{21} & a_{22} & \cdots & a_{2n} \\ \vdots & & & \\ a_{m1} & a_{m2} & \cdots & a_{mn} \end{bmatrix}.$$

We can write this matrix as

$$A = [\mathbf{a}_1 \quad \mathbf{a}_2 \quad \ldots \quad \mathbf{a}_n]$$

where $$\mathbf{a}_i = \begin{bmatrix} a_{1i} \\ a_{2i} \\ \vdots \\ a_{mi} \end{bmatrix} \qquad (i = 1, 2, \ldots, n).$$

The **rank** of the matrix A can be defined as the number of linearly independent vectors in the set $\{\mathbf{a}_1, \mathbf{a}_2, \ldots, \mathbf{a}_n\}$, that is the number of linearly independent columns of the matrix.

If A is a square matrix of order $n \times n$ (say) it can be shown that it is non-singular if the rank of A, written as $r(A)$, is n.

If $r(A) < n$, then A is singular. Conversely, if A is non-singular, then $r(A) = n$ which implies that the columns of A are linearly independent. On the other hand, if $r(A) < n$, the columns of A are linearly dependent.

Solutions to problems

CHAPTER 1

(1) (a) $f(t) = M\dfrac{d^2y}{dt^2}$

(b) $f(t) = ky(t)$

(c) $f(t) = \mu\dfrac{dy(t)}{dt}$.

(2) (a) $M\dfrac{d^2y}{dt^2} + \mu\dfrac{dy}{dt} + ky(t) = 0$

(b) For mass M_1

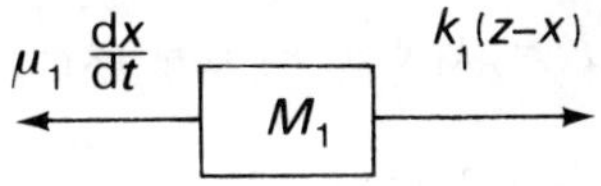

$$M_1\frac{d^2x}{dt^2} = k_1(z-x) - \mu_1\frac{dx}{dt}$$

For mass M_2

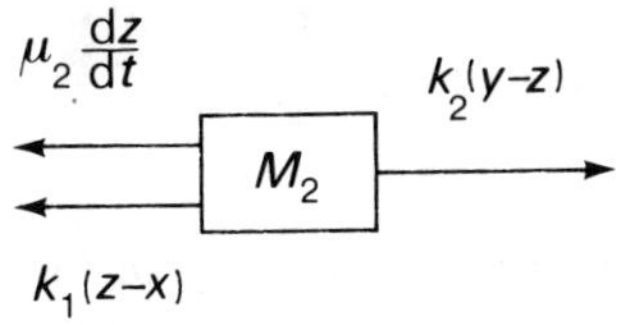

$$M_2\frac{d^2z}{dt^2} = k_2(y-z) - k_1(z-x) - \mu_2\frac{dz}{dt}.$$

On rearranging, we obtain

$$\left.\begin{aligned} M_1\frac{d^2x}{dt^2} + \mu_1\frac{dx}{dt} + k_1x - k_1z &= 0 \\ M_2\frac{d^2z}{dt^2} + \mu_2\frac{dz}{dt} + (k_1+k_2)z - k_1x &= k_2y \end{aligned}\right\}$$

(3)

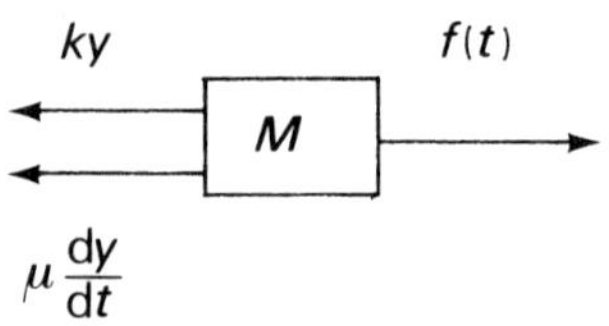

$$M\frac{d^2y}{dt^2} = f(t) - ky - \mu\frac{dy}{dt}$$

so that

$$\frac{d^2y}{dt^2} + \frac{\mu}{M}\frac{dy}{dt} + \frac{k}{M}y = \frac{1}{M}f(t).$$

(4) (i) (a) If the force F is required to deflect the mass a distance x, then both springs are deflected by x, so that

$$F = k_1x + k_2x = (k_1+k_2)x.$$

For the equivalent spring

$$F = Kx.$$

Hence

$$K = k_1 + k_2.$$

(b) If the force F is required to deflect the mass a distance x, the springs are deflected by distances x_1 and x_2 respectively so that

$$x = x_1 + x_2.$$

But $x_1 = \frac{F}{k_1}$ and $x_2 = \frac{F}{k_2}$, so that

$$x = F\left(\frac{1}{k_1} + \frac{1}{k_2}\right).$$

For the equivalent spring

$$x = \frac{F}{K}.$$

Hence $\frac{1}{K} = \frac{1}{k_1} + \frac{1}{k_2}$, so that

$$K = \frac{k_1 k_2}{k_1 + k_2}.$$

(ii) (a)

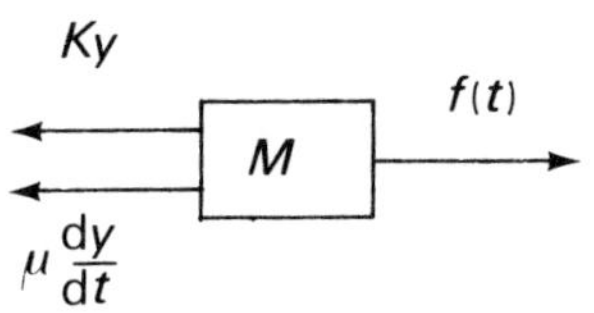

$$M\frac{d^2y}{dt^2} = f(t) - Ky - \mu\frac{dy}{dt}$$

where $K = k_1 + k_2$, hence

$$M\frac{d^2y}{dt^2} + \mu\frac{dy}{dt} + (k_1 + k_2)y = f(t)$$

(b) $$M\frac{d^2y}{dt^2} + \mu\frac{dy}{dt} + \frac{k_1 k_2}{k_1 + k_2}y = f(t).$$

(5) The voltage drop across the resistor is

$$E - V.$$

The current in the circuit is

$$I = \frac{E - V}{R}.$$

Across the capacitor (see equation (1.2))

$$I = C\frac{dV}{dt}.$$

It follows that

$$\frac{E-V}{R} = C\frac{dV}{dt}$$

or

$$\frac{dV}{dt} + \frac{V}{CR} = \frac{E}{CR}.$$

(6) $$\frac{dh}{dt} + \frac{1}{AR}h = \frac{1}{A}q.$$

(7) $$A\frac{dh_1}{dt} = q - \frac{1}{R}h_1$$

$$A\frac{dh_2}{dt} = \frac{1}{R}(h_1 - h_2).$$

CHAPTER 2

(1) (i) $$F(s) = \int_0^\infty te^{-st}dt = -\int_0^\infty \frac{d}{ds}(e^{-st})\,dt.$$

The conditions for the Laplace transform to exist are also sufficient to allow for the interchange of the operations of integration and differentiation, hence

$$F(s) = -\frac{d}{ds}\int_0^\infty e^{-st}dt = -\frac{d}{ds}U(s)$$

where $U(s) = \mathcal{L}$ (unit step function) (see Exercise 2.1)

$$= \frac{1}{s}.$$

It follows that

$$F(s) = \frac{1}{s^2}.$$

(ii) $$\mathcal{L}\{f(t)e^{-\alpha t}\} = \int_0^\infty f(t)e^{-(s+\alpha)t}\,dt = F(s+\alpha).$$

(iii) By (i) above $\mathcal{L}\{t\} = \frac{1}{s^2}$

so that by (ii) above $\mathcal{L}\{te^{-\alpha t}\} = F(s+\alpha)$

$$= \frac{1}{(s+\alpha)^2}.$$

(2) (i) $$\frac{d}{ds}F(s) = \frac{d}{ds}\int f(t)e^{-st}\,dt = \int \frac{d}{ds}[f(t)e^{-st}]\,dt$$

$$= -\quad tf(t)e^{-st}\,dt = -\mathcal{L}\{tf(t)\}.$$

Hence $$\mathcal{L}\{tf(t)\} = -\frac{d}{ds}F(s).$$

(ii) $$\mathcal{L}(t) = -\frac{d}{ds}\left(\frac{1}{s}\right) = \frac{1}{s^2}$$

$$\mathcal{L}(t^2) = -\frac{d}{ds}\left(\frac{1}{s^2}\right) = \frac{2!}{s^3}.$$

Assume that

$$\mathcal{L}(t^{n-1}) = \frac{(n-1)!}{s^n},$$

then $\mathcal{L}\{t^n\} = \mathcal{L}\{tf(t)\}$ where $f(t) = t^{n-1}$

$$= -\frac{d}{ds}F(s) = -\frac{d}{ds}\left[\frac{(n-1)!}{s^n}\right] = \frac{n!}{s^{n+1}}.$$

(3) (i) Using the result of Example 2.3, we find

$$\mathcal{L}\{\sin 2t\} = \frac{2}{s^2+4}.$$

(ii) We found [equation (2.2)], $\mathcal{L}\{f'(t)\} = sF(s) - f(0)$

$$= s\left(\frac{2}{s^2+4}\right) - 0 = \frac{2s}{s^2+4}.$$

(iii) $$\mathcal{L}\left\{\int_0^t f(u)\mathrm{d}u\right\} = \frac{F(s)}{s} = \frac{1}{s}\left(\frac{2}{s^2+4}\right).$$

(4) Use the result of Problem 1, $\mathcal{L}\{f(t)\mathrm{e}^{-\alpha t}\} = F(s+\alpha)$.

(i) When $f(t) = t^3$, $\mathcal{L}\{f(t)\} = \dfrac{3!}{s^4}$, hence

$$\mathcal{L}\{t^3\mathrm{e}^{2t}\} = F(s-2) = \frac{3!}{(s-2)^2}.$$

(ii) $f(t) = \sin 2t$, $F(s) = \dfrac{2}{s^2+4}$, hence

$$\mathcal{L}\{\mathrm{e}^{-3t}\sin 2t\} = F(s+3) = \frac{2}{(s+3)^2+4} = \frac{2}{s^2+6s+13}.$$

(iii) $f(t) = \cos 2t - 3\sin 2t$, $F(s) = \dfrac{s}{s^2+4} - \dfrac{6}{s^2+4}$, hence

$$\mathcal{L}\{\mathrm{e}^{-2t}f(t)\} = F(s+2) = \frac{s+2}{(s+2)^2+4} - \frac{6}{(s+2)^2+4}$$

$$= \frac{s-4}{s^2+4s+8}.$$

(5) (i) Since $\mathcal{L}\left\{\dfrac{t^n}{n!}\right\} = \dfrac{1}{s^{n+1}}$ (see Problem 2)

then $\mathcal{L}^{-1}\left\{\dfrac{1}{s^{n+1}}\right\} = \dfrac{t^n}{n!}$, hence $\mathcal{L}^{-1}\left\{\dfrac{1}{s^3}\right\} = \dfrac{t^2}{2!}$.

(ii) Since $\mathcal{L}\{\cos \alpha t\} = \dfrac{s}{s^2+\alpha^2}$, then $\mathcal{L}^{-1}\left\{\dfrac{s}{s^2+\alpha^2}\right\} = \cos \alpha t$,

hence $\mathcal{L}^{-1}\left\{\dfrac{s}{s^2+4}\right\} = \cos 2t.$

(iii) $$\mathcal{L}^{-1}\left\{\frac{2s+5}{s^2-4s+5}\right\} = \mathcal{L}^{-1}\left\{\frac{2s+5}{(s-2)^2+9}\right\} = \mathcal{L}^{-1}\left\{\frac{2(s-2)+9}{(s-2)^2+9}\right\}$$

$$= 2\mathcal{L}^{-1}\left\{\frac{s-2}{(s-2)^2+9}\right\} + 3\mathcal{L}^{-1}\left\{\frac{3}{(s-2)^2+9}\right\}$$

$$= 2e^{2t}\cos 3t + 3e^{2t}\sin 3t$$

(see table of Laplace transforms)

$$= e^{2t}(2\cos 3t + 3\sin 3t).$$

(iv) By Problem 1 above

$$\mathcal{L}^{-1}\left\{\frac{1}{(s-2)^3}\right\} = e^{2t}\frac{t^2}{2!}.$$

(v) Apply the result

$$\mathcal{L}^{-1}\{e^{-\alpha s}F(s)\} = f(t-\alpha) \quad \text{(see equation 2.6)}$$

with $\alpha = 3$ and $f(t) = e^{2t}\dfrac{t^2}{2!}$.

Then $$f(t-\alpha) = e^{2(t-3)}\frac{(t-3)^2}{2!}.$$

(6) (i) Taking Laplace Theorem of both sides and substituting for initial conditions, we obtain

$$(s^2 - 2s + 2)Y - s + 3 = \frac{3}{s-2},$$

so that

$$Y = \frac{s^2-5s+9}{(s-2)(s^2-2s+2)} = \frac{3/2}{s-2} - \frac{1/2s+3}{s^2-2s+2}$$

$$= \frac{3/2}{s-2} - \frac{1}{2}\frac{s-1}{(s-1)^2+1} - \frac{7}{2}\frac{1}{(s-1)^2+1}.$$

Hence $y(t) = \mathcal{L}^{-1}\{Y\} = \frac{3}{2} e^{2t} - \frac{1}{2} e^{t} (\cos t + 7 \sin t)$.

(ii) $(s^2 - s - 2)Y - s = \frac{s}{s^2+4}$

$$Y = \frac{s^3 + 5s}{(s-2)(s+1)(s^2+4)} = \frac{3/4}{s-2} + \frac{2/5}{s+1} - \frac{1}{20}\frac{3s+2}{s^2+4}$$

$$= \frac{3/4}{s-2} + \frac{2/5}{s+1} - \frac{3}{20}\frac{s}{s^2+4} - \frac{1}{20}\frac{1}{s^2+4}$$

$$y(t) = \frac{3}{4} e^{2t} + \frac{2}{5} e^{-t} - \frac{3}{20} \cos 2t - \frac{1}{20} \sin 2t.$$

(7) On taking the Laplace transform and substituting for initial conditions, we obtain

$$(s^2 - 2s + 1)Y(s) + (s+1)y_0 - y_0' = (s+3)U(s).$$

Hence $$Y(s) = \frac{s+3}{s^2 - 2s + 1} U(s) + \frac{y_0' - (s+1)y_0}{s^2 - 2s + 1}.$$

The first term is the contributions to the response due to the input; the second term is the contribution due to the initial conditions.
The transfer function is

$$G(s) = \frac{s+3}{s^2 - 2s + 1}.$$

(8) (i) In Fig.2.14(a)

$$E(s) = BQ(s) = BA_1R(s), \text{ hence}$$

$$\frac{E(s)}{R(s)} = BA_1.$$

In Fig. 2.14(b)

$$E(s) = \frac{B}{A_2} C(s) = \frac{B}{A_2} A_1A_2R(s) = BA_1R(s), \text{ hence}$$

$$\frac{E(s)}{R(s)} = BA_1.$$

This example indicates how the transfer function must change when moving the starting point of a signal.

(ii) $C(s) = X_1(s) + X_2(s) = AR(s) + BR(s) = (A + B)R(s)$,

hence $G(s) = \dfrac{C(s)}{R(s)} = A + B.$

(9) (i) Apply the equation (2.16) with

$$A(s) = \frac{1}{s+a} \quad \text{and} \quad B(s) = b, \quad \text{to obtain}$$

$$G(s) = \frac{1}{s+a+b}.$$

(ii) Since two feedback loops are involved, both feeding one summing junction, it is recommended to redraw the block diagram, introducing another summing junction as shown in Fig. 2.18.

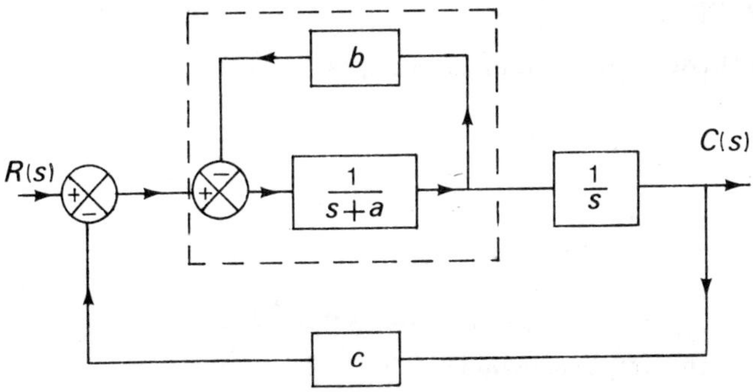

Fig. 2.18

The block diagram can now be reduced using the result of (1) above as shown in Fig. 2.19.

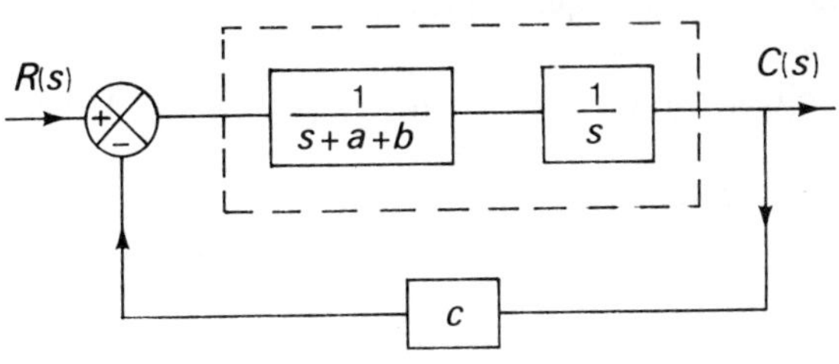

Fig. 2.19

The next reduction, using equation (2.15), results in Fig. 2.20.

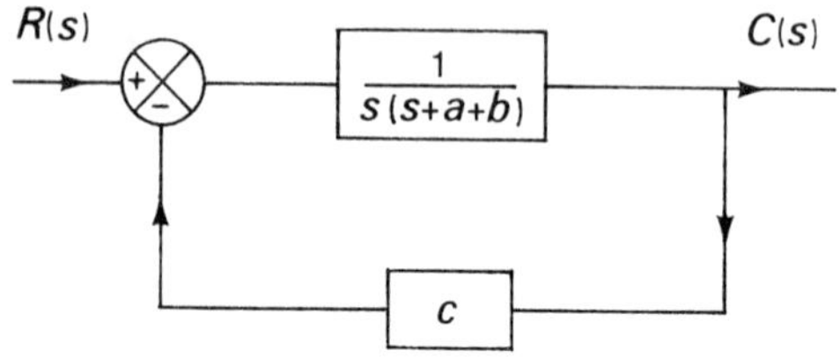

Fig. 2.20

The final reduction, using equation (2.16), results in Fig. 2.21.

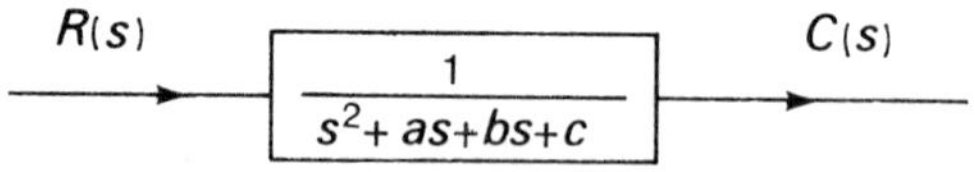

Fig. 2.21

CHAPTER 3

(1) (a) Use transformation equation (3.1)

$x_1 = y, \ x_2 = \dot{y},$ then

$$\begin{bmatrix} \dot{x}_1 \\ \dot{x}_2 \end{bmatrix} = \begin{bmatrix} 0 & 1 \\ 3 & -2 \end{bmatrix} \begin{bmatrix} x_1 \\ x_2 \end{bmatrix} + \begin{bmatrix} 0 \\ 1 \end{bmatrix} u.$$

The output equation is

$$y = [1 \quad 0] \begin{bmatrix} x_1 \\ x_2 \end{bmatrix}$$

(b) Use transformation equation (3.9)

$a_1 = -3, \ a_2 = 1, \ b_0 = 0, \ b_1 = 2$ and $b_2 = 1$, hence

$k_0 = 0$ and $k_1 = 2.$

$$x_1 = y$$

$$x_2 = \dot{x}_1 - 2u.$$

Then $\dot{x}_1 = x_2 + 2u$

and $\dot{x}_2 = \dot{x}_1 - 2\dot{u}$

$$= \ddot{y} - 2\dot{u} = (3\dot{y} - y + 2\dot{u} + u) - 2\dot{u}$$

$$= 3\dot{x}_1 - x_1 + u$$

$$= -x_1 + 3x_2 + 7u, \quad \text{hence}$$

$$\begin{bmatrix} \dot{x}_1 \\ \dot{x}_2 \end{bmatrix} = \begin{bmatrix} 0 & 1 \\ -1 & 3 \end{bmatrix} \begin{bmatrix} x_1 \\ x_2 \end{bmatrix} + \begin{bmatrix} 2 \\ 7 \end{bmatrix} u.$$

$$y = [1 \quad 0] \begin{bmatrix} x_1 \\ x_2 \end{bmatrix}.$$

(2) We must choose

$x_1 = \frac{1}{5}y$, $x_2 = \frac{1}{5}y$, then

$$\begin{bmatrix} \dot{x}_1 \\ \dot{x}_2 \end{bmatrix} = \begin{bmatrix} 0 & 1 \\ 3 & -2 \end{bmatrix} \begin{bmatrix} x_1 \\ x_2 \end{bmatrix} + \begin{bmatrix} 0 \\ 1 \end{bmatrix} u.$$

(3) (i) (a) $G(s) = \dfrac{Y(s)}{U(s)} = \dfrac{3}{s^3 - 2s^2 - s - 2} = \dfrac{\frac{1}{2}}{s-1} - \dfrac{\frac{3}{2}}{s+1} + \dfrac{1}{s+2}.$

Hence $Y(s) = X_1(s) + X_2(s) + X_3(s)$ where

$$X_1(s) = \frac{\frac{1}{2}U(s)}{s-1}, \; X_2(s) = \frac{-\frac{3}{2}U(s)}{s+1} \text{ and } X_3(s) = \frac{U(s)}{s+2}.$$

On taking the inverse Laplace transforms, we obtain the transformations

$$\dot{x}_1 = x_1 + \tfrac{1}{2}u, \; \dot{x}_2 = -x_2 - \tfrac{3}{2}u \text{ and } \dot{x}_3 = -2x_3 + u,$$

hence

$$\begin{bmatrix} \dot{x}_1 \\ \dot{x}_2 \\ \dot{x}_3 \end{bmatrix} = \begin{bmatrix} 1 & 0 & 0 \\ 0 & -1 & 0 \\ 0 & 0 & -2 \end{bmatrix} \begin{bmatrix} x_1 \\ x_2 \\ x_3 \end{bmatrix} + \tfrac{1}{2} \begin{bmatrix} 1 \\ -3 \\ 2 \end{bmatrix} u$$

and $$y = [1 \quad 1 \quad 1] \begin{bmatrix} x_1 \\ x_2 \\ x_3 \end{bmatrix}.$$

(b) $$G(s) = \frac{2s+1}{(s+1)(s+2)} = \frac{3}{s+2} - \frac{1}{s+1}$$

$Y(s) = X_1(s) - X_2(s)$, where

$$X_1(s) = \frac{3U(s)}{s+2} \text{ and } X_2(s) = \frac{U(s)}{s+1}.$$

Hence $\dot{x}_1 = -2x_1 + 3u$

$\dot{x}_2 = -x_2 + u$ or

$$\begin{bmatrix} \dot{x}_1 \\ \dot{x}_2 \end{bmatrix} = \begin{bmatrix} -2 & 0 \\ 0 & -1 \end{bmatrix} \begin{bmatrix} x_1 \\ x_2 \end{bmatrix} + \begin{bmatrix} 3 \\ 1 \end{bmatrix} u$$

and $$y = [1 \quad -1] \begin{bmatrix} x_1 \\ x_2 \end{bmatrix}.$$

(ii) Figs. 3.6 and 3.7 are the block-diagrams representing respectively the state equations for systems (a) and (b).

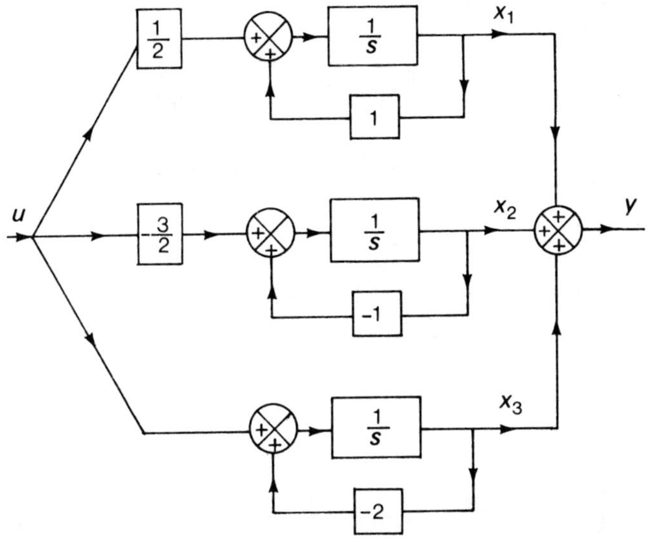

Fig. 3.6

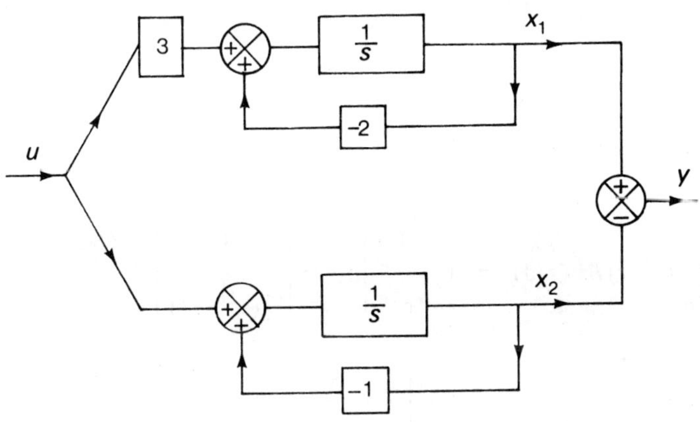

Fig. 3.7

(4) The two equations

$$A_1 \dot{h}_1 = q_1 - \frac{1}{R_1} h_1 - \frac{1}{R_3}(h_1 - h_2)$$

$$A_2 \dot{h}_2 = q_2 + \frac{1}{R_3}(h_1 - h_2) - \frac{1}{R_2} h_2$$

are written in state space form as

$$\begin{bmatrix} \dot{h}_1 \\ \dot{h}_2 \end{bmatrix} = \begin{bmatrix} -\frac{3}{10} & \frac{1}{5} \\ \frac{1}{8} & -\frac{1}{6} \end{bmatrix} \begin{bmatrix} h_1 \\ h_2 \end{bmatrix} + \begin{bmatrix} \frac{1}{5} & 0 \\ 0 & \frac{1}{8} \end{bmatrix} \begin{bmatrix} q_1 \\ q_2 \end{bmatrix}.$$

The output $y = \dfrac{h_1}{R_1} + \dfrac{h_2}{R_2} = [\frac{1}{2} \;\; \frac{1}{3}] \begin{bmatrix} h_1 \\ h_2 \end{bmatrix}.$

(5) (i) $\begin{bmatrix} e^{\lambda_1 t} & 0 \\ 0 & e^{\lambda_2 t} \end{bmatrix}$

(ii) $\begin{bmatrix} e^{t} & e^{2t} - e^{t} \\ 0 & e^{2t} \end{bmatrix}$

(6) $\phi(t) = e^{At} = \begin{bmatrix} e^{-t} & 0 \\ 2e^{-t} - 2e^{-2t} & e^{-2t} \end{bmatrix}$

$$\int_0^t \phi(-\tau)Bu(\tau)d\tau = \int_0^t \begin{bmatrix} 0 \\ e^{2\tau} \end{bmatrix} d\tau = \begin{bmatrix} 0 \\ \frac{1}{2}(e^{2t} - 1) \end{bmatrix}$$

$$\mathbf{x} = -\begin{bmatrix} e^{-t} \\ 2e^{-t} - 4e^{-2t} \end{bmatrix} + \begin{bmatrix} e^{-t} & 0 \\ 2e^{-t} - 2e^{-2t} & e^{-2t} \end{bmatrix} \begin{bmatrix} 0 \\ \frac{1}{2}(e^{2t} - 1 \end{bmatrix}$$

$$= \begin{bmatrix} -e^{-t} \\ \frac{1}{2} - 2e^{-t} + \frac{7}{2}e^{-2t} \end{bmatrix}.$$

(7) (i) The input–output differential equation is

$$\ddot{y} + 5\dot{y} + 2y = u.$$

Let $x_1 = y$ and $x_2 = \dot{y}$, then

$$\begin{bmatrix} \dot{x}_1 \\ \dot{x}_2 \end{bmatrix} = \begin{bmatrix} 0 & 1 \\ -2 & -5 \end{bmatrix} \begin{bmatrix} x_1 \\ x_2 \end{bmatrix} + \begin{bmatrix} 0 \\ 1 \end{bmatrix} u$$

and $y = [1 \quad 0] \begin{bmatrix} x_1 \\ x_2 \end{bmatrix}.$

(ii) Let $x_1 = y$, $x_2 = \dot{y}$ and $x_3 = \ddot{y}$, then

$$\begin{bmatrix} \dot{x}_1 \\ \dot{x}_2 \\ \dot{x}_3 \end{bmatrix} = \begin{bmatrix} 0 & 1 & 0 \\ 0 & 0 & 1 \\ -3 & -5 & -2 \end{bmatrix} \begin{bmatrix} x_1 \\ x_2 \\ x_3 \end{bmatrix} + \begin{bmatrix} 0 \\ 0 \\ 1 \end{bmatrix} u$$

$$y = [1 \quad 0 \quad 0] \begin{bmatrix} x_1 \\ x_2 \\ x_3 \end{bmatrix}.$$

(8) The characteristic roots are $\lambda_1 = -1$ and $\lambda_2 = -2$, so that equation (3.30) becomes

$$V = \begin{bmatrix} 1 & 1 \\ -1 & -2 \end{bmatrix}, \text{ hence } V^{-1} = \begin{bmatrix} 2 & 1 \\ -1 & -1 \end{bmatrix}.$$

It follows that the required transformation is

$$z_1 = 2x_1 + x_2$$

$$z_2 = -x_1 - x_2 \quad \text{(see equation (3.25))}$$

leading to the state equations (see equation (3.26))

$$\begin{bmatrix} \dot{z}_1 \\ \dot{z}_2 \end{bmatrix} = \begin{bmatrix} -1 & 0 \\ 0 & -2 \end{bmatrix} \begin{bmatrix} z_1 \\ z_2 \end{bmatrix} + \begin{bmatrix} 1 \\ -1 \end{bmatrix} u$$

$$y = [1 \quad 1]\begin{bmatrix} z_1 \\ z_2 \end{bmatrix}.$$

(9) Use equation (3.31)

$$\text{(i) } G(s) = [1 \quad 0]\frac{1}{s^2+3s+2}\begin{bmatrix} s+3 & 1 \\ -2 & s \end{bmatrix}\begin{bmatrix} 0 \\ 1 \end{bmatrix} = \frac{1}{s^2+3s+2}.$$

$$\text{(ii) } G(s) = \begin{bmatrix} 0 & 1 & 1 \\ 1 & 1 & 1 \end{bmatrix}\begin{bmatrix} \frac{1}{s+1} & 0 & 0 \\ 0 & \frac{1}{s+2} & 0 \\ 0 & 0 & \frac{1}{s+3} \end{bmatrix}\begin{bmatrix} 0 & 1 \\ 1 & 1 \\ -1 & 0 \end{bmatrix}$$

$$= \begin{bmatrix} \frac{1}{s+2} - \frac{1}{s+3} & \frac{1}{s+2} \\ \frac{1}{s+2} - \frac{1}{s+3} & \frac{1}{s+1} + \frac{1}{s+2} \end{bmatrix}.$$

$$\text{(10) (a) } T(s) = \frac{s+3}{s^4+6s^3+13s^2+12s+4} = \frac{s+3}{(s+1)^2(s+2)^2}.$$

$$\text{(b) } T(s) = \frac{Y(s)}{U(s)} = \frac{K_1}{s+1} + \frac{K_2}{(s+1)^2} + \frac{K_3}{s+2} + \frac{K_4}{(s+2)^2}.$$

$$\text{(c) } K_2 = \left[\frac{s+3}{(s+2)^2}\right]_{s=-1} = 2, \quad K_1 = \left[\frac{\mathrm{d}}{\mathrm{d}s}\left\{\frac{s+3}{(s+2)^2}\right\}\right]_{s=-1} = -3.$$

$$K_4 = \left[\frac{s+3}{(s+1)^2}\right]_{s=-2} = 1, \quad K_3 = \left[\frac{\mathrm{d}}{\mathrm{d}s}\left\{\frac{s+3}{(s+1)^2}\right\}\right]_{s=-2} = 3.$$

$$\begin{bmatrix}\dot{x}_1\\ \dot{x}_2\\ \dot{x}_3\\ \dot{x}_4\end{bmatrix} = \begin{bmatrix}-1 & 1 & 0 & 0\\ 0 & -1 & 0 & 0\\ 0 & 0 & -2 & 1\\ 0 & 0 & 0 & -2\end{bmatrix}\begin{bmatrix}x_1\\ x_2\\ x_3\\ x_4\end{bmatrix} + \begin{bmatrix}0\\ 1\\ 0\\ 1\end{bmatrix}u$$

$$y = [2 \;\; -3 \;\; 1 \;\; 3]\begin{bmatrix}x_1\\ x_2\\ x_3\\ x_4\end{bmatrix}.$$

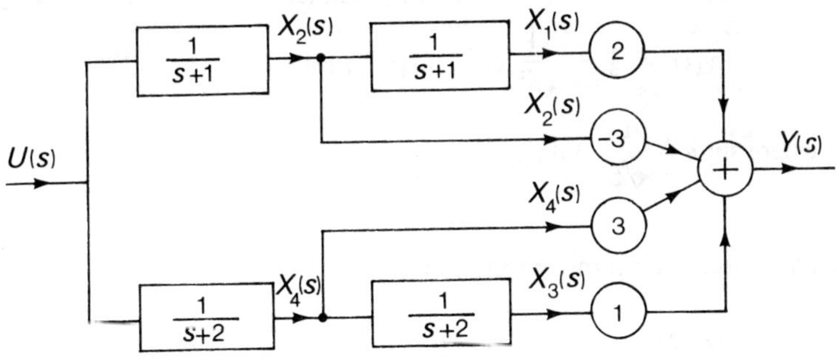

Fig. 3.8

CHAPTER 4

(1) (i) $K_p = \lim_{s\to 0} G(s) = \infty, \quad K_v = \lim_{s\to 0} sG(s) = \frac{1}{3}$

and $K_a = \lim_{s\to 0} s^2G(s) = 0;$

(ii) (a) $R(s) = \dfrac{2}{s}, \; e_{ss} = 0$

(b) $R(s) = \dfrac{2}{s^2}, \; e_{ss} = 6$

(c) $R(s) = \dfrac{3}{2s^3}, \; e_{ss} = \infty$

(d) $R(s) = \dfrac{3}{s} - \dfrac{2}{s^2}, \; e_{ss} = 0 - 6 = -6.$

(Since the system is linear the errors in (d) can be superimposed.)

(2) (i) (a) $y = 1 - e^{-2/3t}$, μ (the time constant) $= \frac{3}{2}$

(b) $y = 1 - e^{-6t}$, $\mu = \frac{1}{6}$.

(ii) (a) $\mu = 3$, (b) $\mu = 2$.

(iii) (a) $G_c(s)$ = closed loop transfer function $= \dfrac{\frac{1}{2}}{1 + \frac{3}{2}s}$, $\mu = \frac{3}{2}$.

(b) $G_c(s) = \dfrac{\frac{1}{2}}{1+s}$, $\mu = 1$.

The time constants have been decreased by a half, hence the responses of both systems have been speeded up.

(3) With the notation of equation (4.6), we have

$a_1 = 4$ and $a_2 = 5$ so that $\alpha = 2$ and $\beta = 1$.

Since $4a_2 > a_1^2$, the system response is underdamped. Equation (4.12) gives

$$y(t) = \tfrac{4}{5} + \frac{4}{\sqrt{5}} e^{-2t} \sin(t + \epsilon)$$

where $\sin \epsilon = \dfrac{1}{\sqrt{5}}$ and $\cos \epsilon = \dfrac{2}{\sqrt{5}}$.

(4) The closed loop transfer function is

$$G_c(s) = \frac{K}{s^2 + 5s + (1+K)}.$$

Hence the output is

$$Y(s) = \frac{K}{s(s^2 + 5s + 1 + K)}.$$

For $K = 1$, $a_1^2 = 25, 4a_2 = 8$, hence the system is overdamped.

$$y(t) = \tfrac{1}{2} + K_2 e^{-p_1 t} + K_3 e^{-p_2 t}$$

where $p_1 = \frac{1}{2}(5 + \sqrt{17})$, $p_2 = \frac{1}{2}(5 - \sqrt{17})$

$$K_2 = \frac{2}{5\sqrt{17} + 17} \quad \text{and} \quad K_3 = \frac{2}{17 - 5\sqrt{17}}.$$

For $K = 5.25$, $a_1^2 = 4a_2$ so the system is critically damped

$$y(t) = \frac{5.25}{6.25}(1 - e^{-2.5t} + 2.5t\, e^{-2.5t}).$$

For $K = 10$, the system is underdamped.

The response is given by equation (4.12) with

$$\alpha = 2.5 \quad \text{and} \quad \beta = \tfrac{1}{2}\sqrt{19}.$$

(5) (a) $Y(s) = \dfrac{\omega_n^2}{s(s^2 + 2\omega_n\tau s + \omega_n^2)}$

(b) $p_1, p_2 = \omega_n\tau \pm \omega_n\sqrt{\tau^2 - 1}.$

(6) On differentiating equation (4.13), we obtain at a turning point

$$\tan(\omega t + \epsilon) = \frac{\omega}{\tau\omega_n} = \frac{\sqrt{1-\tau^2}}{\tau}.$$

Hence at the maximum overshoot when $t = t_1$ (say)

$$\omega t_1 = \tan^{-1}\frac{\sqrt{1-\tau^2}}{\tau} - \tan^{-1}\frac{\sqrt{1-\tau^2}}{\tau} + \pi$$

or

$$t_1 = \frac{\pi}{\omega_n\sqrt{1-\tau^2}}$$

$$\begin{aligned} y(t_1) &= 1 - \frac{1}{\sqrt{1-\tau^2}}\exp\left\{-\frac{\pi\tau}{\sqrt{1-\tau^2}}\right\}\sin(\pi+\epsilon) \\ &= 1 + \frac{1}{\sqrt{1-\tau^2}}\exp\left\{-\frac{\pi\tau}{\sqrt{1-\tau^2}}\right\}\sin\epsilon \\ &= 1 + \exp\left\{-\frac{\pi\tau}{\sqrt{1-\tau^2}}\right\} \end{aligned}$$

since $\sin\epsilon = \sqrt{1-\tau^2}$.

CHAPTER 5

(1) (a) The Routh array is

1	3	1
2	7	0
$\frac{1}{2}$	1	
3		
1		

The system is stable.

(b) The Routh array is

$$\begin{matrix} 1 & 1 \\ 1 & 2 \\ -1 & \\ 2 & \end{matrix}$$

The system is unstable. There are two roots with positive real parts.

(2) The characteristic equation of the closed loop system is

$$s^3 + 3s^2 + 2s + K = 0.$$

From the corresponding Routh array, the conditions for stability are

$$K > 0 \quad \text{and} \quad \frac{6-K}{3} > 0, \quad \text{that is} \quad 0 < K < 6.$$

When $K = 6$ the third row of the Routh array is zero, indicating that the characteristic equation has roots on the imaginary axis - the system is marginally stable.

(3) $P(s-3) = s^3 - 0.25\,s = 0$ is a characteristic equation of an unstable system.
$P(s-2) = s^3 + 3s^2 + 2.75s + 0.75 = 0$ is a characteristic equation of a stable system.
Hence $\sigma = 2$.

(4) V_1 is positive definite.
V_2 and V_3 are not.

(5) (a) In this case the matrix P of equation (5.17) is

$$P = \begin{bmatrix} \frac{1}{4} & \frac{1}{24} \\ \frac{1}{24} & \frac{13}{96} \end{bmatrix}.$$

Since P is positive definite, the system is asymptotically stable.

(b) $$P = \frac{1}{36}\begin{bmatrix} 9 & 1 & 3 \\ 1 & 29 & 11 \\ 3 & 11 & 21 \end{bmatrix}.$$

Since P is positive definite, the system is asymptotically stable.

(6) (i) Let y be the displacement, then

$$m\ddot{y} = -mg.$$

Let $x_1 = y$ and $x_2 = \dot{y}$, then

$$\dot{\mathbf{x}} = \begin{bmatrix} 0 & 1 \\ 0 & 0 \end{bmatrix} \mathbf{x} + \begin{bmatrix} 0 \\ -g \end{bmatrix}.$$

Since $\dfrac{\dot{x}_2}{\dot{x}_1} = \dfrac{\dfrac{dx_2}{dt}}{\dfrac{dx_1}{dt}} = \dfrac{dx_2}{dx_1} = -\dfrac{g}{x_2}$, on integration we obtain the trajectory

$$x_1 = \frac{-1}{2g} x_2^2 + C. \quad (C \text{ is a constant of integration}).$$

This is a family of parabolas.

(ii) $$M\ddot{y} = ky$$

Let $x_1 = y$ and $x_2 = \ddot{y}$, then

$$\dot{\mathbf{x}} = \begin{bmatrix} 0 & 1 \\ -\omega^2 & 0 \end{bmatrix} \mathbf{x} \quad \text{where} \quad \omega^2 = \frac{k}{M}$$

$\dfrac{\dot{x}_2}{\dot{x}_1} = \dfrac{-\omega^2 x_1}{x_2} = \dfrac{dx_2}{dx_1}$, on integration we obtain

$\dfrac{x_1^2}{a^2} + \dfrac{x_2^2}{a^2\omega^2} = 1$ where a^2 is a constant of integration.

This is a family of ellipses.

(7) (i) $$G(s) = \frac{1}{s(s+1)}$$

$$|G(i\omega)| = \frac{1}{\omega(1+\omega^2)^{\frac{1}{2}}} \text{ and } \varphi = -90° - \tan^{-1}\omega.$$

ω	$\|G\|$	φ
0	$-\infty$	-90
1	0.71	-135
5	0.04	-169
10	0.01	-174
⋮	⋮	⋮
∞	0	-180

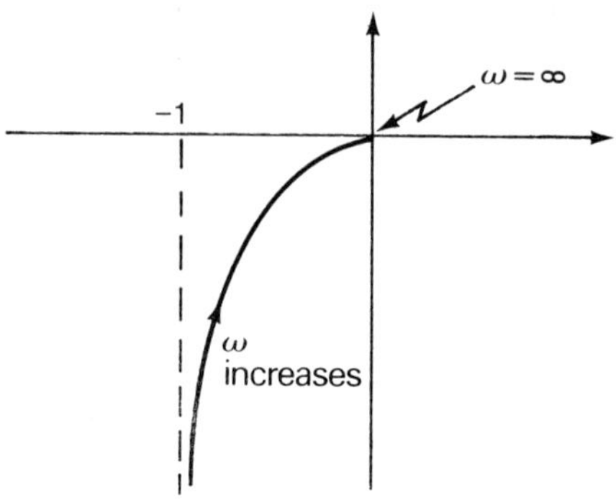

Fig. 5.19

$$\text{Notice that R1 } G(i\omega) = \frac{-1}{1+\omega^2}$$
$$= -1 \text{ for } \omega = 0.$$

(ii) $$G(s) = \frac{2+s}{9+s}$$

$$|G(i\omega)| = \sqrt{\frac{4+\omega^2}{81+\omega^2}} \quad \text{and} \quad \varphi = \tan^{-1}\frac{\omega}{2} - \tan^{-1}\frac{\omega}{9}$$

ω	$\|G\|$	φ
0	0.22	0
5	0.45	39
⋮	⋮	⋮
∞	1	0

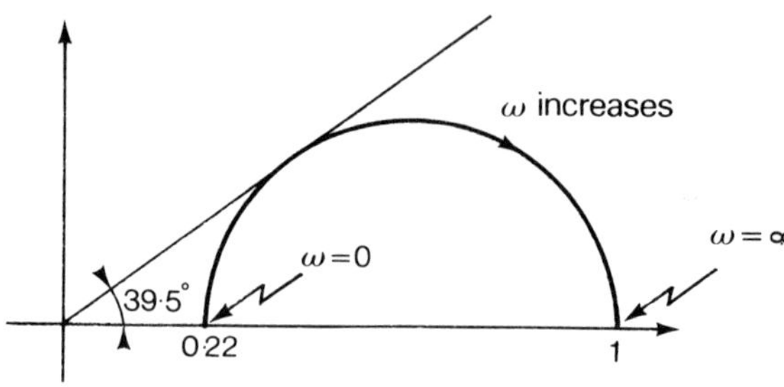

Fig. 5.20

To find $\varphi_{\max}$ use $\dfrac{d}{dx}(\tan^{-1} u) = \dfrac{1}{1+u^2}\dfrac{du}{dx}$ so that

$$\frac{d\varphi}{d\omega} = \frac{2}{4+\omega^2} - \frac{9}{81+\omega^2}, \quad \frac{d\varphi}{d\omega} = 0 \quad \text{for} \quad \omega = 4.24.$$

This corresponds to $\varphi = 39.5°$.

(iii) $G(s) = \dfrac{2}{1 + 3s + s^2}$

$$|G(i\omega)| = \frac{2}{\sqrt{1 + 7\omega^2 + \omega^2}} \quad \text{and} \quad \varphi = -\tan^{-1}\frac{3\omega}{1-\omega^2}.$$

ω	$\lvert G\rvert$	φ
0	2	0
0.5	1.2	−63
1	0.67	−90
10	0.01	−163
⋮	⋮	⋮
∞	0	−180

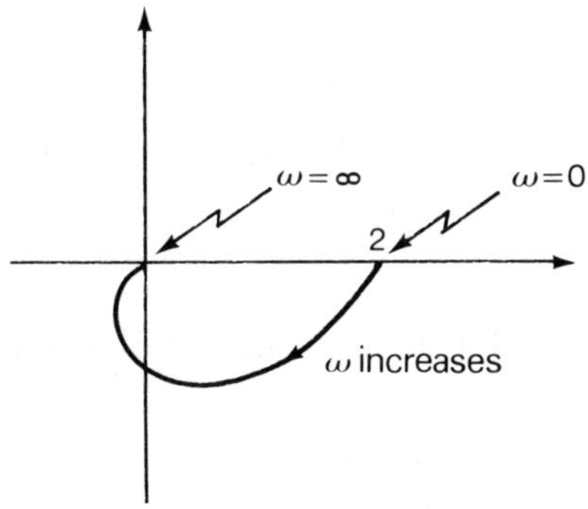

Fig. 5.21

(8) $$\frac{|F(i\omega)|}{K} = \frac{(4+\omega^2)^{\frac{1}{2}}}{\omega(1+\omega^2)^{\frac{1}{2}}(9+\omega^2)^{\frac{1}{2}}(16+\omega^2)^{\frac{1}{2}}}.$$

$$\varphi = \tan^{-1}\omega/2 - 90° - \tan^{-1}\omega - \tan^{-1}\omega/3 - \tan^{-1}\omega/4$$

ω	$\dfrac{\lvert F\rvert}{K}$	φ
0	∞	−90
1	0.12	−141
2	0.04	−168
2.6	0.024	−180
5	0.006	−211
⋮	⋮	⋮
∞	0	−270

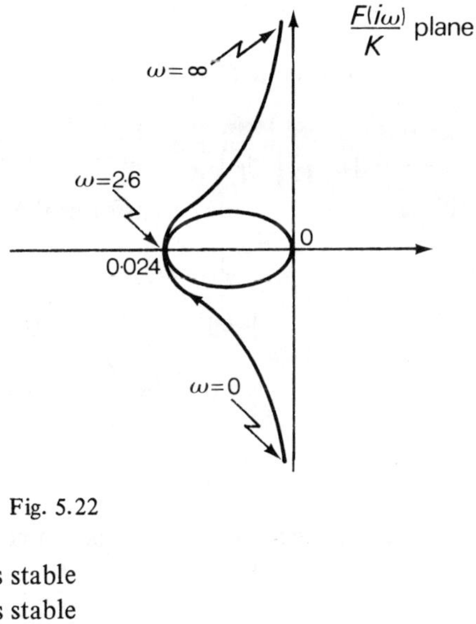

Fig. 5.22

(i) when $K = 10$ the system is stable
(ii) when $K = 30$ the system is stable
(iii) when $K = 60$ the system is unstable.

CHAPTER 6

(1) (i) The transformation $\mathbf{x} = \begin{bmatrix} 2 & 1 \\ -1 & -1 \end{bmatrix}\mathbf{z}$, transforms the system to

$$\dot{\mathbf{z}} = \begin{bmatrix} 2 & 0 \\ 0 & 1 \end{bmatrix}\mathbf{z} = \begin{bmatrix} 2 \\ 0 \end{bmatrix}u$$

$$y = [-2 \quad 0]\mathbf{z}.$$

The second mode ($\lambda = 1$) is both uncontrollable and unobservable.

(ii) $\mathbf{x} = \begin{bmatrix} 3 & -2 \\ -1 & 1 \end{bmatrix}\mathbf{z}$, transforms the system to

$$\dot{\mathbf{z}} = \begin{bmatrix} 3 & 0 \\ 0 & 2 \end{bmatrix}\mathbf{z} + \begin{bmatrix} 0 \\ 1 \end{bmatrix}u$$

$$y = [1 \quad -2]\mathbf{z}.$$

The first mode ($\lambda = 3$) is uncontrollable.

(iii) $\mathbf{x} = \begin{bmatrix} 1 & 1 \\ -1 & -2 \end{bmatrix}\mathbf{z}$, transforms the system to

$$\dot{\mathbf{z}} = \begin{bmatrix} -1 & 0 \\ 0 & 2 \end{bmatrix}\mathbf{z} + \begin{bmatrix} 0 \\ 2 \end{bmatrix}u$$

$$y = [0 \quad 1]\mathbf{z}.$$

The first mode ($\lambda = -1$) is both uncontrollable and unobservable.

(2) (i) $Q = \begin{bmatrix} 4 & 8 \\ -2 & -4 \end{bmatrix}$, $r(Q) = 1$

$R = \begin{bmatrix} -2 & -2 \\ -4 & -4 \end{bmatrix}$, $r(R) = 1$

(ii) $Q = \begin{bmatrix} -2 & -4 \\ 1 & 2 \end{bmatrix}$, $r(Q) = 1$

$R = \begin{bmatrix} -1 & -4 \\ -1 & -6 \end{bmatrix}$, $r(R) = 2$

(iii) $Q = \begin{bmatrix} 2 & 4 \\ -4 & -8 \end{bmatrix}$, $r(Q) = 1$

$R = \begin{bmatrix} -1 & -1 \\ -2 & -2 \end{bmatrix}$, $r(R) = 1.$

(3) (i) $G(s) = -\dfrac{4}{s-2}$

(ii) $G(s) = -\dfrac{2}{s-2}$

(iii) $F(s) = \dfrac{2}{s-2}.$

(4) The controllability matrix is

$$Q = \begin{bmatrix} p & p-12 \\ -1 & p \end{bmatrix}.$$

The system is uncontrollable when $r(Q) = 1$, that is, when $p = -4$ or $p = 3$.

The observability matrix is

$$R = \begin{bmatrix} q & 1 \\ 1+q & 12q \end{bmatrix}.$$

The system is unobservable when $q = \frac{1}{3}$ or $q = -\frac{1}{4}$.

(5)
$$\dot{\mathbf{x}} = \begin{bmatrix} -2 & 0 \\ 0 & -3 \end{bmatrix} \mathbf{x} + \begin{bmatrix} 1 & -2 \\ -1 & 1 \end{bmatrix} \mathbf{u}$$

$$\mathbf{y} = \begin{bmatrix} 1 & 1 \\ -1 & 2 \end{bmatrix} \mathbf{x}.$$

(6)
$$T = \begin{bmatrix} 1 & 0 & 1 \\ 0 & 1 & 0 \\ 0 & 1 & 1 \end{bmatrix}.$$

(7) On differentiating

$$[\mathrm{e}^{-At} b]'\mathbf{w} = \mathbf{0}$$

$(n-1)$ times, we obtain

$$[\mathrm{e}^{-At} A^{i-1}\mathbf{b}]'\mathbf{w} = \mathbf{0} \quad \text{for} \quad 0 \leqslant t \leqslant t_1 \quad (i = 2, \ldots, n).$$

Let $t = 0$ in the above equations, then

$$\mathbf{b}'\mathbf{w} = (A\mathbf{b})'\mathbf{w} = \ldots = (A^{n-1}\mathbf{b})'\mathbf{w} = \mathbf{0}$$

or

$$[\mathbf{b}' \,\vdots\, (A\mathbf{b})' \,\vdots\, \ldots \,\vdots\, (A^{n-1}\mathbf{b})']\mathbf{w} = 0.$$

This is equivalent to

$$Q'\mathbf{w} = \mathbf{0} \quad (Q \text{ is the controllability matrix})$$

and is impossible for a controllable system. The conclusion follows.

(8) With this choice of u, equation (3.20) can be written as

$$e^{-At_1}\mathbf{x}_1 - \mathbf{x}_0 = \left[\int_0^{t_1} e^{-At}\mathbf{b}[e^{-At}\mathbf{b}]'\,dt\right]\mathbf{w}$$

the matrix in the square brackets is non-singular (by the result of Problem 7), so that

$$\mathbf{w} = \left[\int_0^{t_1} e^{-At}\mathbf{b}[e^{-At}\mathbf{b}]'\,dt\right]^{-1}\left[e^{-At_1}\mathbf{x}_1 - \mathbf{x}_0\right].$$

(9) From Problems 7 and 8 we know that we can always find an input function defined in Problem 8, with **w** as above which will transfer the system from $\mathbf{x}_0$ to $\mathbf{x}_1$ in time t_1. This shows that the system is controllable and proves that the criterion is sufficient.

(10)

$$e^{-At}\mathbf{b} = \begin{bmatrix} 1 & -2t \\ 0 & 1 \end{bmatrix}\begin{bmatrix} 0 \\ 1 \end{bmatrix} = \begin{bmatrix} -2t \\ 1 \end{bmatrix}$$

$$\int_0^3 e^{-At}\mathbf{b}[e^{-At}\mathbf{b}]'\,dt = \begin{bmatrix} 36 & -9 \\ -9 & 3 \end{bmatrix}.$$

Hence

$$\mathbf{w} = \begin{bmatrix} \frac{1}{9} & \frac{1}{3} \\ \frac{1}{3} & \frac{4}{3} \end{bmatrix}\left\{\begin{bmatrix} -11 \\ 2 \end{bmatrix} - \begin{bmatrix} 0 \\ 0 \end{bmatrix}\right\} = \begin{bmatrix} -\frac{5}{9} \\ -1 \end{bmatrix}$$

and $u = [-2t \quad 1]\begin{bmatrix} -\frac{5}{9} \\ -1 \end{bmatrix} = \frac{1}{9}(10t - 9)$ for $0 \leqslant t \leqslant 3$.

CHAPTER 7

(1) (i) $\lambda = 0$, $\lambda = -1$.

(ii) $T^{-1} = \frac{1}{30}\begin{bmatrix} -2 & 1 \\ -18 & 24 \end{bmatrix}$ so that $T = \begin{bmatrix} -24 & 1 \\ -18 & 2 \end{bmatrix}$

$\mathbf{k}' = [-6 \quad -4]$ (see equation (7.6)).

Hence $u = \frac{1}{5}[14 \quad -17]\mathbf{x}$.

(2) The modal matrix is

$$T = \begin{bmatrix} 4 & 5 \\ 3 & 4 \end{bmatrix} \quad \text{so that} \quad T^{-1} = \begin{bmatrix} 4 & -5 \\ -3 & 4 \end{bmatrix}$$

and $T^{-1}\mathbf{b} = \begin{bmatrix} -6 \\ 5 \end{bmatrix}$.

Since (equation 7.23) $R_1 = -6,\ R_2 = 2$

$$K_1 = [1 \quad \tfrac{2}{5}] \quad (\text{choosing } f = 1)$$

and $K = \tfrac{1}{5}[14 \quad -17]$.

(3) (a) $K = [-\tfrac{12}{15} \quad 0 \quad -\tfrac{2}{5}] \quad (f = 1)$

(b) $K = \begin{bmatrix} -\tfrac{12}{15} & 0 & -\tfrac{2}{5} \\ 0 & 0 & 0 \end{bmatrix} \quad \left(f = \begin{bmatrix} 1 \\ 0 \end{bmatrix}\right).$

(4) $\dot{\mathbf{z}} = \mathbf{Fz} + \mathbf{Hy} + \mathbf{Bu}$

where $F = A - HC$

(5) $F = \begin{bmatrix} -5 & 1 \\ -2 & -2 \end{bmatrix}$.

CHAPTER 8

(1) (i) $\dfrac{dk}{dt} - \beta k = -\bar{c}, \quad \beta = b - \lambda - \mu$

that is, $k = Ae^{\beta t} + \dfrac{\bar{c}}{\beta} \quad (\rightarrow \dfrac{\bar{c}}{\beta} \text{ if } \beta < 0).$

(ii) For equilibrium solutions, $\dfrac{dk}{dt} = 0$, c = constant, giving

$$bk^{\alpha} - (\lambda + \mu)k = c.$$

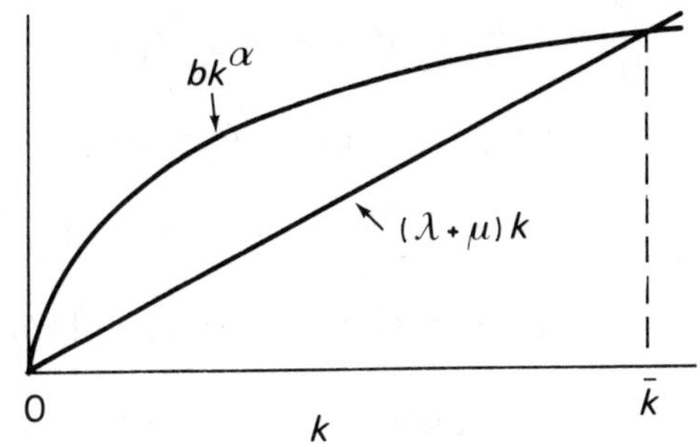

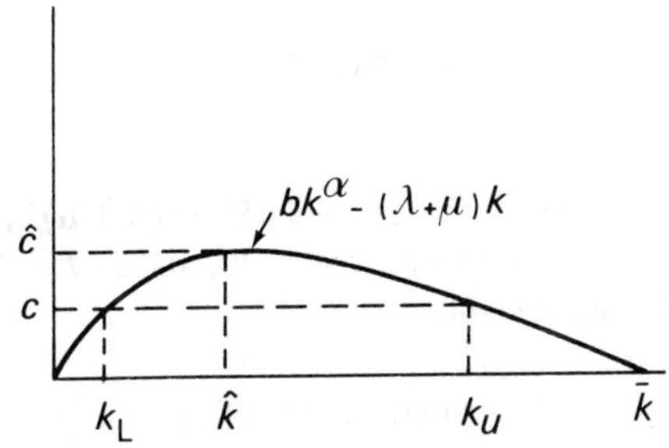

From the sketches, it is clear that provided $0 \leqslant c \leqslant \hat{c}$, there will be two equilibrium values for capital, k_L and k_u. Note that the value of $\hat{c}$, the maximum sustainable value of c, corresponds to $k = \hat{k}$, which is the maximum value of $bk^{\alpha} - (\lambda + \mu)k$, that is, $\alpha b k^{\alpha-1} = (\lambda + \mu)$ or $\hat{k} = [(\lambda + \mu)/\alpha b]^{1/(\alpha-1)}$. Thus

$$\begin{aligned}\hat{c} &= b\hat{k}^{\alpha} - (\lambda + \mu)\hat{k} \\ &= \hat{k}[b\hat{k}^{\alpha-1} - (\lambda + \mu)] \\ &= \hat{k}(\lambda + \mu)(\frac{1}{\alpha} - 1)\end{aligned}$$

that is, $$\hat{c} = \left[\frac{\lambda + \mu}{\alpha}\right]^{\alpha/(\alpha-1)} \frac{(1-\alpha)}{b^{1/(\alpha-1)}}.$$

The stability of k_L and k_u can be determined by substituting

$$k = k_L + x$$

where x is small in

$$\frac{dk}{dt} = bk^{\alpha} - (\lambda + \mu)k - c$$

$$\text{that is,}\quad \frac{dx}{dt} = b(k_L + x)^\alpha - (\lambda+\mu)(k_L + x) - c$$
$$= bk_L^\alpha(1 + x/k_L)^\alpha - (\lambda+\mu)(k_L+x) - c$$
$$= bk_L^\alpha\left(1 + \frac{\alpha x}{k_L} + \ldots\right) - (\lambda+\mu)(k_L+x) - c$$
$$= \{bk_L^\alpha - (\lambda+\mu)k_L - c\} + [\alpha bk_L^{\alpha-1} - (\lambda+\mu)]x + \ldots$$
$$= 0 + [\alpha bk_L^{\alpha-1} - (\lambda+\mu)]x + \ldots$$

$$\text{that is,}\quad \frac{dx}{dt} \approx Ax, \quad A = \alpha bk_L^{\alpha-1} - (\lambda+\mu).$$

But A is the slope of the curve $[bk^\alpha - (\lambda+\mu)k]$, shown above, at $k = k_L$, and so $A > 0$. Hence $x = Be^{At}$, and so $k = k_L$ is unstable. In a similar way $k = k_u$ is stable.

(2) $\dfrac{dx}{dt} = F(x) - h(t) = 0$ when $h = F(x)$.

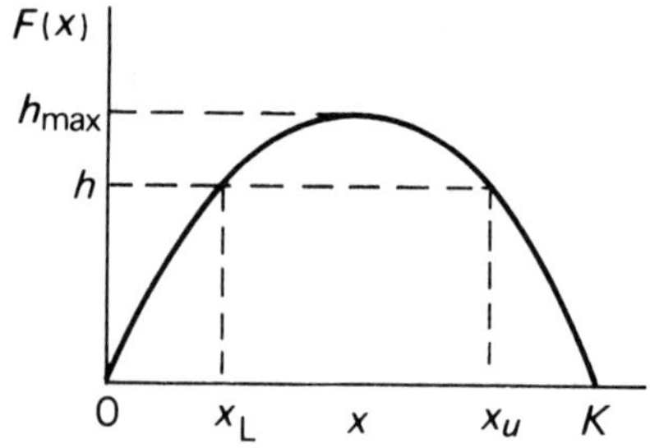

A typical form for $F(x)$ is shown above. For $0 \leqslant h \leqslant h_{max}$, equilibrium is possible at two values of x, x_L and x_u; h_{max} = maximum value of $F(x)$.

For $F(x) = rx(1 - x/K)$, $\dfrac{dF}{dx} = r(1 - 2x/K) = 0$

when $x = K/2$. Hence

$$h_{max} = rK/4.$$

Note that the maximum sustainable yield occurs at $K/2$; that is, at an equilibrium level *half* the saturation level, K. To the layman, it might appear as if we are overexploiting the population at half the saturation level, whereas in reality all is well (provided we do not go above h_{max}!).

Now revenue and costs equated give

$$ph = cE$$

that is, $pEx = cE$.

Thus $x = c/p$ (which will be in equilibrium provided $c/p < K$). The required effort is determined from

$$F(x) = h = Ex$$

that is, $E = pF(c/p)/c$.

Now

$$\text{economic rent} = ph - cE$$

$$= pEx - cE.$$

Also, since we are in equilibrium

$$F(x) = Ex;$$

that is, $\text{economic rent} = pF(x) - c\dfrac{F(x)}{x}$.

For maximum economic rent $pF' - c\dfrac{F'}{x} + c\dfrac{F}{x^2} = 0$,

which determines the equilibrium population level. For the Verhulst population growth, we obtain $x = \frac{1}{2}(K + c/p)$.

(3) $$\frac{ds}{dt} = r\frac{B}{T}(1 - s/M) - \lambda s$$

$$\frac{ds}{dt} + (\lambda + \frac{rB}{TM})s = r\frac{B}{T}.$$

With $a = \lambda + \dfrac{rB}{TM}$, $s = Ke^{-at} + \dfrac{rB}{aT}$, K constant.

At $t = 0$, $s = s_0$, that is, $s_0 = K + \dfrac{rB}{aT}$ giving

$$s = (s_0 - \frac{rB}{aT})e^{-at} + \frac{rB}{aT}.$$

The total sales are given by

$$S = \int_0^T s\,dt = \frac{rB}{a} + \frac{1}{a}\left(s_0 - \frac{rB}{aT}\right)(1 - e^{-aT}).$$

(4) (i) $x = t$ giving

$$J = \int_0^1 (t^2 + 1)dt = (\tfrac{1}{3}t^3 + t)\Big|_0^1 = 4/3.$$

(ii) $x = t^2$ giving

$$J = \int_0^1 (t^4 + 4t^2)dt = (\tfrac{1}{5}t^5 + \tfrac{4}{3}t^3)\Big|_0^1 = 23/15.$$

(iii) $x = \dfrac{(e^t - 1)}{(e - 1)}$ giving

$$J = \int_0^1 \left[\frac{(e^{2t} - 2e^t + 1)}{(e-1)^2} + \frac{e^{2t}}{(e-1)^2}\right] dt$$

$$= \left[\frac{e^{2t} - 2e^t + t}{(e-1)^2}\right]\Big|_0^1 = \frac{e^2 - 2e + 1}{(e-1)^2} - \frac{(-1)}{(e-1)^2}$$

$$= \frac{e^2 - 2e + 2}{(e-1)^2}.$$

(5) (i) $x = t$, $J = \displaystyle\int_0^1 (2t^2 + t^2)dt = t^3\Big|_0^1 = 1.$

(ii) $x = t^2$, $J = \displaystyle\int_0^1 (2t^3 + 2t^3)dt = t^4\Big|_0^1 = 1.$

(iii) $x = \dfrac{(e^t - 1)}{(e - 1)}$,

$$J = \int_0^1 \left[2t\frac{(e^t - 1)}{(e-1)} + \frac{t^2 e^t}{(e-1)}\right] dt$$

$$= \frac{1}{(e-1)}\left\{2te^t\Big|_0^1 - \int_0^1 2e^t dt - t^2\Big|_0^1 + t^2 e^t\Big|_0^1 - \int_0^1 2te^t dt\right\}$$

$$= \frac{1}{(e-1)}\left\{2e - 2e^t\Big|_0^1 - 1 + e - 2te^t\Big|_0^1 + \int_0^1 2e^t dt\right\}$$

$$= \frac{1}{(e-1)}\left\{1 + e - 2e + 2e^t\Big|_0^1\right\} = 1.$$

The value of the integral in all three cases is the same.

This can be seen to be always true, since

$$J = \int_0^1 (2xt + t^2 \frac{dx}{dt})dt = \int_0^1 \frac{d}{dt}(xt^2)dt$$

$$= xt^2\Big|_0^1 = x(1) = 1.$$

So if $x(1)$ is fixed, J has the same value no matter the path taken.

(6) Section 8.2.3: $n=1,\ m=1,\ q=0,\ x_1=x,\ u_1=E,$
$f_0 = e^{-\delta t}(pEx - cE),\ f_1 = F(x) - Ex.$

CHAPTER 9

(1) The integrand $F = (y'^2+1)^2$. Applying Euler's equation (9.10), gives

$$\frac{d}{dx}(2(y'^2+1)2y') = 0.$$

Integrating, $(y'^2+1)y' = K$, constant. This cubic in y' implies that $y' = A$, constant. Thus, on integrating, we obtain optimal paths of the form'

$$y = Ax + B.$$

(i) $y(0)=1, y(1)=2$: this gives $y = 1+x,\ 0 \leqslant x \leqslant 1$, and the corresponding value of J is $J = 4$.

(i) $y(0)=1, y(1)$ not specified: We have $1 = B$, and at $x=1$, from (9.18) we have $\frac{\partial F}{\partial y'} = 0$, that is, $(y'^2+1)y' = 0$, which gives $y' = 0$ at $x=1$; which implies $A = 0$. Hence $y = 1,\ 0 \leqslant x \leqslant 1$ is the optimal path, and the corresponding value of J is $J = 1$.

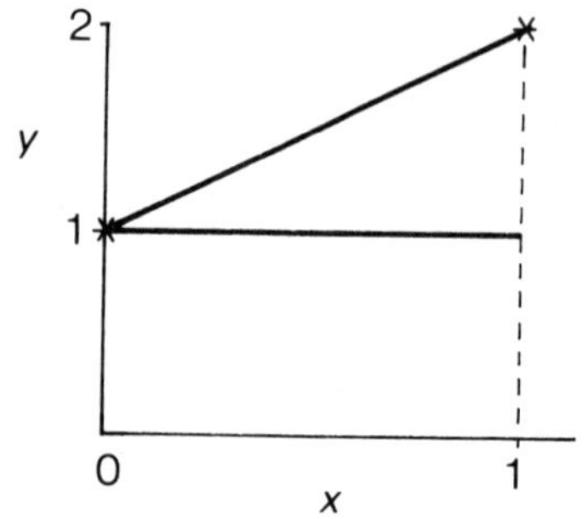

[Clearly the path $y = 1$, $0 \leqslant x \leqslant 1$ gives an overall minimum for J; the path $y = 1 + x$, $0 \leqslant x \leqslant 1$, gives a minimum for J subject to the given boundary conditions, $y(0) = 1$, $y(1) = 2$].

(2) The integrand $F = y^2 + \alpha^2 y'^2$; so that applying Euler's equation,

$$2y - \frac{d}{dx}(2\alpha^2 y') = 0$$

that is, $$y'' - \frac{1}{\alpha^2} y = 0.$$

This has solution $y = A e^{x/\alpha} + B e^{-x/\alpha}$, which we can write as

$$y = C \sinh(x/\alpha) + D \cosh(x/\alpha).$$

Applying the conditions gives

$$y_0 = D;\ 0 = C \cosh(X/\alpha) + D \cosh(X/\alpha).$$

Hence $$y = y_0 \cosh(x/\alpha) - y_0 \frac{\cosh(X/\alpha)}{\sinh(X/\alpha)} \sinh(x/\alpha)$$

that is, $$y = y_0 \frac{\sinh(X - x)/\alpha}{\sinh(X/\alpha)}.$$

Without the restriction $y(X) = 0$, we have $\dfrac{\partial F}{\partial y'} = 0$ at $x = X$, that is $y' = 0$ at $x = X$. This gives

$$\frac{C}{\alpha} \cosh(X/\alpha) + \frac{D}{\alpha} \sinh(X/\alpha) = 0$$

together with

$$y_0 = D.$$

This gives

$$y = y_0 \frac{\cosh(X-x)/\alpha}{\cosh(X/\alpha)}.$$

For the first path, $J_1 = \alpha x_0^2 \coth(X/\alpha)$, and for the second path, $J_2 = \alpha x_0^2 \tanh(X/\alpha)$. We note that $J_2 < J_1$, which is as expected, since there is one less restriction in the second case.

(3) Here the integrand $F = \frac{1}{2}y'^2 + yy' + y' + y$, so that the Euler equation gives

$$y' + 1 - \frac{d}{dx}(y' + y + 1) = 0$$

that is, $y'' = 1$.

Integrating, $y' = x + A$, and again, $y = \frac{1}{2}x^2 + Ax + B$.
The constants, A and B, are determined from the free end point conditions $\frac{\partial F}{\partial y'} = 0$ at $x = 0$ and $x = 2$; that is, $y' + y + 1 = 0$ at $x = 0$ and $x = 2$.

But $y' + y + 1 = (x + A) + (\frac{1}{2}x^2 + Ax + B) + 1 = \frac{1}{2}x^2 + (A+1)x + A + B + 1$.
At

$$x = 0, \quad \text{we have} \quad A + B + 1 = 0;$$

$$x = 2, \quad \text{we have} \quad 2 + 2(A + 1) + A + B + 1 = 0,$$

that is, $A = -2$ and $B = 1$.

Hence the required path is $y = \frac{1}{2}x^2 - 2x + 1$, $0 \leqslant x \leqslant 2$ (which in fact gives a minimum).

(4) $F = \frac{1}{2}y'^2 + y$, so that the Euler equation gives

$$1 - \frac{d}{dx}(y') = 0, \quad \text{that is,} \quad y'' = 1.$$

Integrating twice gives $y' = x + A$ and

$$y = \frac{1}{2}x^2 + Ax + B.$$

Applying $y(0) = 1$ gives $1 = B$, that is,

$$y = \tfrac{1}{2}x^2 + Ax + 1.$$

If, also, $y(1) = 0$, then $0 = \tfrac{1}{2} + A + 1$, that is, $A = -\tfrac{1}{2}$ and

$$y = \tfrac{1}{2}x^2 - \tfrac{3}{2}x + 1.$$

Now, if $y(1)$ is not specified, we must have $\dfrac{\partial F}{\partial y'} = 0$ at $x = 1$, that is, $y'(1) = 0$. This implies that $0 = 1 + A$, that is, $A = -1$ and

$$y = \tfrac{1}{2}x^2 - x + 1.$$

For the first path $J_1 = 23/24$, and the second path $J_2 = 5/6$. As expected, $J_2 < J_1$.

(5) The integrand $F = 2x_1x_2 - 2x_1^2 + \dot{x}_1^2 - \dot{x}_2^2$ is a function of x_1 and x_2, so we have Euler equations for x_1 and x_2, that is,

$$[x_1] \quad \frac{\partial F}{\partial x_1} - \frac{d}{dt}\left(\frac{\partial F}{\partial \dot{x}_1}\right) = 0 \Rightarrow 2x_2 - 4x_1 - 2\ddot{x}_1 = 0$$

$$\text{that is,} \quad x_2 = 2x_1 + \ddot{x}_1$$

$$[x_2] \quad \frac{\partial F}{\partial x_2} - \frac{d}{dt}\left(\frac{\partial F}{\partial \dot{x}_2}\right) = 0 \Rightarrow 2x_1 + 2\ddot{x}_2 = 0$$

$$\text{that is,} \quad x_1 = -\ddot{x}_2.$$

Substituting for x_1 gives

$$x_2 = -2\ddot{x}_2 - \ddddot{x}_2 \qquad \text{that is,} \quad \frac{d^4x_2}{dt^2} + 2\frac{d^2x_2}{dt^2} + x_2 = 0.$$

If $x_2 = e^{mt}$, then $m^4 + 2m^2 + 1 = 0$, that is, $(m^2 + 1)^2 = 0$. This gives $m = \pm i$ (repeated) and so

$$x_2 = (A + Bt)\sin t + (C + Dt)\cos t.$$

Also

$$\dot{x}_2 = (A + Bt + D)\cos t + (B - C - Dt)\sin t$$

$$x_1 = -\ddot{x}_2 = (-A - Bt - 2D)\sin t + (2B - C - Dt)\cos t.$$

The four constants A, B, C, D are determined from the four end point conditions $x_1(0)$, $x_2(0)$, $x_1(\pi/2)$, $x_2(\pi/2)$.

(6) (i) $x = 4 + 2z - 3y$, so we wish to minimise

$$f(y, z) = (4 + 2z - 3y)^2 + y^2 + z^2$$

that is,

$$\frac{\partial f}{\partial y} = \frac{\partial f}{\partial z} = 0 \text{ giving } \begin{cases} 2(4 + 2z - 3y)(-3) + 2y = 0 \\ \qquad\qquad \text{that is, } 10y - 6z = 12 \\ 2(4 + 2z - 3y)(2) + 2z = 0 \\ \qquad\qquad \text{that is, } -6y + 5z = -8. \end{cases}$$

Solving $y = 6/7$, $z = -4/7$ and so $x = 2/7$, giving a minimum value of $f = 8/7$.

(ii) Introducing a Lagrange multiplier λ, and considering the augmented function

$$f^* = x^2 + y^2 + z^2 + \lambda(x + 3y - 2z - 4)$$

For extremum values of f^*, we require

$$\frac{\partial f^*}{\partial x} = 0 \Rightarrow 2x + \lambda = 0$$

$$\frac{\partial f^*}{\partial y} = 0 \Rightarrow 2y + 3\lambda = 0$$

$$\frac{\partial f^*}{\partial z} = 0 \Rightarrow 2z - 2\lambda = 0.$$

Hence $x = -\lambda/2$, $y = -3\lambda/2$, $z = \lambda$; substituting into the constraint yields $\lambda = -4/7$, so that, as before,

$$x = 2/7,\ y = 6/7,\ z = -4/7.$$

(7)

We need to minimise the surface area

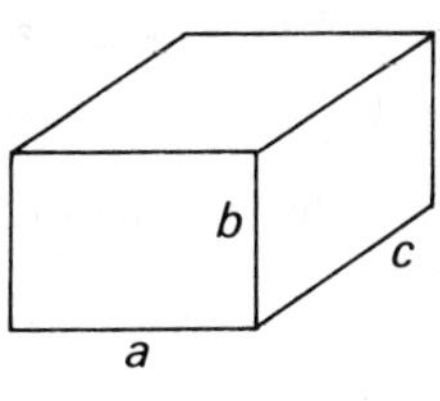

$$S = 2ab + 2bc + 2ca$$

subject to the constraint,

$$V = abc.$$

So we introduce the Lagrange multiplier λ and form the augmented functional

$$S^* = 2ab + 2bc + 2ca + \lambda(V - abc).$$

For a minimum

$$\frac{\partial S^*}{\partial a} = 0 \Rightarrow 2(b+c) - \lambda bc = 0$$

$$\frac{\partial S^*}{\partial b} = 0 \Rightarrow 2(a+c) - \lambda ac = 0$$

$$\frac{\partial S^*}{\partial c} = 0 \Rightarrow 2(b+a) - \lambda ab = 0.$$

Multiplying the first equation by a, the second by b and subtracting gives $ac - bc = 0$, that is, $a = b$. By symmetry $a = b = c = V^{\frac{1}{3}}$, that is, a cube.

(8) (i) Minimise

$$J = \int_0^1 (y'')^2\,\mathrm{d}x \quad \text{subject to} \quad y(0) = 0, \frac{\mathrm{d}y}{\mathrm{d}x}(0) = a$$

$$y(1) = 0, \frac{\mathrm{d}y}{\mathrm{d}x}(1) = b$$

$$(y'' = \frac{\mathrm{d}^2y}{\mathrm{d}x^2}).$$

Following the method of Example 9, put

$$x_1 = y, \quad x_2 = \frac{\mathrm{d}y}{\mathrm{d}x}$$

so that the problem is to minimise $J = \int_0^1 (x_2')^2\,\mathrm{d}x$ ($'$ means $\frac{\mathrm{d}}{\mathrm{d}x}$) subject to $x_1(0) = 0$, $x_2(0) = a$; $x_1(1) = 0$, $x_2(1) = b$, and constraint $x_1' - x_2 = 0$. We form the augmented functional

$$J^* = \int_0^1 [(x_2')^2 + \lambda(x_2 - x_1')]\,\mathrm{d}x.$$

Following the analysis of Example 9, we obtain

$$x_1 = Ax^3/_3 + Bx^2/_2 + Cx + D$$

$$x_2 = Ax^2 + Bx + C.$$

Fitting the conditions, gives $D=0, C=a, B=-2(b+2a), A=3(a+b)$. Hence

$$y = x_1 = (a+b)x^3 - (b+2a)x^2 + ax.$$

(iii) As before we have

$$x_1 = Ax^3/_3 + Bx^2/_2 + Cx + D$$

$$x_2 = Ax^2 + Bx + C.$$

Now $x_1(0)=0$, $x_1(1)=0$, but there are no given conditions on x_2 at $x=0,1$.

Hence $\dfrac{\partial F}{\partial x_2'} = 0$ at $x=0,1$,

that is, $x_2' = 0$ at $x=0,1$; $x_2'(0)=x_2'(1)=0$.

Fitting these conditions, gives $A=B=C=D=0$, that is, $y=0$, $0 \leqslant x \leqslant 1$ (and so $J=0$ – which is clearly a minimum).

(9) As before, we introduce the variables

$$x_1 = x, \quad x_2 = \dot{x},$$

so that we seek the stationary path for

$$J = \int_0^1 (1+\dot{x}_2^2)\mathrm{d}t$$

subject to $\dot{x}_1 - x_2 = 0$; $x_1(0)=0$, $x_2(0)=1$, and $x_1(1)=1$, $x_2(1)=1$.

We form the augmented functional

$$J^* = \int_0^1 [1 + \dot{x}_2^2 + \lambda(\dot{x}_1 - x_2)]\mathrm{d}t$$

and form the Euler equations for x_1 and x_2:

$$[x_1] \quad \frac{\partial F}{\partial x_1} - \frac{\mathrm{d}}{\mathrm{d}t}\left(\frac{\partial F}{\partial \dot{x}_1}\right) = 0, \quad \text{that is,} \quad 0 - \frac{\mathrm{d}}{\mathrm{d}t}(\lambda) = 0$$

$$[x_2] \quad \frac{\partial F}{\partial x_2} - \frac{\mathrm{d}}{\mathrm{d}t}\left(\frac{\partial F}{\partial \dot{x}_2}\right) = 0, \quad \text{that is,} \quad -\lambda - \frac{\mathrm{d}}{\mathrm{d}t}(2\dot{x}_2) = 0$$

$$\Rightarrow \ddot{x}_2 = -\tfrac{1}{2}\lambda.$$

Hence $\ddot{x}_2 = 0,$

so $x_2 = At^2 + Bt + C,$ and integrating again

$$x_1 = At^3/_3 + Bt^2/_2 + Ct + D.$$

Applying the end conditions gives $A = 3$, $B = -2$, $C = 1$, $D = 0$. Hence the optimal curve is

$$x = t^3 - t^2 + t.$$

(10) As in Exercise 8, we introduce $x_1 = y$, $x_2 = \dfrac{dy}{dx}$, so that we look for extremals of the functional

$$J = \int_0^{\pi/2} (x_2'^2 - x_1^2 + x^2)dx$$

subject to $x_1' = x_2$; $x_1(0) = 1$, $x_2(0) = 0$; $x_1(\pi/2) = 0$, $x_2(\pi/2) = 1$.

We form the augmented functional

$$J^* = \int_0^{\pi/2} [x_2'^2 - x_1^2 + x^2 + \lambda(x_2 - x_1')]dx$$

and form Euler equations for x_1 and x_2;

$[x_1]$ $\quad -2x_1 - \dfrac{d}{dx}(-\lambda)$ $\quad$ that is, $\quad \lambda' = 2x_1$

$[x_2]$ $\quad \lambda - \dfrac{d}{dx}(2x_2')$ $\quad$ that is, $\quad x_2'' = \frac{1}{2}\lambda.$

Hence $x_2''' = \frac{1}{2}\lambda' = x_1$ $\quad$ that is, $\quad \dfrac{d^4x_2}{dx^4} = x_1' = x_2.$

Thus $x_2 = A\sin x + B\cos x + C\sinh x + D\cosh x$

and $x_1 = x_2''' = -A\cos x + B\sin x + C\cosh x + D\sinh x.$

Applying the conditions gives

$$A = -\sinh\pi/2,\ B = -\cosh\pi/2,\ C = 1 - \sinh\pi/2,\ D = \cosh\pi/2.$$

That is,

$$y = \sinh\frac{\pi}{2}\cos x - \cosh\frac{\pi}{2}\sin x + (1 - \sinh\pi/2)\cosh x + \cosh\pi/2\ \sinh x.$$

CHAPTER 10

(1) Eliminating u gives

$$J = \int_0^T (2x^2 + 2x\dot{x} + \dot{x}^2)\mathrm{d}t$$

so that the Eluer equation is

$$\frac{\partial F}{\partial x} - \frac{\mathrm{d}}{\mathrm{d}t}\left(\frac{\partial F}{\partial \dot{x}}\right) = 0, \quad \text{that is,} \quad 4x + 2\dot{x} - \frac{\mathrm{d}}{\mathrm{d}t}(2x + 2\dot{x}) = 0.$$

Thus $\ddot{x} - 2x = 0$, and the solution follows as in section 10.1.

(2) The Hamiltonian for this system is

$$H = (Q - x)^2 + u^2 + \lambda(u - ax).$$

We have $\dfrac{\partial H}{\partial u} = 0$, that is, $2u + \lambda = 0$

$$\dot{\lambda} = -\frac{\partial H}{\partial x} \quad \text{that is,} \quad \dot{\lambda} = +2(Q - x) + \lambda a.$$

Eliminating λ gives $\dot{u} = au + x - Q$; and eliminating u from $u = \dot{x} + ax$ gives

$$\ddot{x} - (1 + a^2)x = -Q.$$

Writing $\alpha^2 = 1 + a^2$, we have solution

$$x = A\sinh\alpha t + B\cosh\alpha t + Q/\alpha^2.$$

Fitting the boundary conditions, gives

$$x = \frac{Q(a^2 + \cosh\alpha T)}{\alpha^2 \sinh\alpha T}\sinh\alpha t + \frac{Q}{\alpha^2}(1 - \cosh\alpha t)$$

and the optimal control u is determined from $u = \dot{x} + ax$.

(3) Using the method of section 10.4, we form the Hamiltonian

$$H = \tfrac{1}{2}u^2 + p_1 x_2 + p_2(-x_2 + u)$$

where the adjoint variables satisfy

$$\dot{p}_1 = -\frac{\partial H}{\partial x_1} = 0 \qquad \text{that is,} \quad p_1 = A, \text{ constant}$$

$$\dot{p}_2 = -\frac{\partial H}{\partial x_2} = -p_1 + p_2 \quad \text{that is,} \quad \dot{p}_2 - p_2 = -A \Rightarrow p_2 = Be^t - A.$$

Also $\dfrac{\partial H}{\partial u} = 0$, that is, $u + p_2 = 0$. Hence $u = A - Be^t$, and so

$$\dot{x}_2 + x_2 = A - Be^t.$$

Solving,

$$x_2 = Ce^{-t} + A - \frac{B}{2}e^t$$

and

$$\dot{x}_1 = x_2 \quad \text{gives} \quad x_1 = -Ce^{-t} + At - \frac{B}{2}e^t + D.$$

Applying the boundary conditions gives the appropriate values for A, B, C and D.

(4) This is a special case of the problem in section 10.3, with $\alpha = \frac{16}{3}$; so we obtain (10.20), that is,

$$\frac{d^4 x_1}{dt^4} - \frac{d^2 x_1}{dt^2} + \tfrac{3}{16}x_1 = 0.$$

For solutions of the form $x_1 = e^{mt}$, we require

$$16m^4 - 16m^2 + 3 = 0$$

that is, $(4m^2 - 1)(4m^2 - 3) = 0.$

Hence $m = \pm\frac{1}{2}, \pm\dfrac{\sqrt{3}}{2}$ and

$$x_1 = Ae^{t/2} + Be^{-t/2} + Ce^{\sqrt{3}t/2} + De^{-\sqrt{3}t/2}$$

and $$x_2 = \dot{x}_1 = \frac{A}{2}e^{t/2} - \frac{B}{2}e^{-t/2} + \frac{\sqrt{3}}{2}Ce^{\sqrt{3}\,t/2} - \frac{\sqrt{3}D}{2}e^{-\sqrt{3}\,t/2}.$$

Since $x_1, x_2 \to 0$ as $t \to \infty$, we see that $A = C = 0$; and applying the initial conditions gives

$$B = \frac{\sqrt{3}a + 2b}{\sqrt{3} - 1}, \quad D = \frac{a + 2b}{1 - \sqrt{3}}.$$

Now $$u = \dot{x}_2 + x_2 = \left(\frac{B}{4} - \frac{B}{2}\right)e^{-t/2} + \left(\frac{3}{4}D - \frac{\sqrt{3}}{2}D\right)e^{-\sqrt{3}\,t/2}.$$

that is, $$u = -\frac{(\sqrt{3}a + 2b)}{4(\sqrt{3} - 1)}e^{-t/2} + \frac{\sqrt{3}}{4}\frac{(2 - \sqrt{3})(a + 2b)}{(\sqrt{3} - 1)}e^{-\sqrt{3}t/2}.$$

(5) (i) Here $$H = \tfrac{1}{2}(3x^2 + u^2) + \lambda(x + u)$$

where
$$\dot{\lambda} = -\frac{\partial H}{\partial x} = -(3x + \lambda)$$

and
$$\frac{\partial H}{\partial u} = 0, \quad \text{that is,} \quad u + \lambda = 0.$$

Thus eliminating λ gives $\dot{u} = 3x - u$, and since $u = \dot{x} - x$, we have, $\ddot{x} - \dot{x} = 3x - (\dot{x} - x)$; that is,

$$\ddot{x} - 4x = 0.$$

Hence $x = Ae^{2t} + Be^{-2t}$. Now $x(0) = x_0$, given; that is, $x_0 = A + B$, and $x = Ae^{2t} + (x_0 - A)e^{-2t}$. Since $x(T)$ is free, $\lambda = 0$ at $t = T$, that is, $u = 0$ at $t = T$. But

$$u = \dot{x} - x = 3Ae^{2t} - (x_0 - A)e^{-2t}.$$

At $t = T$, $u(T) = 0 = 3Ae^{2T} - (x_0 - A)e^{-2T}$.
This gives $A = x_0e^{-2T}/(3e^{2T} + e^{-2T})$; that is,

$$u(t) = \frac{3x_0e^{2(t-T)}}{(3e^{2T} + e^{-2T})} - \frac{3e^{2(T-t)}x_0}{(3e^{2T} + e^{-2T})}.$$

(ii) As $T \to \infty$, $A \to 0$, and $u(t) \approx -x_0e^{-2t}$, and $x \approx x_0e^{-2t}$.
Thus $u \approx -x$, that is, $F = -1$.

(6) (i) $H = u^2 + \lambda(-ax + u)$ and

$$\dot{\lambda} = -\frac{\partial H}{\partial x} = +a\lambda$$

and $\dfrac{\partial H}{\partial u} = 0$, that is, $2u + \lambda = 0$. Eliminating λ gives $\dot{u} - au = 0$.

Hence $u = Ae^{at}$ and $\dot{x} + ax = Ae^{at}$. Solving

$$x = Be^{-at} + \frac{A}{2a}e^{at}.$$

Now at $t = 0$, $x_0 = B + \dfrac{A}{2a}$, and at $t = T$, $0 = Be^{-aT} + \dfrac{A}{2a}e^{at}$.

Thus $A = 2ax_0/(1 - e^{2aT})$ and $B = x_0e^{2aT}/(e^{2aT} - 1)$. Hence

$$x = \frac{(e^{a(T-t)} - e^{a(t-T)})x_0}{(e^{aT} - e^{-aT})}.$$

(ii) $$u = \frac{2ax_0e^{a(t-T)}}{(e^{-aT} - e^{aT})}.$$

(iii) We see that $$x = Be^{-2at}e^{at} + \frac{A}{2a}e^{at}$$

$$= \left(\frac{B}{A}e^{-2at} + \frac{1}{2a}\right)u$$

that is, $$x = \frac{1}{2a}(1 - e^{2a(T-t)})u.$$

Hence $$u = \frac{2ax}{[1 - e^{2a(T-t)}]}$$ giving $$F = \frac{2a}{[1 - e^{2a(T-t)}]}.$$

CHAPTER 11

(1) We form the Hamiltonian

$$H = 1 + p_1(-2x_1 + u) + p_2u$$

where the adjoint variables satisfy

$$\dot{p}_1 = -\frac{\partial H}{\partial x_1} = 2p_1 \Rightarrow p_1 = Ae^{2t}$$

$$\dot{p}_2 = -\frac{\partial H}{\partial x_2} = 0 \Rightarrow p_2 = B.$$

Now $H = (p_1 + p_2)u + 1 - p_1 x_1$, so the maximum of H with respect to u, $|u| \leqslant 1$, occurs when

$$u = \begin{cases} +1 \text{ if } p_1 + p_2 > 0 \\ -1 \text{ if } p_1 + p_2 < 0. \end{cases}$$

But $p_1 + p_2 = Ae^{2t} + B$, and this passes through zero at most once; so the control u is bang-bang, with at most one switch. Now if $u = k = \pm 1$, then

$$\dot{x}_1 + 2x_1 = k \Rightarrow x_1 = Ce^{-2t} + k$$

$$\dot{x}_2 = k \Rightarrow x_2 = D + kt.$$

Eliminating t, gives

$$x_2 + E = -\frac{k}{2}\log|x_1 - k|, \quad E \text{ constant.}$$

So we have for

(i) $k = 1$: $x_2 + E = -\frac{1}{2}\log|x_1 - 1|$, which is illustrated below

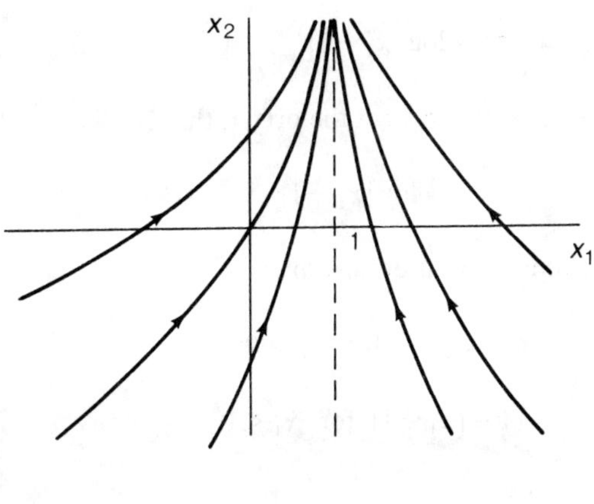

(ii) $k = -1$: $x_2 + E = \frac{1}{2}\log|x_1 + 1|$.

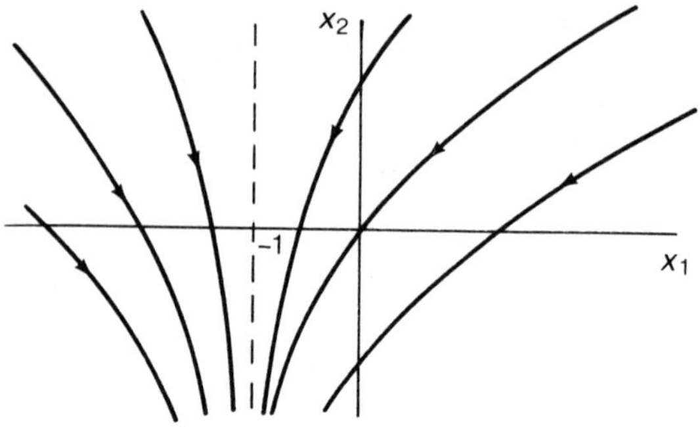

Combining these two diagrams gives

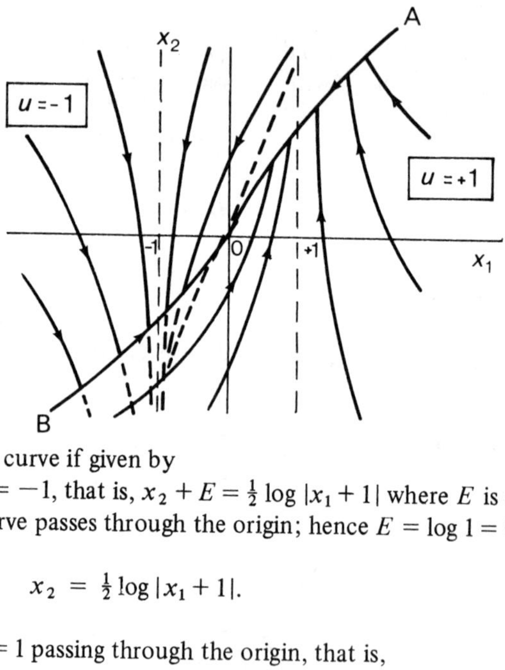

The switching curve if given by

(i) $x_1 > 0$: $u = -1$, that is, $x_2 + E = \frac{1}{2}\log|x_1 + 1|$ where E is such that the curve passes through the origin; hence $E = \log 1 = 0$ and

$$x_2 = \tfrac{1}{2}\log|x_1 + 1|.$$

(ii) $x_1 < 0$: $u = 1$ passing through the origin, that is,

$$x_2 = -\tfrac{1}{2}\log|x_1 - 1|.$$

Hence the switching curve is defined by

$$x_2 = \begin{cases} \frac{1}{2}\log|x_1 + 1| & \text{for } x_1 > 0 \\ -\frac{1}{2}\log|x_1 - 1| & \text{for } x_1 < 0. \end{cases}$$

(2) The Hamiltonian is given by

$$H = 1 + p_1(-x_1 - u) + p_2(-2x_2 - 2u)$$

where

$$\dot{p}_1 = -\frac{\partial H}{\partial x_1} = p_1 \Rightarrow p_1 = Ae^t$$

$$\dot{p}_2 = -\frac{\partial H}{\partial x_2} = 2p_2 \Rightarrow p_2 = Be^{2t}.$$

Now $H = -(p_1 + 2p_2)u + 1 - x_1 p_1 - 2x_2 p_2$, so by the maximum principle,

$$u = \begin{cases} +1 & \text{if } p_1 + 2p_2 < 0 \\ -1 & \text{if } p_1 + 2p_2 > 0. \end{cases}$$

Now $p_1 + 2p_2 = Ae^t + 2Be^{2t} = Ae^t(1 + 2Be^t/A)$, and this has at most one zero. Hence there is at most one switch in the control u.

If $u = k = \pm 1$, $\dfrac{dx_1}{dt} + x_1 = -k$ that is, $x_1 = Ae^{-t} - k$

$\dfrac{dx_2}{dt} + 2x_2 = -2k$ that is, $x_2 = Be^{-2t} - k.$

Eliminating t gives $(x_1 + k)^2 = C(x_2 + k)$. Thus

(i) $k = 1$: $(x_1 + 1)^2 = C(x_2 + 1)$ which is illustrated below.

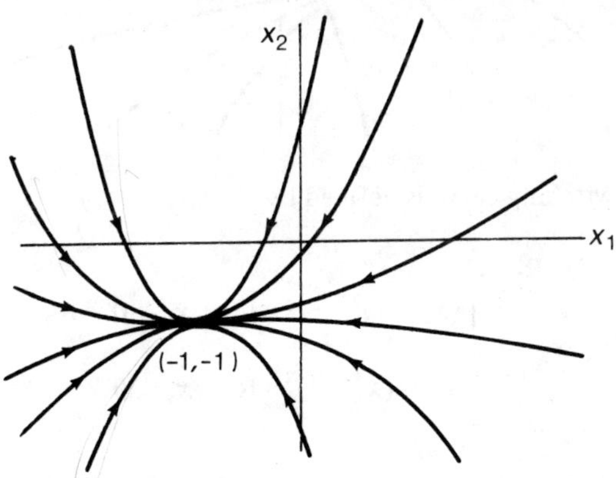

[NB: $\dot{x}_1 > 0$ if $-x_1 - 1 > 0$, that is, $x_1 < -1$]

(ii) $k=-1$: $(x_1-1)^2 = C(x_2-1)$

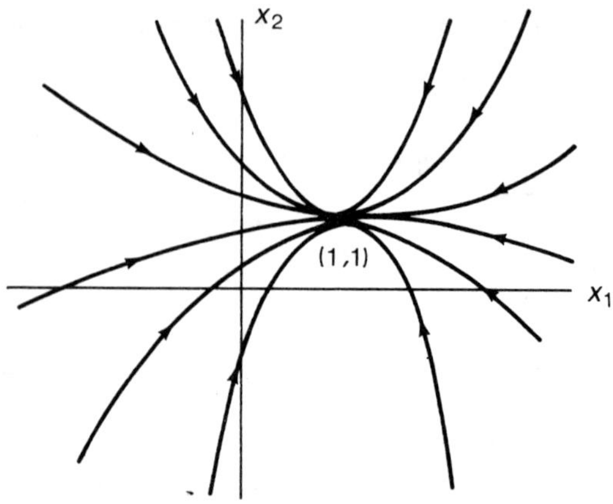

[NB: $\dot{x}_1 > 0$ if $-x_1+1>0$, that is, $x_1<1$]

Combining the diagrams gives

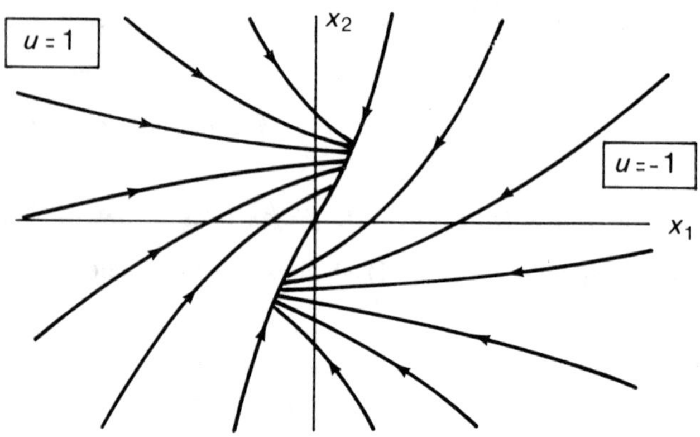

The switching curve is defined by

$$x_2 = \begin{cases} (x_1+1)^2 - 1 & \text{for } x_1 > 0 \\ 1-(x_1-1)^2 & \text{for } x_1 < 0. \end{cases}$$

(3) This system has already been studied in section 11.4. Following that analysis, we obtain $u = \pm 1$, and

$$x_2^2 = 2k(x_1 - B), \quad k = \pm 1$$

as the optimal trajectories. Also $u = \pm 1$ according to the sign of $p_2 = -At + B$, giving at *most* one switch in the optimal control. Combining the diagrams in Fig. 11.7, we obtain the picture below.

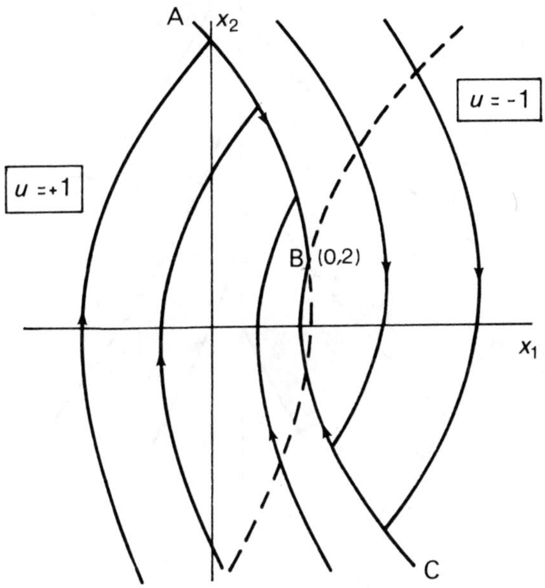

ABC is the switching curve

(4) This is the problem considered in section 11.3, except for the terminal condition on x_2, which is now free. As in section 11.3, we obtain bang-bang control, that is, $u = \pm 1$, with optimal trajectories

$$x_2 - B = -k \log|(x_1 - k)/A|, \quad k = \pm 1$$

$$= -k \log|x_1 - k| + k \log|A|$$

that is, $x_2 - C = \mp \log|x_1 \mp 1|$.

Now $u = \pm 1$, depending on the sign of $p_1 + p_2$, where $p_1 = Ae^t$ and $p_2 = B$. Since x_2 is not specified at the final time, the transversality condition gives $p_2 = 0$ at the final time. Hence $B = 0$ and $p_2 = 0$ for all time. Thus $p_1 + p_2 = Ae^t$ does *not* change sign, and there is *no* switch in the optimal control. This means that we obtain optimal trajectories as illustrated below.

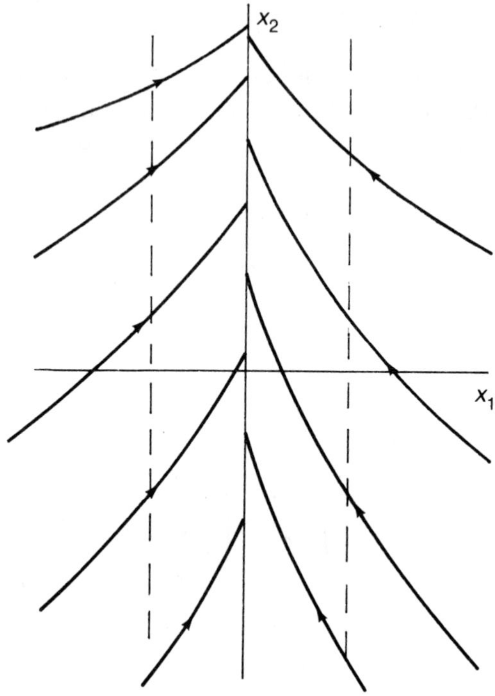

(5) Here $H = 1 + p(2x + u)$ and so for optimal control

$$u = \begin{cases} +1 & \text{if } p > 0 \\ -1 & \text{if } p < 0 \end{cases}$$

That is, $u = k$, constant $(= \pm 1)$. Also $\dot{p} = -\partial H/\partial x = -2p$, that is, $p = Ae^{-2t}$, and so there is no zero for p and hence *no* switch in u. We have $a\dot{x} = 2x + k$, and solving

$$x = (x_0 + k/2)e^{2t} - k/2.$$

Now, at the end time, say $t = T$, we have

$$e^{2T} = k/(2x_0 + k).$$

But for this to have a real positive solution for T, we must have $e^{2T} > 1$, that is, $\dfrac{k}{2x_0 + k} > 1$. For $k = 1$, we require $\dfrac{1}{2x_0 + 1} > 1$. That is,

$-\frac{1}{2} < x_0 < 0$; and for $k = -1$, we require $\dfrac{1}{1-2x_0} > 1$, that is, $0 < x_0 < \frac{1}{2}$. Thus we can reach the origin if $-\frac{1}{2} < x_0 < \frac{1}{2}$, and there is no time optimal control to the origin if

$$\{x : x < -\tfrac{1}{2} \text{ or } x > \tfrac{1}{2}\}.$$

CHAPTER 12

(1) We have

$$\dot{k} = f(k) - (\lambda + \mu)k - c$$

and

$$\dot{c} = \eta[f'(k) - \beta]c$$

we can eliminate the consumption c. This gives

$$f'(k)\dot{k} - (\lambda + \mu)\dot{k} - \ddot{k} = \eta[f'(k) - \beta][f(k) - (\lambda + \mu)k - \dot{k}]$$

that is, $\ddot{k} + \dot{k}\{\lambda + \mu - \eta(f'(k) - \beta) - f'(k)\}$

$$+ \eta\{f'(k) - \beta\}\{f(k) - (\lambda + \mu)k\} = 0.$$

This is the non-linear second order differential equation which the optimal capital path must satisfy.

(2) Here we wish to maximise

$$\text{P.V.} = \int_0^\infty e^{-\delta t}(px^\alpha E^\beta - cE^\gamma)dt$$

subject to $\dot{x} = F(x) - x^\alpha E^\beta$. Hence

$$H = -e^{-\delta t}(px^\alpha E^\beta - cE^\gamma) + \lambda(F(x) - x^\alpha E^\beta)$$

where the adjoint variable, λ, satisfies

$$\dot{\lambda} = -\frac{\partial H}{\partial x} = \alpha p e^{-\delta t} x^{\alpha-1} E^{\beta-1} - \lambda(F'(x) - \alpha x^{\alpha-1} E^\beta)$$

and from the maximum principle

$$\frac{\partial H}{\partial E} = 0, \text{ that is, } -p\beta e^{-\delta t} x^\alpha E^{\beta-1} + e^{-\delta t} c\gamma E^{\gamma-1} - \beta\lambda x^\alpha E^{\beta-1} = 0.$$

Hence $\lambda = e^{-\delta t}\{c\gamma x^{-\alpha}E^{\gamma-\beta}/\beta - p\}$

and so, eliminating λ gives, in the case $\beta = \gamma$,

$$\delta = \frac{\alpha c\gamma x^{-\alpha-1}F(x)}{(p\beta - c\gamma x^{-\alpha})} + F'.$$

Since there is no direct dependence on t, we have an equation for an equilibrium solution for the optimal population.

(3) The first part of this question is dealt with in section 12.7, with $a = 0$ and $\omega^2 = 1$.

From Fig. 12.13 we see that the optimal trajectory from (1,1) has *one* switch in the control (starting with $u = -1$, and finishing with $u = 1$). On the other the circle, ending with OM_1, also passes through (1,1). Hence taking $u = -1$ with *no* switches will achieve the desired goal, but this would not be the time-optimal path.

(4) Here we wish to maximise $h(T) = \int_0^T v\,dt$, subject to the state equations

$$\dot{v} = -g + c\beta/m$$

$$\dot{m} = -\beta$$

and at $t = 0$, $m = m_0$, $v = 0$, and $0 \leqslant \beta \leqslant \bar{\beta}$. Thus

$$H = -v + p_1(-g + c\beta/m) + p_2(-\beta)$$

where the adjoint variables satisfy

$$\dot{p}_1 = -\frac{\partial H}{\partial v} = 1$$

$$\dot{p}_2 = -\frac{\partial H}{\partial m} = c\beta p_1/m^2.$$

Now H is linear in the control β, so we have bang-bang control with;

$$\beta = \begin{cases} 0 & \text{if} \quad p_1 c/m - p_2 < 0 \\ \bar{\beta} & \text{if} \quad p_1 c/m - p_2 > 0, \end{cases}$$

provided $p_1\dfrac{c}{m} - p_2 \neq 0$ on a non-zero sub-interval $[t_1, t_2]$. Now suppose $p_1\dfrac{c}{m} - p_2$ is identically equal to zero on a non-zero sub-interval; that is,

$$p_1 \frac{c}{m} = p_2$$

and differentiating with respect to time gives

$$\dot{p}_1 \frac{c}{m} - p_1 \frac{c}{m^2} \dot{m} = \dot{p}_2 \Rightarrow \frac{c}{m} = 0,$$

having substituted for $\dot{p}_1$, $\dot{m}$ and $\dot{p}_2$. Clearly this cannot be true, so we conclude that $(p_1 \frac{c}{m} - p_2)$ cannot be zero on a non-zero sub-interval and so we do indeed have bang-bang control.

Assuming only one switch in the control, we must clearly start with $\beta = \bar{\beta}$, and finish with $\beta = 0$. The switch will occur when all the fuel has been burnt. If the rocket's mass is m_1 when this occurs, then form $\dot{m} = -\bar{\beta}$, we have $m = m_0 - \bar{\beta} t$, and the switching time is given by $t_1 = (m_0 - m_1)/\beta$. Thus the optimal control is given by

$$\beta = \begin{cases} \bar{\beta} & 0 \leqslant t \leqslant t_1 \\ 0 & t_1 \leqslant t \leqslant T. \end{cases}$$

(The final time T is, of course, determined by $v = 0$ at $t = T$.)

(5) Let $x_1 = \theta$, $x_2 = \dot{\theta}$, so that

$$\dot{x}_1 = x_2$$

$$\dot{x}_2 = p - x_2$$

with $|p| \leqslant 1$ and at $t = 0$, $x_1 = \alpha$, $x_2 = 0$ whilst at the final time $x_1 = x_2 = 0$.

Now

$$H = 1 + \lambda_1 x_2 + \lambda_2 (p - x_2)$$

so that the control is bang-bang with

$$p = \begin{cases} +1 & \text{if } \lambda_2 > 0 \\ -1 & \text{if } \lambda_2 < 0. \end{cases}$$

The adjoint variables satisfy

$$\dot{\lambda}_1 = -\frac{\partial H}{\partial x_1} = 0 \qquad \text{that is,} \quad \lambda_1 = A, \text{ constant}$$

$$\dot{\lambda}_2 = -\frac{\partial H}{\partial x_2} = -\lambda_1 + \lambda_2 \quad \text{that is,} \quad \lambda_2 = -A + B\mathrm{e}^{+t}.$$

We see that λ_2 changes sign at most once, and so we have at most one switch in the control.

If $p = k(= \pm 1)$, then

$$x_2 = C\mathrm{e}^{-t} + k$$

and

$$x_1 = -C\mathrm{e}^{-t} + kt + D$$

where C and D are arbitrary. Eliminating t gives

$$x_1 = -x_2 + E - k\log(x_2 - k).$$

We can complete the picture by considering $k = +1$ and $k = -1$. For the starting conditions stated we take $p = -1$ followed by $p = +1$.

CHAPTER 13

(1) Optimal plan: Newcastle - Doncaster - Nottingham - Leicester - Oxford; 12 additions; 16 possible paths; 48 additions.

(2)

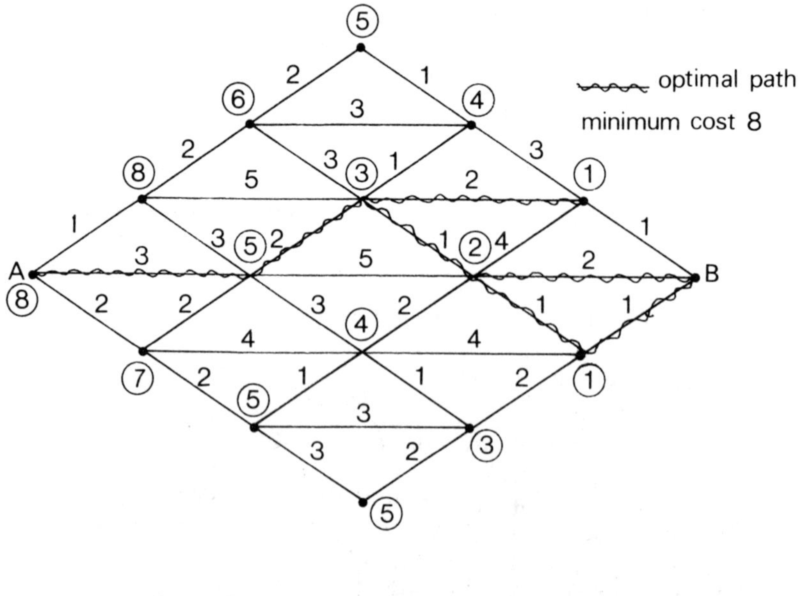

(3)

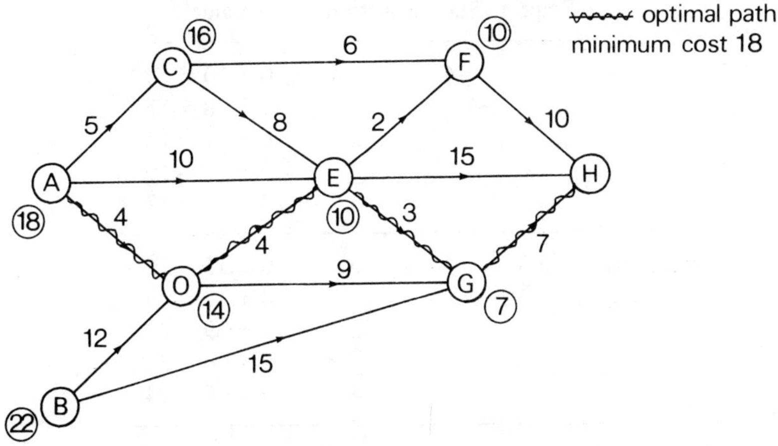

(4)

	Stage	State	Action	Value
(Product C)	1	0	0	0
		1	1	6
		2	2	7
		3	3	8
		4	4	10
(Product B)	2	0	0	0
		1	<u>0</u> 1	0 + 6 = <u>6</u> 3 + 0 = 3
		2	0 <u>1</u> 2	0 + 7 = 7 3 + 6 = <u>9</u> 6 + 0 = 6
		3	0 1 <u>2</u> 3	0 + 8 = 8 3 + 7 = 10 6 + 6 = <u>12</u> 9 + 0 = 9

	Stage	State	Action	Value
		4	0	0 + 10 = 10
			1	3 + 8 = 11
			2	6 + 7 = 13
			<u>3</u>	9 + 6 = <u>15</u>
			4	12 + 0 = 12
(Product A)	3	4	0	0 + 15 = 15
			1	8 + 12 = 20
			2	18 + 9 = 27
			<u>3</u>	22 + 6 = <u>28</u>
			4	24 + 0 = 24

Optimal allocation: 3 units to A, 0 to B, 1 to C.

(5) This is similar to the allocation problem of section 13.4.1, except that the number of tons of raw material available can be 2, 4, 6 or 8. We first follow through the usual method, remembering that the final quantity of raw material to be allocated is not necessarily 8.

Stage, n — Allocation of raw material to a product, n being the number of products remaining (A, B, C will be considered at stages 3, 2, 1, respectively).

State, (n, i) — i tons remain to be allocated, and n products remain to be considered.

Action — Allocate k tons to current product.

Return — Return from allocation of k tons to product n.

	Stage	State	Action	Value
(Product C)	1	0	0	0
		1	<u>1</u>	<u>8</u>
		2	<u>2</u>	<u>10</u>
		3	<u>3</u>	<u>11</u>
(Product B)	2	0	0	0
		1	<u>0</u>	0 + 8 = <u>8</u>
			1	6 + 0 = 6

Stage	State	Action	Value	
	2	0	0 + 10 = 10	
		1	6 + 8 = 14	
		2	12 + 0 = 12	
	3	0	0 + 11 = 11	
		1	6 + 10 = 16	
		2	12 + 8 = 20	
		3	21 + 0 = 21	
	4	1	6 + 11 = 17	
		2	12 + 10 = 22	
		3	21 + 8 = 29	
	5	2	12 + 11 = 23	
		3	21 + 10 = 31	
	6	3	21 + 11 = 32	
(Product A) 3	0	0	0	Net Profit (returns - cost)
	2	0	0 + 14 = 14	
		1	9 + 8 = 17	
		2	18 + 0 = 18	18 − 15 = £3
	4	0	0 + 29 = 29	
		1	9 + 21 = 30	
		2	18 + 14 = 32	32 − 28 = £4
		3	23 + 8 = 31	
	6	0	0 + 32 = 32	
		1	9 + 31 = 40	
		2	18 + 29 = 47	47 − 40 = £7
		3	23 + 21 = 44	
	8	2	18 + 32 = 50	
		3	23 + 31 = 54	54 − 50 = £4

Thus 6 tons raw material should be used and allocated
2 tons for A; 3 tons for B; 1 ton for C.

(6) STAGE: Completion of Stripping, Repairing and Rebuilding
(1) (2) (3)

STATE: Number of days to end of overhaul

ACTION: 1 Complete next stage in 2 days
2 ,, ,, ,, ,, 4 ,,
3 ,, ,, ,, ,, 6 ,,

RETURN: Cost associated with action.

Stage	State	Action	Value	
3	0	–	0	
2	2	1	20	
	4	2	15	
	6	3	9	←
1	4	1	11 + 20 = 31	
	6	1	11 + 15 = 26	
		2	8 + 20 = 28	
	8	1	11 + 9 = 20	←
		2	8 + 15 = 23	
		3	7 + 20 = 27	
0	10	1	19 + 20 = 38	←
		2	17 + 26 = 43	
		3	8 + 31 = 39	

Optimum plan - Stripping 2 days
Repairing 2 days
Rebuilding 6 days – cost £38 000.

(7) Stage n number of months until end of planning period

State i length of fish in stock at start of the month, corresponding to stage n.

Action k feed at rate k where $k = 1, 2$ or 3

Return revenue received for fish at the end of the month less the feeding cost.

Note that there are **five** times as many fish in the first month compared with the last and corresponding months similarly. Thus unit feeding costs must be multiplied by stage variable n for month labelled n. Details of the calculation are shown below.

We see that the optimum feeding rate is

Jan	Feb	Mar	Apl	May
1	1	2	2	3

and the maximum return is then 24 units.

Stage	State	Action	Value	
1	8	1	0 − 1 = −1	
		2	2 − 2 = 0	
		3	8 − 5 = 3	
	9	1	2 − 3 = −1	
		2	8 − 4 = 4	
		3	14 − 7 = 7	
	10	1	8 − 5 = 3	
		2	14 − 6 = 8	
		3	22 − 7 = 15	←
	11	1	14 − 8 = 6	
		2	22 − 9 = 13	
	12	1	22 − 10 = 12	
2	8	1	2 − 2 + 3 = 3	
		2	4 − 4 + 7 = 7	
		3	8 − 10 + 15 = 13	
	9	1	4 − 6 + 7 = 5	
		2	8 − 8 + 15 = 15	←
		3	14 − 14 + 13 = 13	
	10	1	8 − 10 + 15 = 13	
		2	14 − 12 + 13 = 15	
		3	20 − 14 + 12 = 18	

Stage	State	Action	Value	
	11	1	14 − 16 + 13 = 11	
		2	20 − 18 + 12 = 14	
	12	1	20 − 20 + 12 = 12	
3	8	1	4 − 3 + 13 = 14	
		2	6 − 6 + 15 = 15	←
		3	10 − 15 + 18 = 13	
	9	1	6 − 9 + 15 = 12	
		2	10 − 12 + 18 = 16	
		3	14 − 21 + 14 = 7	
	10	1	10 − 15 + 18 = 13	
		2	14 − 18 + 14 = 10	
		3	20 − 21 + 12 = 11	
	11	1	14 − 24 + 14 = 4	
		2	20 − 27 + 12 = 5	
	12	1	20 − 30 + 12 = 2	
4	8	1	8 − 4 + 15 = 19	←
		2	10 − 8 + 16 = 18	
		3	14 − 20 + 13 = 7	
	9	1	10 − 12 + 16 = 14	
		2	14 − 16 + 13 = 11	
		3	20 − 28 + 5 = −3	
	10	1	14 − 20 + 13 = 7	
		2	20 − 24 + 5 = 1	
		3	24 − 28 + 2 = −2	
5	8	1	10 − 5 + 19 = 24	←
		2	15 − 10 + 14 = 19	
		3	20 − 25 + 7 = 2	

(8) Using the variational calculus method, we have

$$\frac{\partial F}{\partial x} = \frac{\mathrm{d}}{\mathrm{d}t}\left(\frac{\partial F}{\partial \dot{x}}\right), \quad \text{where} \quad F = \dot{x}^2 - x^2.$$

Thus $\quad -2x = 2\ddot{x}, \quad$ that is, $\quad \ddot{x} + x = 0, \quad$ giving

$$x = A\sin x + B\cos x.$$

Since $x(0) = 0$, $B = 0$ and $x(1) = 1$ gives $1 = A\sin 1$. Hence the optimal path is $x = \sin x/\sin 1$, and the corresponding value of the functional is

$$\int_0^1 \left(\frac{\cos^2 x}{\sin^2 1} - \frac{\sin^2 x}{\sin^2 1}\right) \mathrm{d}t = \cot 1 = 0.642$$

Bibliography

[1] Anand, D. K. (1974), *Introduction to Control Systems*, Pergamon Press.
[2] Auslander, D. M. Takahashi, Y. and Rabins, M. J. (1974), *Introducing Systems and Control*, McGraw-Hill.
[3] Bailey, N. T. J. (1975), *The Mathematical Theory of Infectious Diseases*, Griffin.
[4] Barnett, S. (1975), *Introductions to Mathematical Control Theory*, Clarendon Press.
[5] Barnett, S. and Storey, C. (1968), Some Applications of Liapunov Matrix Equations, *J. Inst. Applics.*, **4**, 33-42.
[6] Berkovitz, L. D. (1974), *Optimal Control Theory*, Springer-Verlag.
[7] Blackman, P. F. (1977), *State-Variable Analysis*, Macmillan.
[8] Brewer, J. W. (1974), *Control Systems Analysis, Design and Simulation*, Prentice-Hall.
[9] Brogan, W. L. (1974), *Modern Control Theory*, Quantum Publishers.
[10] Brown, B. M. (1965), *The Mathematical Theory of Linear Systems*, Science Paperbacks.
[11] Burghes, D. N. and Downs, A. M. (1975), *Classical Mechanics and Control*, Ellis Horwood.
[12] Burley, D. (1974), *Studies in Optimization*, Intertext.
[13] Clark, R. N. (1966), *Introduction to Automatic Control Systems*, Wiley.
[14] Clarke, C. W. (1976), *Mathematical Bio-economics*, Wiley.
[15] Close, M. C. and Frederick, D. K. (1978), *Modelling and Analysis of Dynamic Systems*, Houghton Mifflin.
[16] Craggs, J. W. (1973), *Calculus of Variations*, George Allen and Unwin.
[17] Dasgupta, P, and Heal, G. M. (1974), The Optimal Depletion of Exhaustible Resources, *Rev. Econ. Studies Symposium*, pp. 3-28.
[18] Davison, R. B. (1978), *Rev. Econ. Studies*,
[19] Desoer, C. A. (1970), *Notes for a Second Course on Linear Systems*, Van Nostrand.
[20] Director, S. W. and Rohrer, R. A. (1972), *Introduction to Systems Theory*, McGraw-Hill.

[21] Dransfield, P. (1968), *Engineering Systems and Automatic Control*, Prentice–Hall.

[22] Flemming, W. H. and Richel, R. W. (1975), *Deterministic and Stochastic Optimal Control*, Springer–Verlag.

[23] Fortmann, T. E. and Hitz, K. L. (1977), *An Introduction to Linear Control Systems*, Marcel Dekker.

[24] Fossard, A. J. (1977), *Multivariable System Control*, North–Holland.

[25] Gelfano, I. M. and Fomin, S. V. (1963), *Calculus of Variations*, Prentice–Hall.

[26] Graham, A. (1979), *Matrix Theory and Applications for Engineers and Mathematicians*, Ellis Horwood.

[27] Hadley, G. and Kemp, M. C. (1971), *Variational Methods in Economics*, North–Holland.

[28] Hastings, N. A. J. (1972), *Dynamic Programming with Management Applications*, Butterworths.

[29] Intrilligator, M. D. (1971), *Mathematical Optimization and Economic Analysis*, Prentice–Hall.

[30] Jacobs, O. L. R. (1974), *Introduction to Control Theory*, Clarendon Press.

[31] Kirk, D.E. (1970), *Optimal Control Theory: An Introduction*, Prentice-Hall.

[32] Lasalle, J. P. and Lefschetz, S. (1961), *Stability by Liapunov's Direct Method*, Academic Press.

[33] Lawden, D. F. (1963), *Optimal Trajectories for Space Navigation*, Butterworths.

[34] Lawden, D. F. (1975), *Analytical Methods of Optimization*, Scottish Academic Press.

[35] Layton, J. M. (1976), *Multivariable Control Theory*, Peter Peregrinus.

[36] Lee, E. B. and Markus, L. (1967), *Foundations of Optimal Control Theory*, Wiley.

[37] Lootsma, F. A. (Ed.) (1972), *Numerical Methods for Non-Linear Optimization*, Wiley.

[38] MacFarlane, A. G. J. (1970), *Dynamical System Models*, Harrap.

[39] MacFarlane, A. G. J. (1972), A Survey of some Recent Results in Linear Multivariable Feedback Theory, *Automatica*, **8**, 455–492.

[40] Marshall, S. A. (1978), *Introduction to Control Theory*, Macmillan.

[41] Melsa, J. L. and Schultz, D.G. (1969), *Linear Control Systems*, McGraw-Hill.

[42] Ogata, K. (1970), *Modern Control Engineering*, Prentice–Hall.

[43] Owens, D. H. (1978), *Feedback and Multivariable Systems*, Peter Peregrinus.

[44] Padulo, L. and Arbib, M. A. (1974), *System Theory*, Saunders.

[45] Parks, P. C. (1964), Stability Analysis for Linear and Non-Linear Systems, *Progress in Control Eng.* **2**, 29–64.

[46] Pars, L. A. (1962), *Introduction to the Calculus of Variations*, Heinemann.

[47] Pearce, D. W. (Ed.) (1975), *The Economics of Natural Resource Depletion*, Macmillan.

[48] Polak, E. and Wong, E. (1970), *Notes for a First Course On Linear Systems*, Van Nostrand.

[49] Pontryagin. L. S., Boltyanskii, V. G., Gamkrelidze, R. V. and Mishchenko, E. F. (1962), *The Mathematical Theory of Optimal Processes*, Wiley.

[50] Porter, B. (1967), *Stability Criteria for Linear Dynamical Systems*, Oliver & Boyd.

[51] Porter, B. and Crossley, T. R. (1972), *Modal Control: Theory and Applications*, Taylor & Francis.

[52] Porter, W. A. (1966), *Modern Foundations of System Engineering*, Macmillan.

[53] Power, H. M. and Simpson, R. J. (1978), *Introduction to Dynamics and Control*, McGraw-Hill.

[54] Prime, H. A. (1969), *Modern Concepts in Control Theory*, McGraw-Hill.

[55] Sage, A. P. (1968), *Optimum System Control*, Prentice-Hall.

[56] Sánchez. D. A. (1968), *Ordinary Differential Equations Stability Theory: an Introduction*, Freeman.

[57] Schwarzenbach, J. and Gill, K. F. (1978), *System Modelling and Control*, Edward Arnold.

[58] Sethi, P. S. (1973), Optimal Control of the Vidale-Wolfe Advertising Model, *Operational Res.*, **21**, 998-1013.

[59] Stephenson, G. (1971), *Inequalities and Optimal Problems in Mathematics and the Sciences*, Longman.

[60] Timothy, L. K. and Bona, B. E. (1968), *State Space Analysis*, McGraw-Hill.

[61] Truxal, J. G. (1972), *Introductory System Engineering*, McGraw-Hill.

[62] Wiberg, D. M. (1971), *State Space and Linear Systems*, Schaum.

[63] Wickwire, K. (1977), Mathematical Models for the Control of Pests and Infectious Diseases: a survey, *Theor. Pop. Biol.*, **11**, 182-238.

[64] Willems, J. L. (1970), *Stability Theory of Dynamical Systems*, Nelson.

[65] Wolovich, W. A. (1974), *Linear Multivariable Systems*, Springer-Verlag.

[66] Zadeh, L. A. and Desoer, C. A. (1963), *Linear System Theory: the State Space Approach*, McGraw-Hill.

Index

A

abscissa of convergence, 28
acceleration error constant, 97
action, 290
adjoint, 321
– equation, 239
– variable, 239
admissible region, 207, 249

B

bang-bang control, 245, 248
Bellman's equations, 300
block-diagram, 41
Boltza problem, 261
brachistochrone, 209, 216, 226

C

capacitor, 13
capacitance, 13
capital stock, 200
catenary, 219
Cauchy's theorem, 141
Cayley–Hamilton theorem, 64, 325
characteristic equation, 108, 323
– frequency, 179
classical control, 23
closed-loop, 21, 99
– transfer function, 137
cofactor, 320
coil, 13
companion form, 51, 73, 165, 325
complementary function, 62
conformal mapping, 138, 141, 142
constraints, 222
consumption path, 200
control, 19
– optimal, 197
– vector, 206
controllability, 149
– criterion, 154
– matrix, 154
controller, 100
critical point, 119
critically damped system, 87
current, 13
cycloid, 217

D

damping ratio, 89
decoupled variables, 51
delay function, 32
depletion, 200
depletion rate, 273
depreciation, 198
diagonal form, 52, 73
Dirac delta function, 33
direct method, 116
– solution, 62
discount, 199
dynamic programming, 289

E

economic growth, 197, 267
– rent, 201
eigenvalue, 323
eigenvector, 323
elasticity of marginal utility, 268
electromagnetic, 13
electromotive, 13
electrostatic, 13

equilibrium point, 119
error, 20
– coefficients, 96
– function, 95
signal, 42
steady state, 83
Euler equation, 210
Euler–Lagrange, 214
exploited population, 200, 274
extremum, 210, 214, 231

F

feed-back, 99
– control, 20, 21, 99
– law, 172
– signal, 42
feed forward, 99
final value theorem, 36
first order system, 84
frequency response, 131
functional, 204

G

gain, 98
golden-rule level, 207

H

Hamiltonian, 233
harvesting, 201
head level, 14

I

impulse response, 41
indefinite form, 124
inequality constraints, 246
initial value theorem, 36
integrating factor, 62
inverse transform, 28

J

Jordan canonical form, 61

L

Lagrange multiplier, 223, 231
Laplace transform, 27
Liapunov's function, 123, 128
– method, 116
– theorem, 124
liquid tank, 14
linear differential equation, 22
– dependence, 328
linearisation, 15

M

marginal stability, 108
mathematical modelling, 22
matrix, companion, 51, 165, 325
– eigenvalue, 324
– input, 57
– modal, 324
– partitioned, 175
– output, 57
– similar, 74, 325
– singular, 323
– system, 57
– Vandermonde, 76
maximum principle, 307
minimal realisation, 163
– polynomial, 328
modal form, 66
mode, 66
modern control, 23
monic polynomial, 180
multiplicity, 93
multivariable feedback, 171
– system, 56, 175

N

negative definite, 124
– feedback, 43
nonlinear differential equation, 23
Nyquist, 131
– criterion, 144
– path, 142

O

observability, 149, 155
– criterion, 157
– matrix, 157
observer, 188
– identity, 190
– reduced order, 190
open-loop, 21
– transfer function, 98, 137
optimal control, 197, 247, 258, 267
– plan, 299
– solution, 239
ordinary differential equation, 22
overdamped system, 87
overshoot, 90

P

partial function, 315
particular integral, 62

performance index, 91
piston, 14
phase lag, 135
– lead, 135
– plane, 117, 248
plan, 290
plant, 19
pole, 66, 86, 98
– location, 171
polynomial, matrix, 326
Pontryagin, 249
– principle, 252
position error constant, 96
positive definite form, 124
production function, 197
– planning, 297
proper system, 39, 56
proportional-gain matrix, 178

Q

quadratic form, 126

R

rank of a matrix, 154, 328
realisation, 163
recurrence relation, 294
reference signal, 42
relative stability, 147
residue, 93, 181
resistance, 16
resolvent matrix, 71
resource allocation, 295
– depletion, 200, 270
– stock, 200
response, impulse, 41
– steady state, 83
– transient, 83
return, 290
– function, 292
rise time, 91
rocket trajectory, 203, 279
Routh's criterion, 110
routing problem, 289

S

sales rate, 201
saturation level, 201
sensing device, 42
servomechanism, 83, 204, 281
settling time, 91
signal, error, 42
– reference, 42
singular control, 278
spectral form, 66
stability, 100, 107, 120
stage, 290
state, 23, 290
– feedback, 171
– space, 49, 51
– transformation, 74
– transition matrix, 70
– variable, 57
– vector, 206
steady-state, 83
– analysis, 83
– error, 83, 91, 95
– terms, 132
subsystem, 19
superposition, 22
sustainable yield, 208
switching curve, 254, 257
– time, 247
Sylvester's criterion, 127
system, 19, 87, 92
– autonomous, 49
– decomposition, 157
– error, 91
– first order, 84
– frequency response, 131
– improper, 39, 94
– MIMO, 56
– proper, 39, 56
– second order, 86
– SISO, 56
– transfer function, 42

T

Taylor series, 302
time constant, 84
– delay, 90
– invariant, 22
– varying, 22
trajectory, 117, 125, 203, 248, 256, 259
transducer, 21
transfer function, 23, 39, 57, 78, 98
transient response, 83
– terms, 132
transition function, 292
– equation, 294
transversality condition, 240, 250, 258
trivial solution, 328

U

unbounded controls, 231
uncontrollable, 151
undamped frequency, 89
underdamped system, 87
unit impulse, 33
– step, 29
unity feedback, 44
unobservable, 151
unstable, 100, 121
utility function, 199

V

Vandermonde matrix, 76
variational calculus, 209
velocity error constant, 96
Verhulst model, 201, 275
viscous, 14

W

weighting function, 41, 107
welfare integral, 199, 200

Z

zero, 93